Sandy Beaches as Endangered Ecosystems

Environmental Problems, Possible Assessment and Management Solutions

Editors
Sílvia C. Gonçalves
and
Susana M.F. Ferreira
MARE, Marine and Environmental Sciences Centre
ESTM – IPLeiria, School of Tourism and Maritime Technology
Polytechnic of Leiria
Peniche, Portugal

CRC Press
Taylor & Francis Group
Boca Raton London New York

CRC Press is an imprint of the
Taylor & Francis Group, an **informa** business
A SCIENCE PUBLISHERS BOOK

Cover credit: Top photograph: Praia da Gralha, São Martinho do Porto, Alcobaça, Portugal (photo by Susana M.F. Ferreira). Down photograph: Playa Las Galeras, Samaná, Dominican Republic (photo by Sílvia C. Gonçalves)

First edition published 2022
by CRC Press
6000 Broken Sound Parkway NW, Suite 300, Boca Raton, FL 33487-2742

and by CRC Press
2 Park Square, Milton Park, Abingdon, Oxon, OX14 4RN

Library of Congress Cataloging-in-Publication Data

Names: Gonçalves, Sílvia C., 1976- editor.
Title: Sandy beaches as endangered ecosystems : environmental problems and possible assessment and management solutions / editors, Sílvia C. Gonçalves and Susana M.F. Ferreira, MARE, Marine and Environmental Sciences Centre ESTM - IPLeiria, School of Tourism and Maritime Technology, Polytechnic of Leiria, Peniche, Portugal.
Description: First edition. | Boca Raton : CRC Press, 2021. | Includes bibliographical references and index.
Identifiers: LCCN 2021029969 | ISBN 9780367147495 (hardcover)
Subjects: LCSH: Seashore ecology. | Coastal zone management. | Coastal biodiversity conservation. | Endangered ecosystems.
Classification: LCC QH541.5.S35 S26 2021 | DDC 577.69/9--dc23
LC record available at https://lccn.loc.gov/2021029969

ISBN: 978-0-367-14749-5 (hbk)
ISBN: 978-1-032-17258-3 (pbk)
ISBN: 978-0-429-05325-2 (ebk)

DOI: 10.1201/9780429053252

Typeset in Times New Roman
by Shubham Creation

Preface

Who has never gazed at the idyllic sight of a sandy beach? For those who have not, we hope that some day you will have the opportunity. For those who did... have you ever wonder: *Do we want our children and their progenies to experience this same soothing feeling?* These questions were lingering in our thoughts and were our main drivers into preparing this book...

Sandy beaches are complex, dynamic, but also sensitive interface ecosystems that fringe the largest proportion of the world's coastlines. The physical environment of the beach, characterized by the permanent interactions between sand, waves, tides and winds, always in motion, has a profound effect on its inhabitants. Due to this dynamism, living on these ecosystems is a very complex task, notably in open oceanic beaches, where the exposure to environmental changes is more pronounced. Sandy beach biota, especially the fauna, must keep pace and fully adapt to harsh and complex cycles of changing environmental conditions, displaying a wide array of adaptations, if they are to endure. The low diversity of organisms observed in most sandy beaches, when compared for instance to rocky shores, should therefore not be surprising. Another relevant feature of the beach environment is the absence of large primary producers, such as macroalgae and plants. Instead, benthic microflora and surf-zone phytoplankton are the main primary producers, when present, and often represent small primary production values. As such, these ecosystems are strongly dependent of irregular inputs of allochthonous organic debris from both marine and terrestrial origins, which frequently act as food subsidies, by fuelling their food webs.

Despite their barren appearance for the most oblivious, beaches have a high natural value and represent unique habitats for several macrofaunal and meiofaunal species, as well as feeding and/or reproduction grounds for some iconic animals, such as birds, marine turtles and fish. Beaches offer diversified habitat possibilities and may shelter a considerable diversity of organisms, even if in low biomass. In fact, only a few species exhibit abundant populations in sandy beaches, especially among macrofauna (e.g., sandhoppers and ocypodid crabs). These ecosystems also have several other important ecological functions, such as their crucial contribution in the decomposition of organic matter and in nutrients recycling, being functional links between the terrestrial and the marine domains. The comprehension of these features is extremely important to help decision makers opting for the most adequate solution when problems arise, as well as raising awareness among common citizens to adopt more ecological and sustainable behaviours. The example of the 2020's COVID-19 pandemic, during which some municipalities around the world executed a sand disinfection with sodium hypochlorite (bleach) in public beaches. Unaware that a sandy beach is a living entity, those who employed this strategy with good intentions, contributed to an ecological crime. It is hoped that the information provided by this book, as well as other similar publications, may reach public opinion to avoid similar situations in the future.

From an anthropocentric perspective, beaches offer several goods and services to humans. They are fishing areas for fish, shellfish and bait harvesting; offer coastal protection from extreme storm surges; and keep a multitude of touristic and recreational activities. Beaches are vital for tourism-based economies. In many regions of the world (for instance, in the Mediterranean), these ecosystems are the main asset that allows maintaining the tourism industry. Hence, tourism in coastal zones and recreational activities on the beach have drastically intensified in the last decades, acting as extra stressors, besides the ones derived from increased urbanization in the coastline, caused by the growth and expansion of human populations.

Presently, sandy beaches are also one of the most endangered ecosystems on our planet. A vast array of human activities and infrastructures severely affect these environments, especially in the most populated coastal areas. In fact, most countries that have a straight connection with the sea, present a higher population density near the coast than on their inland territory. For instance, the introduction of urbanistic and touristic facilities on the beach and its surrounding areas, often causes habitat degradation and fragmentation, disturbance of the ecological balance in these natural areas, as well as simultaneously promoting beach erosion. On the other hand, global climate changes and their consequences, such as retrieval of the coastline driven by sea level rise, are already an imminent threat in several coastal areas around the world.

In this context, monitoring and assessing the environmental disturbances occurring in sandy beaches, and the responses of their different components (biotic and abiotic), has become a priority, if adequate management and conservation strategies are to be developed and implemented. Specific tools encompassing the physical environment and the biota, at different levels of ecological organization, as well as suitable management, conservation actions and programmes where the ecologic, economic and social dimensions are comprehensively integrated, are therefore pivotal. Relying on the vast expertise and understanding of a team with internationally renowned scientists, who have joined us on this adventure, this book aims to consolidate knowledge about these issues. It also intends to point out future directions that can guide researchers, managers, stakeholders, policy decision makers, plus graduation and post-graduation students, towards a more sustainable path while working with sandy beaches. Only this way will we continue to enjoy nature and the environment that surround us, hopefully in a conscious and respectful way.

Meanwhile, we still can have the privilege of putting our bare feet firmly in the sand, feel its grains caress our soles, running in between our toes. We may scrutinize the horizon, stare at the tranquilizing beauty of a sunset or a sunrise on the beach (depending on where we stand) and be cradled by the sound of the waves crashing into the shore. As we close our eyes and take a deep breath, we may linger on the idea that tomorrow will be another day.

Editors
Sílvia C. Gonçalves
Peniche, Portugal
Susana M.F. Ferreira
Coimbra, Portugal

Acknowledgements

The editors would like to thank all the authors for their fruitful collaboration in preparing this book. The authors' notable contributory chapters, sharing of expertise knowledge and diligent efforts to prepare them on the most comprehensive way, were mainly appreciated.

In individual terms, SCG would like to thank her family for their encouragement and unconditional support, whenever she predisposes to do something new, and all her students interested in marine ecology for the soulful strength they give her, which allowed to keep this flame alive. SMFF feels overwhelmed by the invitation of her colleague SCG to get on board of this adventure, but also by the authors' kind readiness to embrace the idea of materializing this book and compile current information on sandy beaches. She is grateful to all of those (family, friends, students, colleagues and peers) that during her life have contributed in some way to carry on her passion for the sea and its life forms.

Finally, to all of those who will read or consult this book, the editors and authors express their gratitude for your interest and hope that somehow the information provided helps to contribute for the conservation of coastal environments.

Editors
Sílvia C. Gonçalves
Susana M.F. Ferreira

Contents

Chapter 1

Sandy Beach Heterogeneity: Intertidal and Supralittoral Communities

Leonardo L. Costa[1]*, Abílio Soares-Gomes[2] and Ilana R. Zalmon[1]

[1]Universidade Estadual do Norte Fluminense Darcy Ribeiro, Laboratory of Environmental Sciences, Av. Alberto Lamego, 2000, 28013-602, Campos dos Goytacazes, Rio de Janeiro, Brazil (costa.ecomar@gmail.com, ilana@uenf.br).

[2]Fluminense Federal University, Laboratory of Sediment Ecology, Marine Biology Deparment, Niterói, Rio de Janeiro, 24001-970, Brazil (abiliosg@id.uff.br).

INTRODUCTION

At first glance, the homogeneous appearance of sandy beaches might provide a false notion that these are systems devoid of life. Nevertheless, wherever you look in the beach interface there are live creatures. How many and which organisms one can detect depend on specific conditions on several spatial-temporal dimensions among and within habitats: substrate depths, distances from the waterline, tide height, disturbance level, and some luck and a careful inspection. Beaches have particular characteristics following a continuum of morphodynamics types, from microtidal reflective beaches (steep slopes, coarser grains, narrow surf zones and harsher hydrodynamics) to macrotidal dissipative beaches (gentle slopes, finer grains, wide

*Corresponding author: costa.ecomar@gmail.com

surf zones and benign swash climate), with intermediate morphodynamic stages between these extremes.

A gradient of conditions according to the distance from the water line is a common characteristic of intertidal systems, as well as in all sandy beach types. As an ecotone in the interface between aquatic and continental environments, marine, terrestrial and semi-terrestrial organisms coexist. For this reason, it is apparent that discrete assemblages of species occurring together in space and time, herein defined as communities can be found on beaches. The farther from the water, the harsher are physical conditions for marine organisms, but more feasible is the habitat for terrestrial organisms and vice versa.

In general terms, species that assemble to make up a community are determined by dispersal and environmental constraints and internal dynamics (Belyea and Lancaster 1999). In other words, a set of populations of different species coexist spatially and temporally because these species can disperse for the same place, they share ecological tolerances to environmental conditions and, in most cases, they interact with each other to ensure their resources requirements. Till date, ecologists debate whether communities are defined as fortuity assemblages of species by means of their ecological tolerances (individualistic-reductionist view) or whether communities are tightly integrated and functionally organized 'superorganisms' with recognizable borders and strong inter-dependence among species (organicist–holistic view) (Dussault 2020).

In the context of sandy beaches, both concepts can be partially merged. In the last three decades ago, it has been postulated that high energy sandy beaches and their adjacent surf zones may function as a discrete and self-sustainable ecosystem (McLachlan 1980). In this sense, beach communities have at least recognizable physical borders. Obviously, organisms transit among boundaries and can connect marine and terrestrial food webs co-occurring in adjacent habitats along the Littoral Active Zone (LAZ) (Fanini et al. 2021). The open nature of LAZ becomes the exchange of species being prominent, while it is frequent to find species from adjacent habitats on the beach. This includes mobile species that use beaches as foraging and roosting sites and sedentary organisms transported passively by waves and tides. However, in general, each entity of LAZ (surf zone-beach-dunes) is characterized by distinct assemblages, tightly related to the physical constrains and habitats they inhabit.

While communities on the beach itself, disregarding surf zones and dunes, have discrete assemblages, they are not structured by ecological interactions as postulated by the organicist–holistic ecological views. Actually, scientific evidence points out that biological communities of sandy beaches are structured by independent responses of individual species to the physical environment (McLachlan and Dorvlo 2005, Barboza and Defeo 2015), following the Auto-Ecological Hypothesis (AEH), as expected for homogeneous systems (Noy-Meir 1979). Therefore, assuming that species are only weakly dependent upon one another on sandy beaches, it is unlikely that assemblages were subject to community-level natural selection. In this context, beach communities can be viewed as discrete assemblages of species tolerant to the hydrodynamics harshness and bordered by physical constrains along the LAZ. The beach, as a single unity of LAZ, has physical constrains over a gradient of distance from the waterline as the main source of heterogeneity. The sand characteristics, swash movement, periodic inundation, time of air exposure, temperature and moisture are the main habitat conditions following this fine-scale gradient (Celentano et al. 2019).

These wide range of conditions determine which species of macroinvertebrates will inhabit each beach zone by means of their ability to burrow, feeding requirements, sensitivity to desiccation and life traits. Among beach zones, intertidal and supralittoral have been usually considered discrete communities. The former remains exposed during low tide, except for the swash zone, and are submerged during high tide. The supralittoral zone is the dry portion of the beach located above the high-tide limit, being only sporadically inundated during storms. While the intertidal zone has assemblages of water-breather and mobile air-breather species, supralittoral is inhabited exclusively by air-breathers.

Community ecology seeks to understand how species assemblages are distributed in the environment at various scales, and how these species can be influenced by their abiotic environment and ecological interactions. On sandy beaches, several hypotheses have emerged and are being tested by correlational or manipulative experiments to explain the main community patterns constantly related to the physical environment and consistent along with the AEH. This chapter provides an overview of the main characteristics, taxonomic composition, patterns and processes regulating sandy beach biological communities and ecological interactions. The main drivers of intertidal and supralittoral communities' structure are depicted at several spatial scales based in scientific literature. And finally, the main future perspectives of studies in community ecology of sandy beaches are stated.

COMPOSITION

In general, sandy beaches support abundant life, albeit of relatively low diversity compared to heterogeneous ecosystems. Sandy beaches present heterogeneity in physical conditions following morphodynamic types, tide range and distance from the waterline, but this is not as prominent as found in rocky shores and coral reefs, where a high diversity of microhabitats increases niche partitioning. Most invertebrate phyla are represented on sandy beaches, but the relatively low diversity becomes feasible to enumerate a general pattern of dominant families, particularly of macroinvertebrates, in which special attention is given because they are by far the most well-known inhabitants of this ecosystem.

Crustaceans, molluscs and polychaetes usually make up more than 90% of communities on sandy beaches. It is estimated that crustacean usually represent at least a half of species pool on sandy beaches and dominate abundance on most beaches, whilst molluscs are also important in terms of biomass (McLachlan and Brown 2006). Notoriously, dominance patterns and number of species vary according to beach types (see later), thus, taxonomic and trophic composition must be deconstructed by means of morphodynamics.

Crustaceans are common across all beach types, being dominant on reflective beaches, especially cirolanid isopods, talitrid amphipods and hippid mole crabs

(Table 1.1). These macroinvertebrates are particularly mobile and have robust morphological structure, being able to cope with a harsh environment or inhabiting upper parts of the beach. Ghost crabs are also representative of all beach types, but they are usually disregarded in studies on community ecology mainly because of specific sampling methods (burrow counts). Among polychaetes, spionids are usually dominant on dissipative beaches (Table 1.1). Some families have been more representative in terms of relative density in temperate latitudes and on macro and mesotidal intermediate beaches, such as the amphipod Oedicerotidae compared to microtidal beaches where it has been usually absent (Table 1.1).

Table 1.1 SIMPER analysis showing the contribution of macroinvertebrates families to community similarity of intertidal-supralittoral interface of sandy beaches with distinct morphodynamics. Only studies that sampled intertidal and supralittoral fringe are included in the analysis. The calculation of Bray-Curtis similarity relied in fourth-transformed relative abundance values of macroinvertebrates families included as variables of a multivariate matrix. AB = Average relative Abundance; IC = Individual Contribution; CC = Cumulated Contribution. Data was gathered from the meta-analysis by Costa and Zalmon (2021).

Taxa	Families	AB(%)	IC(%)	CC(%)
	Dissipative beaches – Average similarity: 25.15			
Polychaete	Spionidae	0.53	37.67	37.67
Crustacea	Cirolanidae	0.34	17.24	54.91
Mollusca	Donacidae	0.28	17.21	72.12
Crustacea	Hippidae	0.21	12.24	84.36
Polychaete	Opheliidae	0.24	7.09	91.44
	Reflective beaches – Average similarity: 27.87			
Crustacea	Cirolanidae	0.59	37.81	37.81
Crustacea	Hippidae	0.49	31.82	69.64
Mollusca	Donacidae	0.24	11.02	80.65
Crustacea	Talitridae	0.34	8.87	89.52
Polychaete	Glyceridae	0.21	6.15	95.67
	Intermediate beaches – Average similarity: 26.22			
Crustacea	Cirolanidae	0.53	45.02	45.02
Polychaete	Spionidae	0.29	12.88	57.90
Crustacea	Talitridae	0.30	12.40	70.30
Polychaete	Glyceridae	0.24	8.53	78.84
Crustacea	Hippidae	0.23	4.36	83.20
Mollusca	Donacidae	0.16	4.18	87.38
Crustacea	Mysidae	0.14	3.07	90.45

The same macroinvertebrates are typical dominants from dissipative to reflective sandy beaches globally, acting as ecological equivalents and playing similar functions in the ecosystem. Generally, beaches are dominated by scavengers in upper intertidal and supralittoral zones and by suspension-feeders in low intertidal zones (see later). The dominance patterns of trophic guilds are very close to morphodynamics: filter-feeders are usually the most representative groups in dissipative high-energy beaches, where surf zones present high primary productivity as the main trophic subsidy, whilst scavengers are prevalent in reflective extremes (Lercari et al. 2010).

When supralittoral is included in sampling campaigns, insects (adults and larvae) become also representative of beach communities. Insects, mainly Coleoptera, as *Phaleria* spp., represent an average of 10% of communities in terms of relative abundance when only the supralittoral is sampled (Fig. 1.1). Nevertheless, in studies restricted to intertidal zones or including only supralittoral fringe, insects have less than 3% of relative abundance (Fig. 1.1). Apparently, insect counts are underestimated by sediment sampling tools, usually efficient for collecting only buried invertebrates, whilst several insects, as well as spiders, are mobile surface-active species on beaches. Thus, even studies of beach communities relying on samplings over the entire intertidal-supralittoral interface must include more than one sampling tool to represent accurately the biodiversity composition, including sediment samplers, pitfall traps, burrow counts and visual census (Fanini and Lowry 2016, Costa et al. 2020). This is particularly important on beaches subject to large wrack inputs. The supralittoral fauna associated with wrack is typically dominated by insects in terms of species—with beetles, ants and flies as most common (Colombini and Chelazzi 2003).

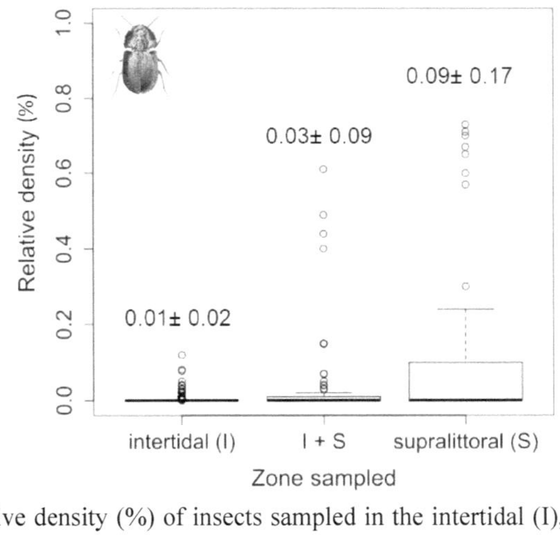

Figure 1.1 Relative density (%) of insects sampled in the intertidal (I), in supralittoral (S) and in intertidal + supralittoral fringe (I+S). Data was gathered from the meta-analysis by Costa and Zalmon (2021). Values above each box are mean ± standard deviation. Horizontal lines are median values, boxes are the interquartile ranges and vertical dashed lines are the lower/upper limits of non-outliers. Dots are outliers.

PATTERNS AND PROCESSES

Morphodynamics and Tide Regime

Despite the natural variability of beach communities, clear trends of richness, abundance and biomass occur on several spatial scales. The most conspicuous macroscale pattern is the reduction of species richness from dissipative fine-grained beaches towards reflective coarse-grained beaches, and this apparently does not depend on the human impact level (Fig. 1.2). Virtually, all typical beach macroinvertebrates from intertidal and supralittoral are successful on dissipative beaches, where colonization by surf zone species is also easier than in harshness reflective extremes. The Swash Exclusion Hypothesis (SEH) states that this pattern results from the exclusion of species toward harsher swash climate and coarser sediments of reflective beaches (McLachlan et al. 1993). The harsher swash climate towards reflective conditions determines the progressive exclusion of less tolerant species until, in the extreme reflective situation, no intertidal species occur and only supralittoral forms remain. Indeed, the swash climate that macroinvertebrates experience on the beach face is closely coupled to beach type; dissipative beaches present wider and lasting swashes with finer grains (Mcardle and McLachlan 1992).

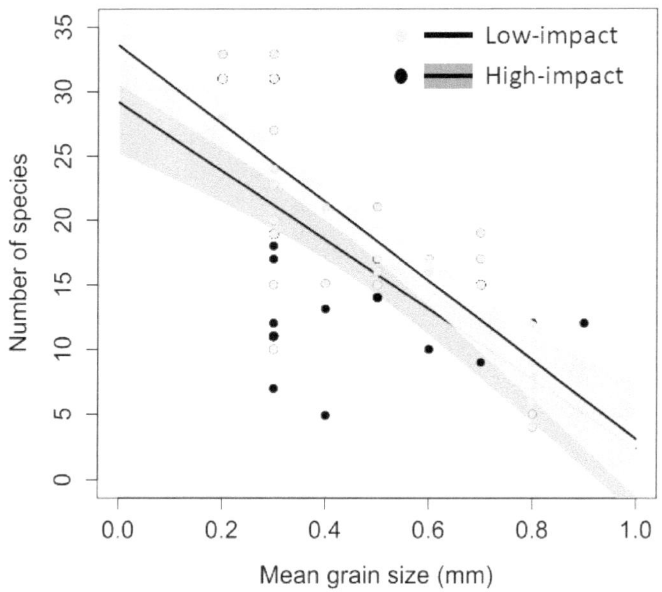

Figure 1.2 The number of species on the intertidal-supralittoral interface reduces toward coarser sediments in low-impacted ($R^2 = 0.57$) and high-impacted beaches ($R^2 = 0.40$) in temperate latitudes. Analysis of Covariance indicated that macrofauna richness is higher on non-urbanized beaches than on urbanized beaches ($F_{impact\text{-}level} = 15.987$, $p < 0.001$), but both respond equally to grain size ($F_{interaction} = 0.464$, $p = 0.497$). N = 120 sampling units gathered from five articles of impact assessments at community level (Dugan et al. 2003, De La Huz et al. 2005, Junoy et al. 2005, Bessa et al. 2013, 2014).

The SEH is well-accepted among beach ecologists, but mechanisms (e.g., larvae settlement, trophic subsidies and tolerance ranges) still remain low-tested, probably because of the difficulty of deconstructing covariating environmental drivers (grain size, slope, accretion-erosion dynamic, retention of organic matter, primary productivity) acting distinctly according to beach types (Brazeiro 2001). For example, if the macrofauna richness increases towards gentler slopes, finer grains, longer swash periods, higher primary productivity and retention of organic matter, simultaneously, all typical characteristics of dissipative beaches, which are the actual drivers or relevant interactions producing mechanisms at community level? Do they have independent influences on different biological processes or species dynamics? Which and how many specific processes are affected? Perhaps only manipulative experiments can solve these issues. For instance, there is evidence that some cirolanids are able to select actively certain sediment features, probably because grain size is an environmental filter for their burrowing and locomotion performance (Defeo et al. 1997). Indeed, the burrowing ability of several macroinvertebrates can be directly affected by changes in sediment, although contrasting results (disrupted burrowing in coarse and fine sands) have been produced in manipulative experiments (Brazeiro 1999; Viola et al. 2014).

Actually, most studies have incorporated environmental covariables into morphodynamic indexes instead of deconstructing the role of each one. This limitation is similar to impact assessments that have not been able to depict the relative importance of each stressor disrupting indicator species on urbanized beaches, where multiple disturbances co-occur (Costa et al. 2020). Nevertheless, in general it is assumed that in reflective beaches the macroinvertebrates have lower feeding time through frequent but short swashes, increased risk of animals being displaced from the substratum with increasing swash frequency, being also more exposed to predation, and increased risk of animals being swept and stranded above the effluent line, where they are unable to burrow (McLachlan et al. 1993). All of these mechanistic conjectures are restricted to species from the swash zone.

The challenge that covariates pose to beach ecologists becomes pronounced when tide range is included in predictive models of macrofauna species richness at macroscales. Clearly, species richness increases from microtidal reflective beaches to macrotidal dissipative beaches (Barboza and Defeo 2015; Defeo et al. 2017). This leads to confusing effects, since increasing tide range makes beaches more dissipative in some degree. As stated earlier, long-lasting and wide swash, fine grains and gentle slope are well-known characteristics of dissipative beaches that provide less harsh hydrodynamics and support higher macrofauna richness. Also similar to dissipative conditions, higher tide ranges create a wider exposure gradient along which intertidal species establish themselves (Fig. 1.3). Exposure gradient is closed related with specific-species tolerances, resources availability and ecological interactions, producing multidimensional ecological niches. Thus, the higher the exposure gradient, as in macrotidal dissipative beaches, the higher is the niches availability and, consequently higher is the species richness.

This is part of the Hypothesis of Macroscale Physical Control (HMPC) (McLachlan and Dorvlo 2005) stating that beach communities are primarily controlled by (1) tide range, which defines the dimensions of intertidal habitat and the number of niches that can be accommodated (HAH– Habitat Availability Hypothesis) and (2) latitude, because of evolutionary processes producing a higher pool of potential settler species in the tropics (see next). According to HMPC, swash and sediment features determine species richness at local scales, similar to premises of almost all ecological hypotheses regarding beach communities. The HAH requires further investigations beyond dimensions of water-land gradient, since the role of environmental heterogeneity for beach diversity is not well described and tested. Predominantly, sandy beaches are homogeneous systems and the gradient of distance from the water line is practically the only prominent source of heterogeneity. However, beach cast can be composed of a heterogeneous quantity of wood fragments, fruits, seeds and carrion that can provide additional microhabitats, mainly for supralittoral species.

Figure 1.3 Aerial photography showing the physical constrains (evidenced by sand colours and immersion/emersion patterns) along the intertidal-supralittoral interface following the distance from the waterline of a dissipative beach in Southeastern Brazil. According to the Habitat Availability Hypothesis macrotidal beaches present the highest number of species because there are more dimensions of intertidal habitat following the higher gradient of distance from the waterline, which increase the number of niches that can be accommodated. Picture by Victor Rangel.

Not surprisingly, macrofauna communities of sandy beaches are inhabited by species with different life-history traits, mainly when considering intertidal and supralittoral (i.e., beach) as a unique entity of LAZ. Taking into account the large interspecific variability of marine, semi-terrestrial and terrestrial organisms and the many physical conditions covariating with morphodynamics, the idea that swash harshness is the unique key factor limiting the distribution of all marine species along the morphodynamic gradient is not convincing. For this reason, Brazeiro (2001) proposed the Multicausal Environmental Severity Hypothesis (MESH), which states that different species are limited by specific ecological factors, mainly physical variables, linked to morphodynamics (Fig. 1.4).

Figure 1.4 shows a schematic representation of the rationale behind the MESH. The squares with their letters and colours represent different species, with distinct habitat requirements. For example, dissipative beaches have in general, a predominance of fine grains, gentle slope and long-lasting swash, which are all habitat requirements of the species A. The species B requires both sediments with a predominance of fine grains and gentle slopes. All other species co-occurring on dissipative beaches have only one limiting factor among grain size, slope and swash climate as decisive for their occurrence only in dissipative beaches and not in reflective beaches. Thus, the MESH allows that species A, B and C are controlled by different drivers, but such variables covariate toward the reflective-dissipative continuum, producing convergent cross-taxon patterns caused by multiple drivers. Finally, species J, K and L are habitat generalists that colonize beaches regardless of morphodynamics characteristics, forming less complex communities on reflective beaches composed by species with high tolerance range to physical factors. The MESH represents a simplified tri-dimensional physical niche (sediment type, swash climate and accretion-erosion dynamics) of beach species. The presented niche breath of each fictitious species is, therefore: J = K = L > C = D = E = F = G = H = I > B > A. SEH simplifies physical constrains among beach morphodynamics into a swash climate as the unique limiting factor. SEH is, therefore, part of MESH, with both following the AEH, since communities are structured by individual responses of species to the physical environment.

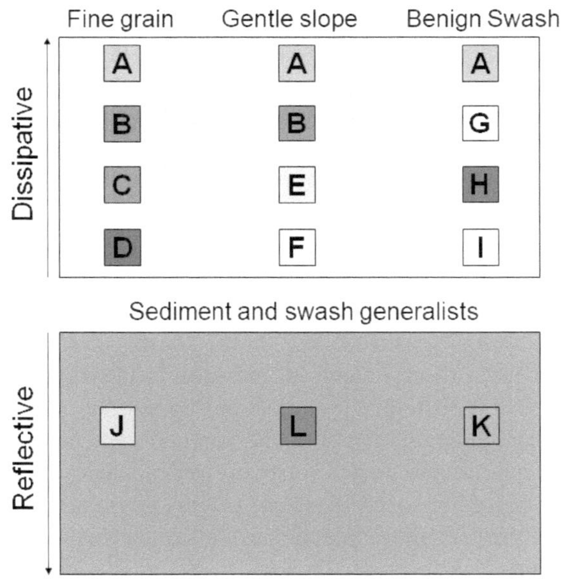

Figure 1.4 Schematic representation of the rationale behind the Multicausal Environmental Severity Hypothesis (MESH), which states that species with different life traits are controlled by different limiting factors (co variates) interrelated and linked with morpho dynamics. Each letter inside the squares represents a different species.

Abundance and biomass do not respond as consistently to beach morphodynamics and tide range as species richness, but similar trends have been found. In general, beaches of fine sand and flat slopes support high abundance and biomass, even though there is scatter in secondary datasets compared to species richness. The Habitat Harshness Hypothesis (HHH) is an attempt to explain community patterns through populational parameters of beach species (Defeo et al. 2003). This hypothesis postulates that in reflective beaches the harsh environment poses a trade-off for beach species, which have to divert more energy towards survival, resulting in lower fecundity and higher mortality. Thus, post-settlement processes may prevent some species from establishing populations under harsh swash of reflective beaches, contributing for the lower species richness. Even species that are able to colonize harsher environments tend to be less abundant than in dissipative beaches, where they can invest more energy for reproduction. Thus, communities usually have increasing abundance from reflective toward dissipative beaches.

HHH is the most intuitive hypothesis for explaining population patterns, and consequently community descriptors, according to the physical environment. Populations of species living in the swash and those immediately landwards of reflective beaches are expected to have lower abundance, individual sizes, growth rates, longevity, weight, fecundity, reproductive output, higher mortality and burrowing rates than those from dissipative beaches (Defeo et al. 2001). Predictions of HHH have been corroborated for intertidal species, such as *Emerita brasiliensis* (Crustacea: Hippidae) and *Donax hanleyanus* (Mollusca: Donacidae). However, the response of species that co-occur at beaches with contrasting morphodynamics do not always act as predicted by HHH. The HHH was not robust for explaining the reproductive cycle of *Anomalocardia brasiliensis* (Mollusca: Veneridae) when considering tidal flats as a more benign habitat, indicating that only physical variables may not be always adequate for describing populations' dynamics on beaches or similar systems (Corte et al. 2014). Similarly, Herrmann et al. (2010) found contradictory results, with many reproductive responses of *D. hanleyanus* to physical variables opposing those predicted by the HHH, although only three beaches were sampled.

The SEH and HHH are apparently powerful to the intertidal fringe species, but this is not necessarily valid for supralittoral and upper intertidal air-breathing crustaceans, which usually increase in richness and abundance from dissipative to reflective conditions, a reverse trend to that seen in intertidal crustaceans (Defeo and McLachlan 2011). Life-history traits of supralittoral forms, such as direct development, aplanktonic larval phase and parental care with eggs/embryos turn the majority of supralittoral forms relatively independent of the swash climate throughout their entire life span. These are additional evidences that intertidal and supralittoral communities may be regulated by independent factors and present discrete communities. In this way, the Habitat Safety Hypothesis (HSH) emerged and predicts higher abundance of supralittoral species under harsher environments, assuming that the combination of narrow swashes and steep slopes makes reflective beaches a more stable and safer environment for supralittoral species by means of the lower risk of inundation.

The HSH has been less tested, particularly because supralittoral assemblages are neglected compared to intertidal ones in beach ecology. This is surprising, because easy and low-cost sampling methods (e.g., pitfalls) could be applied to sampling supralittoral organisms. Interesting study models, such as ghost crabs, have provide quite different results regarding both HSH and HHH. Ghost crabs are supralittoral crustacean at first independent of swash. However, they have oceanic larvae and thus, larvae settlement is possibly affected by a swash climate. Using burrow counts, some authors have observed an increasing density toward dissipative conditions corroborating HHH (Lucrezi 2015; Pombo et al. 2017), whilst others have found more burrows on reflective and coarser-grained beaches, following the HSH (e.g., Defeo and McLachlan 2011). At macroscale, Barboza et al. (2021) performed a meta-analysis and found that the grain size did not consistently explain the variability of burrow density of *Ocypode quadrata*. However, dataset of grain size was considered limited, with some authors considering the entire intertidal zone, but others sampling only supralittoral, biasing the potential of grain size as a truly proxy of morphodynamics and habitat harshness. Ghost crabs are an important link between intertidal and supralittoral communities and have been prioritized so far in studies considering the supralittoral macrofauna.

Sandy beaches are assumed to have low *in situ* primary production, although some evidence points to the grazing of benthic diatoms by macro- and meiofaunal species (Maria et al. 2011). Despite this, subsidies of macroalgae cast ashore provide food and shelter for various species. Intertidal scavengers are also benefitted from a feeding opportunity, but supralittoral communities are enriched by the diverse wrack-associated species (mainly insects and crustaceans). Thus, whilst intertidal communities are mainly controlled by the individual responses of marine species to sand features and a swash climate, and slope seems to drive the risk of inundation of terrestrial and semi-terrestrial organisms (see Habitat Safety Hypothesis), wrack availability is an important structuring factor of supralittoral communities (Colombini and Chelazzi 2003). In some cases, wrack also washes around in the swash on the lower shore, and then the associated species can be entirely marine, but supralittoral communities have been the focus of wrack-associated assemblages on beaches.

Wrack quantity and deposition frequency on the beach usually have a positive influence on several faunal descriptors such as density and number of species of various functional groups (Gonçalves and Marques 2011). This illustrates the role of wrack subsidy on supralittoral community organization, not only scavengers from the entire beach face, but mainly because of the simple relationship between microhabitat availability and occurrence of obligate wrack-associated species. The removal of wrack to make beaches more attractive to tourists has disrupted beach communities, with reducing species richness being more prominent for wrack-associated assemblages than intertidal ones (Schooler et al. 2019). In addition, studies have found that wrack subsidies not only provide more diverse supralittoral invertebrates assemblages, but have been proved as pivotal for food provisioning for birds that prey on wrack-associated species (Dugan et al. 2003; Schlacher et al. 2017). The potential of wrack subsidy as major drivers of supralittoral communities have still not been extensively tested regarding ecological theories, perhaps because beaches subject to large wrack inputs are a special situation occurring at low spatial scales, precluding large extrapolations and paradigms. However, the availability of wrack can be viewed in the scope of the HAH.

Latitude

Scale is a paramount feature in discussions of patterns of species richness in any ecosystem, as well on sandy beaches. As described in this chapter, the number of species living on a beach reflects mainly local influences regarding morphodynamics, tide regime, the range of habitats (heterogeneity) and apparently weaker ecological interactions. These local influences provide various ecological theories to explain patterns and processes of the biodiversity of sandy beaches. However, there are large scale drivers of both spatial and temporal patterns of macrofauna communities which are also important to assess.

Sandy beach macrofauna communities respond to latitude as those found in other ecosystems: greater diversity toward the tropics, particularly if the beaches with similar morphodynamics are compared. If species richness changes with latitude, then there must be some causative agent or covariate also changing with the latitude. Firstly, the evolutionary process has admittedly led to the development of greater species pools in the tropics. Thus, the magnitude of the differences between dissipative and reflective beaches driven by local physical processes might be higher in the tropics than in the temperate areas because the number of potential colonizers of beaches regardless of morphodynamics is higher in the first. Meta-analyses have endorsed that differential historical factors in tropical, sub-tropical and temperate latitudes related to geographic isolation and coastal complexity, have been considered important drivers by themselves of species richness on sandy beaches (Barboza and Defeo 2015; Defeo et al. 2017).

Even including other possible controlling factors at large spatial scales, beach slope, tidal range, and grain size, which are proxies of habitat harshness and habitat availability (see earlier) have an important role in structuring beach communities across latitudes. Schooler et al. (2017) argued, by comparing datasets of three decades apart, that local scale processes, including human disturbances, can exert a stronger influence on intertidal biodiversity on beaches than regional processes. This partially explains the regulating role of morphodynamic variables in species richness at large spatial scales. Nevertheless, when latitudinal trends are evaluated considering morphodynamic types separately, temperature becomes the most important predictor of species richness (Barboza and Defeo 2015). This is consistent with the Kinetic Energy or Temperature Hypothesis stating that higher metabolic rates or relaxed thermal constraints promote diversity through increasing mutation and speciation rates, mainly in ectotherms that are most sensitive to temperature variations and constitute the vast majority of obligate sandy beach species (Clarke and Gaston 2006, Barboza and Defeo 2015).

Contrasting results have emerged when considering the role of primary productivity as an explanatory variable in structuring latitudinal trends of species richness on beaches. At first, higher productivity can be correlated with a wider range of available resources (and available niches), then this is likely to lead to an increase in species richness. This was found by Defeo et al. (2017) assessing the macrofauna richness along the Atlantic and Pacific coasts of South America. At this regional scale, the productivity-richness hypothesis was, thus, corroborated. However, when a global scale was considered in the meta-analysis proposed by Barboza and Defeo (2015), primary productivity has a minor effect on species richness, evidencing the scale-dependence of ecological studies on sandy beaches.

A more productive environment may have a higher rate of supply but not necessarily a greater variety of resources. This might lead to more individuals per species rather than more species by means of the increased environmental carrying capacity. Previously, McLachlan and Dorvlo (2005) had provided insights from a meta-analysis that abundance (in gentle slope beaches) and biomass (over the whole range of beach slopes) are higher at temperate sites (opposing its species richness), because wave action is greater at a higher latitude, especially for crustaceans. High wave energy provides the accumulation of diatoms in their surf zones and increases wrack input to the beach, providing higher carrying capacity and higher macrofauna abundance and biomass.

Other factors probably vary geographically but quite independently of latitude. They tend to counteract relationships between beach community descriptors and richness at some level. Among them, the amount of natural and anthropogenic disturbance that beaches experience, connectivity among beach arcs that affect the input of larvae, and heterogeneity have not been considered at macroscales. They are a challenge to control in statistical models targeting the disentanglement of ecological and geographical patterns, but they may explain some spatial-temporal variation of communities.

Disturbances

Disturbance is defined as a discrete event that suddenly disorders communities by influencing environmental conditions, resources or ecological interactions. This is considered a factor whose variation is primarily temporal, different from spatial drivers discussed until now. Patterns of abundance of species can vary following disturbances because they may (1) provide opportunities for organisms reaching new locations (e.g., by passive transport); (2) change conditions and resources, and (3) promote ecological releases. Natural and anthropogenic disturbances have been shown to follow some of these premises, influencing macroinvertebrates communities on sandy beaches.

In general, there are emergent properties that appear in the face of natural disturbances on sandy beaches. Storms Wave Events (SWEs) are the main sources of natural disturbances in this ecosystem. These disturbances at moderate levels have been followed by a short-term increasing of macrofauna richness and abundance rather than apparent negative effects. SWEs usually affect the variety of food resources, by increasing the wrack stranded on the beach surface usually removed from adjacent rocky shores, benefitting scavengers (Alves and Pezzuto 2009). In addition, waves may transport surf zone species to intertidal areas (Machado et al. 2016). Changes in morphodynamic parameters following storms are also pivotal for restructuring beach communities (see earlier). These lead to changes in species richness and distribution, but the beach community usually recovers within weeks to month after storms. Various authors have shown that anthropogenic disturbances may reduce resilience of beach communities to natural disturbances, thus constituting the main threat under scenarios of increasing frequency and intensity of storms predicted as consequences of climate change (Harris et al. 2011, Witmer and Roelke 2014, Machado et al. 2016).

Human disturbances have been considered one of the most threating for beach communities worldwide (Defeo et al. 2009). However, whilst the negative effects for macroinvertebrates at population level are evident so far, community aggregate descriptors are not as responsive as species abundance in detecting impacts (Costa et al. 2020). Community changes result mainly from species lost that is apparently prominent for some taxa without oceanic larval stage (Costa and Zalmon 2019). Thus, similar to the response of the abiotic environment, responses of beach communities to human disturbances are apparently a result of individual sensitivity of species to stressors.

The nature of most natural communities is considered more than just the sum of its constituent species. However, it would be contradictory to state that disturbances enable the colonization of species after ecological release from their predators or competitors, since ecological interactions are rare in controlling factors of beach communities. This differs from typical dominance-controlled communities, where some species are competitively superior to others so that an initial colonizer of an opening left by a disturbance cannot necessarily maintain its presence there. Evidences that human disturbances pose cascade effects or secondary successions on beach communities are truly rare (Costa et al. 2017). It is also unknown if disturbance level shapes beach communities at broader spatial-temporal scales.

Metacommunity and Connectivity

The Source and Sink Hypothesis offers an explanation for communities' structure based on the population's level. The hypothesis is based on the concept of meta-population: a group of subpopulations of the same species that are spatially separated, but that interact at some level by the dispersion of individuals (Levins 1969). According to the hypothesis, meta-populations from beaches could be a source or sink of new individuals that are dispersed mainly during the larval pelagic phase. Sink meta-populations are subsided by larvae coming from other meta-populations: the sources. The set of meta-populations in a region can be considered meta-communities.

The predictable increase in macroinvertebrates species richness from reflective to dissipative beaches suggests that the latter act as sources of biodiversity and the former as sinks. As organisms from dissipative beaches live in a more benign environment, their populations would be more abundant and stable, with individuals breeding more efficiently. The harsher environment of reflective beaches constrains the abundance and reproduction, thus their populations would be kept by larval inputs from meta-populations from dissipative beaches. Testing this 'source-sink hypothesis' requires investigations in a meta-community framework, but this knowledge is still lacking on sandy beaches. This is particularly relevant because most beach species have oceanic larval stages, which allows dispersion among beaches with contrasting characteristics and even human disturbance levels.

Yet, few studies were made to evaluate the source-sink hypothesis, particularly using species nestedness patterns (Checon et al. 2018) as a proxy. A nested pattern of sandy beaches communities implies that the set of species inhabiting reflective beaches is a subset of the set of species inhabiting dissipative beaches. Thus, the number of exclusive species must be higher in dissipative beaches than in reflective ones at regional scales. While Brazeiro (1999) found exclusive species occurring mainly in dissipative beaches in Chile, Checon et al. (2018) observed non-nested communities with a similar number of exclusive species among nine beaches with distinct morphodynamics in Brazil. The last authors also found that spatial distance was a more decisive variable than morphodynamics to explain assemblages' dissimilarities, reinforcing the importance of larval dispersion as structuring factors of beach communities.

The metacommunity approach has several implications for impact assessments and management. Firstly, as pointed by Checon et al. (2018), if dispersal processes are more determinant for fitting beach communities than local characteristics and then exclusive species occur in beaches regardless of morphodynamic, conservation practices across large spatial scales should consider all beach types. Secondly, if beach communities are meta-communities connected by larval dispersion, landscape features, such as the percentage of low-urban beaches adjacent to disturbed ones become pivotal as drivers of community structure of the former. For example, studies have shown that species without oceanic larval dispersion are prone to be locally extinct from disturbed areas because individuals cannot breed successfully and the population do not receive larvae from adjacent low-impacted areas (Costa and Zalmon 2019). Nevertheless, if urban beaches are adjacent to low-urban areas, where populations are able to reproduce and release larvae, species with pelagic larval stage will persist even under some disturbance level. Thus, the higher is the length of low-impacted beaches in the coastal landscape, the milder the human disturbances and coastal tourism will be for beach communities at regional scales. These insights regarding larval dispersion, pre-recruitment processes, connectivity patterns and settlement of beach species are, however, not known enough.

Zonation

Species coexisting on beaches use different sources of energy, have different reproductive and development modes, and occupy, by means of their habitat requirements and life history, different zones in the intertidal-supralittoral interface. The relative importance of each requirement is hard to generalize, due to the frequently high number of potential factors involved and mainly, due to their complex interactions. Different from intertidal rocky coasts where zonation of organisms is a conspicuous pattern worldwide, on sandy beaches this phenomenon is subtle because the majority of their dwellers is hidden inside the substrate: endobenthos predominate in soft bottoms while epibenthos predominate in hard bottoms (Raffaelli and Hawkins 1996). In addition, the constant movement of sediments and organisms precludes static zonation patterns on sandy beaches.

The first causes suggested for sandy beaches zonation were based on the physical drivers related to tides and the air-sea interface stresses, similar to the pioneering ideas on rocky shores. However, different from rocky shores, where posterior studies demonstrated the important role of biological interactions on zonation, there is a strong correspondence between physical and biological zones on sandy shores. Celentano et al. (2019) conducted an intensive across-shore sampling in a dissipative beach and verified that sediment temperature decreased towards the swash zone and at deeper sediments. Moisture, grain size and organic content increased seaward. A significant vertical segregation of body size was also observed, suggesting different microhabitat preferences and burrowing abilities. Consequently, species spatial patterns vary according to life history traits and differential susceptibility to variations in environmental conditions.

Conflicting views on the zonation of beach macroinvertebrates were pointed by MacLachlan and Jaramillo (1995): (1) no clear zonation; (2) two zones above and below the drift line; (3) three zones described by Dahl (1952) based on crustaceans; (4) four zones described by Salvat (1964) based on changes of sand moisture across the shore. Trying to conciliate the existing conflicts, McLachlan and Jaramillo (1995) proposed a zonation pattern similar to the pioneering scheme of Dahl (1952), composed by three zones: (1) supralittoral; (2) littoral or intertidal; and (3) sublittoral. According to them, a fourth zone between swash and high-tide line emerged during low tide (retention zone) proposed by Salvat (1964) is not a general rule. However, they warned that the nature and number of zones is a complex issue and they could change depending to the beach morphodynamics and temporal variation in wave energy and sediment dynamics. For example, when tide rises, zones tend to get closer, whilst some organisms migrate, resulting in rearrangements of mobile fauna and almost indistinguishable biological zones. The lack of clear zonation patterns is due partly to the manner that some dominant organisms moving up and down the beach for feeding or by the fact that most organisms of sandy beaches display escape behaviours to avoid dangerous conditions (e.g., risk of desiccation or inundation).

Schlacher and Thompson (2013) tested the hypothesis that the instability and variability of habitats, combined with plastic behaviour and mobility of beach macroinvertebrates result in indistinct or not well-established zonation pattern on sandy beaches. Among the 52 cross-shore fauna data they analyzed, however, 94% demonstrated distinct biological zones. However, they warned that they found considerable variation in the number of zones and the way they are distributed across the intertidal gradient. The mean number of zones they found (2.56) was similar to the global average (2.94). In their results, beach dimensions, sediment properties, and morphodynamic indexes were not correlated to the number of zones.

Although some variation had been reported in the literature, it seems plausible that a three-zone pattern (sublittoral-intertidal-supralittoral) is the most frequent on sandy beaches. This reinforces the view that intertidal and supralittoral are discrete communities driven by particular environmental forcing, the first strongly related with to a swash climate and the last affected by still little-known factors. Increasing environmental stress has historically been related to lower diversity, richness and higher species dominance, since the physical factors such as heat and desiccation are more severe due to the longer exposure time. However, tide oscillations at the intermediate levels of the substrate turn the environment into a less stressful one and allow several organisms to inhabit intertidal zone rather than supralittoral zone (Table 1.2).

Table 1.2 Main characteristics of the there zones observed on sandy beaches, and the more frequent macroinvertebrates taxa found.

Zones	**Main Features**	**Typical Macroinvertebrates**
Supralittoral	Surficial dry sand most of the time; water table very far from the surface; dunes and vegetation; urban infrastructure; air-breather macroinvertebrates	Ocypodidade crabs (ghost crabs); talitrid amphipods; oniscid and cirolanid isopods
Intertidal/ Littoral	Periodical water level oscillation; water table in intermediary depth of sediment column; high hydrodynamics; effluent line; swash; water retention; association of water- and air breather' species	Cirolanid isopods (*Excirolana*); haustorids and other amphipods; spionid and ophelid polychaetes (e.g., *Scolelepis*, *Euzonus* and Glyceridae); thalassinids decapods (deep-burrowing ghost-shrimps); Hippid crabs (e.g., *Emerita*); donacid bivalve molluscs
Sublittoral	High hydrodynamics; extension of the surf zone; saturated sand; water-breather macroinvertebrates	Hippid crabs (e.g., Emerita, Hippa, Albunea); mysid crustaceans; donacid bivalve molluscs; nephtyid and glycerid polychaetes; idoteid, oedicerotid, and haustoriid amphipods; echinoderms (Mellita and Encope)

ECOLOGICAL INTERACTIONS

Predation

Apparently, beach communities are organized according to the individualistic-reductionist view, becoming practicably a sum of species tolerant to similar environmental conditions. Not surprisingly, as shown in this chapter, various paradigms exist in beach literature regarding general patterns of community structure and physical factors relationships. Most hypotheses point out the individual responses of species to sand, swash and slope at several spatial scales. As a result, interspecific responses to various physical covariates converge to more diverse communities in dissipative beaches, where environmental conditions are less harsh than in reflective beaches. In this context, ecological interactions were rarely tested as important controlling factors of intertidal-supralittoral communities. However, ecological interactions might not be totally disregarded as important factors at some level, particularly under dissipative conditions. Thus, even though evolutionary interactions do not intensify the community characteristics, beaches present at least an ephemeral mosaic of interacting intra- and interspecific individuals. Therefore this must be considered to explain at least partially ecological patterns and their importance for the ecosystem functioning.

Resource-consumer relationships constitute the more fundamental ecological interaction, shaping the energy flow and nutrient cycling of any ecosystem. The main top predators of sandy beaches are fishes, birds and crabs, but some terrestrial mammals are also consumers in marine intertidal communities (Carlton and Hodder 2003). Except for ghost crabs, some polychaetes, gastropods and insects, predators of sandy beaches are mostly transient species that feed on beach macroinvertebrates. Many studies have shown the importance of crustacean, molluscs and polychaetes in the diet and for the habitat selection of birds (e.g., Dugan et al. 2003, Lunardi et al. 2012) and fishes (e.g., Takahashi et al. 1999, Costa and Zalmon 2017). Reductions in beach clams and mole crabs due to human disturbances may be especially problematic given their high caloric content and pivotal as feeding resources of resident and migratory species (Gül and Griffen 2018). Energetic demands are particularly critical during nesting and chick rearing phases for resident birds, for instance (Schlacher et al. 2017). For migratory birds, beach invertebrates can become further critical prey in the face of urban occupation and habitat loss on low-energy shores (Hubbard and Dugan 2003).

Accordingly, macroinvertebrates communities of the intertidal-supralittoral interface are subject to intense predation, but regulation of macroinvertebrates communities by predators (top-down control) on sandy beaches has not been consistently demonstrated. Wolcott (1978) estimated that the Atlantic ghost crab *O. quadrata* crop the majority of mole crabs and donacid clams production on a North Carolina barrier beach. However, the author estimated prey biomass relying on dataset from other regions and different years, precluding reliable conclusions and extrapolations. More recently, Van Tomme et al. (2014) endorsed that very high consumption rates and preference of a dominant beach polychaete and crustaceans by a flatfish and a shrimp in mesocosm experiments are evidences of the controlling role of macrofauna community of dissipative beaches by epibenthic predators. However, the authors claimed for caution when extrapolating these laboratory results to the field, where predators and prey experience a three-dimensional dynamic habitat, and prey can hide in sediment and be replaced by other resources.

Indeed, exclusion experiments (e.g., caging) are challenging on exposed beaches, so that most insights about predation effects on beach communities come from studies performed in tidal flats. Thus, in general, beach ecologists assume that keystones prey or predators are absent on sandy beaches, and predation is not an important structuring driver of intertidal and supralittoral communities. However, this hypothesis needs further investigations, including specificities at local scales, where predation of macroinvertebrates by ghost crabs, portunid crabs, cicindelid insects, fishes or birds are intense. It has been postulated using trophic models, that predation on non-urban beaches is a more determinant driver than in urban beaches (Reyes-Martínez et al. 2014, Costa et al. 2017). Nevertheless, it is necessary to contrast demographic dynamics of interacting populations, as well to estimate feeding rates of predators *in situ*. Even though predation does not have dramatic cascade effects for a physically controlled ecosystem, it is plausible that populations are significantly regulated by specific or diffuse predation-prey interactions, but this is still little known. At least for now, scientific literature has indicated only bottom-up effects, since predators usually select beaches with high food availability, and primary productivity is an important driver of latitudinal patterns of species richness (Barboza and Defeo 2015).

Competition

One of the most robust empirical evidence that beach communities might be partially fitted by biotic factors was brought by Defeo et al. (1997). The authors performed field sampling and laboratory experiments and raised evidence that cirolanids populations can be driven by interspecific competition on Uruguayan beaches. The evidences pointed were: (1) *Excirolana armata* was restricted to fine-grained beaches, whilst the congeneric *Excirolana brasiliensis* was more abundant in coarse-grained beaches, even the last displays selectivity by fine sediments; (2) allopatric populations of *E. braziliensis* occur in higher densities in fine-grained beaches than those observed when co-occurred with *E. armata*; (3) body size of both species was lower in sympatry than in allopatry; (4) species were spatially segregated in intertidal zones, with *E. braziliensis* being restricted to upper intertidal where it co-occurred with *E. armata.*

Survival experiments performed by Defeo et al. (1997), in which populations of cirolanid cogenerics were monitored in isolation and together in fine and coarse-grained mesocosms failed to detect competition. Thus, doubts still exist if different covariates produced the observed correlational patterns from field sampling. Other correlational studies also raised the competition as a structuring component of beach populations' dynamics. Defeo and Alava (1995) observed that the clams *Mesodesma mactroides* and *Donax hanleyanus* had opposing abundance patterns, whilst body size of the first was inversely related with density of the second one. These clams explore similar feeding resources and live at similar beach strata and sediment depth. In supralittoral, Jaramillo et al. (2003) found that a coleopteran insect and a terrestrial crustacean change their distributional and activity patterns resulting in spatial and temporal partitioning along with a sympatric population of talitrids; all these species are scavengers, and these correlational insights were interpreted as avoiding competition .

In fact, there are numerous evidences of spatial segregation among sympatric interespecifics with niche similarities and potential overlap, hence being interpreted as competition, particularly in denser communities toward dissipative conditions. However, all scientific evidences have been raised on fine scales, apparently affecting species distribution or relative abundance, but not community structure. Thus, competition can exist on beaches, but these do not override at first the physical control of beach communities. The extremely crowded conditions for organisms as observed on rocky shores are not found on sandy beaches. The fauna is sparse and does not occupy all available space, suggesting that competition for space is not a major driver of community or populational patterns. Competition for food also appears negligible, since the sparse populations and abundant plankton and detritus make competition unlikely. However, specific hypothesis regarding competition must be properly tested, particularly in the supralittoral in which knowledge is even less.

Commensalism, Amensalism and Mutualism

There are examples of other interactions, such as commensalisms, mutualisms and amensalisms affecting species distribution and behaviour on beaches. Pinnotherid pea crabs are typical commensals during their adult life, for instance, living in facultative or obligate association with clams, sand-dollars (Echinodermata: Melitidae) and into ghost shrimps (Crustacea: Callianassidae) burrows. The unique evidence of mutualism comes from Manning and Lindquist (2003), who exposed clams with and without epibiontic hydroid colonies to multiple predators, and observed that the hydroid defended the clam against Florida pompano *Trachinotus carolinus* in mesocosms. Otherwise, the hydroid projects above the sand surface and allows ghost crabs to readily detect clams, which poses doubt whether hydroid-clam interaction is actually a mutualism or an amensalism.

Ghost shrimps have proved to significantly affect surrounding communities by turning over large quantities of sediment when they are in huge densities. Berkenbusch et al. (2000) showed that the number of species and abundance of co-occurring macroinvertebrates were usually lower in sites with high density of the ghost shrimp *Callianassa filholi* on an intertidal sandflat in Otago Harbour, southeast New Zealand. Deleterious effects of sediment turnover are prominent for tube-building and suspension-feeder organisms in particular. Considering that ghost shrimps do not take clear benefits from bioturbation per se, and various filter-feeders are impaired, this activity can be interpreted as an amensalism on swash zones of sheltered beaches where ghost shrimps are more common.

PERSPECTIVES IN COMMUNITY ECOLOGY OF BEACHES

In addition to the changes imposed by beach morphodynamics that are very well-known in the scientific literature, salinity has been identified as a critical variable that affects coastal biodiversity patterns. Salinity range, rather than salinity itself, drives variations in abundance and distribution of macrofaunal components along estuarine gradients. In this context, differential susceptibility of species with distinct development modes to salinity variations could cloud the understanding of observed patterns. According to Barboza et al. (2012), the reduced species richness in outer estuary beaches with high salinity ranges, suggests that sandy beach macrofauna has relatively low tolerance to large salinity variations. Both salinity and salinity range are recognized as key factors in structuring macrofaunal communities. Nevertheless, the studies designed to set global patterns of beach communities do not consider estuarine beaches. Macroscale patterns and the importance of salinity as drivers of community descriptors could change drastically if sandy beaches from transitional estuarine systems are included in predictive models.

In conclusion it has been suggested in various meta-analyses and manipulative experiments (Defeo et al. 1997, McLachlan and Dorvlo 2005, Defeo and McLachlan 2005, Defeo and McLachlan 2013, Barboza and Defeo 2015) that: (a) biological interactions are not major controlling factors of beach communities, even though they are more important regulatory agents than previously thought, particularly in benign dissipative beaches or undisturbed sites; (b) supralittoral forms are relatively independent of the swash regime and show no clear response to beach type; this compartment is neglected as compared to the intertidal zone, and studies at community level about patterns and processes are still lacking, including variables such as wrack availability and habitat heterogeneity; (c) fluctuations are noticeable in species with planktonic larvae structured as meta-populations, due to environmental disturbances and stochasticity in reproduction and recruitment. Studies focussed on larvae dispersion and connectivity among beaches with distinct morphodynamic are scant. In synthesis, mainly physical factors govern community and population features of sandy beach macrofauna in different scales. Human induced impacts are increasingly imposing variability in population demography, structure and dynamics and will probably remain the main target of community ecology studies on sandy beaches in the next decades.

REFERENCES

Alves, E.S. and P.R. Pezzuto. 2009. Effect of cold fronts on the benthic macrofauna of exposed sandy beaches with contrasting morphodynamics. Brazilian J. Oceanogr. 57: 73–96.

Barboza, F.R., J. Gómez, D. Lercari and O. Defeo. 2012. Disentangling diversity patterns in sandy beaches along environmental gradients. PLoS One. 7(7): e40468.

Barboza, R.F. and O. Defeo. 2015. Global diversity patterns in sandy beach macrofauna: A biogeographic analysis. Sci. Rep. 5: 1–9.

Barboza, C.A.M., G. Mattos, A. Soares-Gomes, I.R. Zalmon and L.L. Costa. 2021. Low densities of the ghost crab *Ocypode quadrata* related to large scale human modification of sandy shores. Front. Mar. Sci. (in press).

Belyea, L.R. and J. Lancaster. 1999. Assembly rules within a contingent ecology. Oikos. 86: 402–416.

Berkenbusch, K., A.A. Rowden and P.K. Probert. 2000. Temporal and spatial variation in macrofauna community composition imposed by ghost shrimp *Callianassa filholi* bioturbation. Mar. Ecol. Prog. Ser. 192: 249–257.

Bessa, F., D. Cunha, S.C. Gonçalves and J.C. Marques. 2013. Sandy beach macrofaunal assemblages as indicators of anthropogenic impacts on coastal dunes. Ecol. Indic. 30, 196–204.

Bessa, F., S.C. Gonçalves, J.N. Franco, J.N. André, P.P. Cunha and J.C. Marques 2014. Temporal changes in macrofauna as response indicator to potential human pressures on sandy beaches. Ecol. Indic. 41: 49–57.

Brazeiro, A. 1999. Patrones comunitarios en ensambles de macroinvertebrados intermareales de playas arenosas expuestas de Chile: exploración de procesos y mecanismos subyacentes. Ph.D. thesis, Pontificia Universidad Católica de Chile, Santiago, Chile.

Brazeiro, A. 2001. Relationship between species richness and morphodynamics in sandy beaches: What are the underlying factors? Mar. Ecol. Prog. Ser. 224: 35–44.

Carlton, J.T. and J. Hodder. 2003. Maritime mammals: Terrestrial mammals as consumers in marine intertidal communities. Mar. Ecol. Prog. Ser. 256: 271–286.

Celentano, E., D. Lercari, P. Maneiro, P. Rodríguez, I. Gianelli, L. Ortega, et al. 2019. The forgotten dimension in sandy beach ecology: Vertical distribution of the macrofauna and its environment. Estuar. Coast. Shelf Sci. 217: 165–172.

Checon, H.H., G.N. Corte, Y.M.L Shah-Esmaeili and A.C.Z. Amaral. 2018. Nestedness patterns and the role of morphodynamics and spatial distance on sandy beach fauna: ecological hypotheses and conservation strategies. Sci. Rep. 8(1): 3759.

Clarke, A. and K.J. Gaston. 2006. Climate, energy and diversity. Proc. R. Soc. B Biol. Sci. 273: 2257–2266.

Colombini, I. and L. Chelazzi. 2003. Influence of marine allochthonous input on sandy beach communities. Oceanogr. Mar. Biol. an Annu. Rev. 41: 115–159.

Corte, G.N., L.Q. Yokoyama and A.C.Z. Amaral. 2014. An attempt to extend the Habitat Harshness Hypothesis to tidal flats: A case study of *Anomalocardia brasiliana* (Bivalvia: Veneridae) reproductive biology. Estuar. Coast. Shelf Sci. 150, 136–141.

Costa, L.L. and I.R. Zalmon. 2017. Surf zone fish diet as an indicator of environmental and anthropogenic influences. J. Sea Res. 128: 61–75.

Costa, L.L., D.C. Tavares, M.C. Suciu, D.F. Rangel and I.R. Zalmon. 2017. Human-induced changes in the trophic functioning of sandy beaches. Ecol. Indic. 82: 304–315.

Costa, L.L. and I.R. Zalmon 2019. Sensitivity of macroinvertebrates to human impacts on sandy beaches: A case study with tiger beetles (Insecta, Cicindelidae). Estuar. Coast. Shelf Sci. 220: 142–151.

Costa, L.L., L. Fanini, I.R. Zalmon and O. Defeo. 2020. Macroinvertebrates as indicators of human disturbances: a global review. Ecol. Indic. 118: 106764.

Costa, L.L. and I.R. Zalmon. 2021. Macroinvertebrates as umbrella species on sandy beaches. Biol. Conserv. 253: 108922.

Dahl, E. 1952. Some aspects of the ecology and zonation of the fauna of sandy beaches. Oikos: 4: 1–27.

De La Huz, R., M. Lastra, J. Junoy, C. Castellanos and J.M. Viéitez. 2005. Biological impacts of oil pollution and cleaning in the intertidal zone of exposed sandy beaches: Preliminary study of the "prestige" oil spill. Estuar. Coast. Shelf Sci. 65: 19–29.

Defeo, O. and A. de Alava. 1995. Effects of human activities on long-term trends in sandy beach populations: the wedge clam *Donax hanleyanus* in Uruguay. Mar. Ecol. Prog. Ser. 123: 73–82.

Defeo, O., A. Brazeiro, A. de Alava and G. Riestra. 1997. Is sandy beach macrofauna only physically controlled? Role of substrate and competition in isopods. Estuar. Coast. Shelf Sci. 45: 453–462.

Defeo, O., J. Gomez and D. Lercari. 2001. Testing the swash exclusion hypothesis in sandy beach populations: The mole crab *Emerita brasiliensis* in Uruguay. Mar. Ecol. Prog. Ser. 212: 159–170.

Defeo, O., D. Lercari and G. Júlio. 2003. The role of morphodynamics in structuring sandy beach populations and communities: What should be expected? J. Coast. Res. 35: 352–362.

Defeo, O. and A. McLachlan. 2005. Patterns, processes and regulatory mechanisms in sandy beach macrofauna: A multi-scale analysis. Mar. Ecol. Prog. Ser. 295: 1–20.

Defeo, O., A. McLachlan, D.S. Schoeman, T.A. Schlacher, J. Dugan, A. Jones, et al. 2009. Threats to sandy beach ecosystems: A review. Estuar. Coast. Shelf Sci. 81: 1–12.

Defeo, O. and A. McLachlan. 2011. Coupling between macrofauna community structure and beach type: A deconstructive meta-analysis. Mar. Ecol. Prog. Ser. 433: 29–41.

Defeo, O. and A. McLachlan. 2013. Global patterns in sandy beach macrofauna: Species richness, abundance, biomass and body size. Geomorphology 199: 106–114.

Defeo, O., C.A.M. Barboza, F.R. Barboza, W.H. Aeberhard, T.M.B. Cabrini, R.S. Cardoso, et al. 2017. Aggregate patterns of macrofaunal diversity: An interocean comparison Glob. Ecol. Biogeogr. 26(7): 823–834.

Dugan, J.E., D.M. Hubbard, M.D. McCrary and M.O. Pierson. 2003. The response of macrofauna communities and shorebirds to macrophyte wrack subsidies on exposed sandy beaches of southern California. Estuar. Coast. Shelf Sci. 58: 25–40.

Dussault, A.C. 2020. Neither superorganisms nor mere species aggregates: Charles Elton's sociological analogies and his moderate holism about ecological communities. Hist. Philos. Life Sci. 42: 1–27.

Fanini, L. and J.K. Lowry. 2016. Comparing methods used in estimating biodiversity on sandy beaches: Pitfall vs. quadrat sampling. Ecol. Indic. 60: 358–366.

Fanini, L., C. Piscart, E. Pranzini, C. Kerbiriou, I. Le Viol and Pétillon, J. 2021. The extended concept of littoral active zone considering soft sediment shores as social-ecological systems, and an application to Brittany (North-Western France). Estuar. Coast. Shelf Sci. 250: 107148.

Gonçalves, S.C. and J.C Marques. 2011. The effects of season and wrack subsidy on the community functioning of exposed sandy beaches. Estuar. Coast. Shelf Sci. 95: 165–177.

Gül, M.R. and B.D. Griffen. 2018. Impacts of human disturbance on ghost crab burrow morphology and distribution on sandy shores. PLoS One. 13: e0209977.

Harris, L., R. Nel, M. Smale and D. Schoeman. 2011. Swashed away? Storm impacts on sandy beach macrofaunal communities. Estuar. Coast. Shelf Sci. 94: 210–221.

Herrmann, M., C.R.B. de Almeida, W.E. Arntz, J. Laudien and P.E. Penchaszadeh. 2010. Testing the habitat harshness hypothesis: Reproductive biology of the wedge clam *Donax hanleyanus* (Bivalvia: Donacidae) on three Argentinean sandy beaches with contrasting morphodynamics. J. Molluscan Stud. 76: 33–47.

Hubbard, D.M. and J.E. Dugan. 2003. Shorebird use of an exposed sandy beach in southern California. Estuar. Coast. Shelf Sci. 7714: 41–54.

Jaramillo, E., H. Contreras, C. Duarte and M.H. Avellanal. 2003. Locomotor activity and zonation of upper shore arthropods in a sandy beach of north central Chile. Estuar. Coast. Shelf Sci. 58: 177–197.

Junoy, J., C. Castellanos, J.M. Vie. 2005. The macroinfauna of the Galician sandy beaches (NW Spain) affected by the Prestige oil-spill. Mar. Pollut. Bull. 50: 526–536.

Lercari, D., L. Bergamino and O. Defeo. 2010. Trophic models in sandy beaches with contrasting morphodynamics: Comparing ecosystem structure and biomass flow. Ecol. Modell. 221: 2751–2759.

Levins, R. 1969. Some demographic and genetic consequences of environmental heterogeneity for biological control. Am. Entomol. 15: 237–240.

Lucrezi, S. 2015. Ghost crab populations respond to changing morphodynamic and habitat properties on sandy beaches. Acta Oecol. 62: 18–31.

Lunardi, V.O., R.H.F. Macedo, J.P. Granadeiro and J.M. Palmeirim. 2012. Migratory flows and foraging habitat selection by shorebirds along the northeastern coast of Brazil: The case of Baia de Todos os Santos. Estuar. Coast. Shelf Sci. 96: 179–187.

Machado, P.M., L.L. Costa, M.C. Suciu, D.C. Tavares, D.C and I.R. Zalmon. 2016. Extreme storm wave influence on sandy beach macrofauna with distinct human pressures. Mar. Pollut. Bull. 107: 125–135.

Manning, L.M. and N. Lindquist. 2003. Helpful habitant or pernicious passenger: Interactions between an infaunal bivalve, an epifaunal hydroid and three potential predators. Oecologia. 134: 415–422.

Maria, T.F., M. de Troch, J. Vanaverbeke, A.M. Esteves and A. Vanreusel. 2011. Use of benthic vs planktonic organic matter by sandy-beach organisms: A food tracing experiment with 13C labelled diatoms. J. Exp. Mar. Bio. Ecol. 407: 309–314.

Mcardle, S.B. and A. McLachlan. 1992. Sand beach ecology: swash features relevant to the macrofauna. J. Coast. Res. 8, 398–407.

McLachlan, A. 1980. Exposed Sandy Beaches as Semi-Closed Ecosystems. Mar. Environ. Res. 4: 59–63.

McLachlan, A., E. Jaramillo, T.E. Donn and F. Wessels. 1993. Sandy Beach macrofauna communities and their control by the physical environment: A Geographical Comparison. J. Coast. Res. 15: 27–38.

McLachlan, A. and E. Jaramillo. 1995. Zonation on sandy beaches. Oceanogr. Mar. Biol. Ann. Rev. 33: 305–335.

McLachlan, A. and A. Dorvlo. 2005. Global Patterns in sandy beach macrobenthic communities. J. Coast. Res. 214: 674–687.

McLachlan, A. and A.C. Brown. 2006. The Ecology of Sandy Shores, 2nd Ed. Academic Press, Burlington, Massachusetts.

Noy-Meir, I., 1979. Structure and function of desert ecosystems. Isr. J. Bot. 28: 1–19.

Pombo, M., A.L. Oliveira, L.Y. Xavier, E. Siegle and A. Turra. 2017. Natural drivers of distribution of ghost crabs *Ocypode quadrata* and the implications of estimates from burrows. Mar. Ecol. Prog. Ser. 565: 131–147.

Raffaelli, D. and S. Hawkins. 1996. Intertidal Ecology. Kluwer Academic Press, Dordrecht, Boston, London.

Reyes-Martínez, M.J., D. Lercari, M.C. Ruíz-delgado, J.E. Sánchez-Moyano, A. Jiménez-Rodríguez, A. Pérez-hurtado, et al. 2014. Human pressure on sandy beaches: Implications for trophic functioning. Estuaries Coasts. 38: 1782–1796.

Salvat, B. 1964. Les conditions hydrodynamiques interstitielles des sediments muebles intertidaux et la repartition vertical de la fauna endogee. C.R. Acad. Sci. Paris, 259: 1576–1579.

Schlacher, T.A. and L. Thompson. 2013. Spatial structure on ocean-exposed sandy beaches: Faunal zonation metrics and their variability. Mar. Ecol. Prog. Ser. 478: 43–55.

Schlacher, T.A., B.M. Hutton, B.L. Gilby, N. Porch, G.S. Maguire, B. Maslo, et al. 2017. Algal subsidies enhance invertebrate prey for threatened shorebirds: A novel conservation tool on ocean beaches? Estuar. Coast. Shelf Sci. 191: 28–38.

Schooler, N.K., J.E. Dugan, D.M. Hubbard and D. Straughan. 2017. Local scale processes drive long-term change in biodiversity of sandy beach ecosystems. Ecol. Evol. 7: 4822–4834.

Schooler, N.K., J.E. Dugan and D.M. Hubbard. 2019. No lines in the sand: Impacts of intense mechanized maintenance regimes on sandy beach ecosystems span the intertidal zone on urban coasts. Ecol. Indic. 106: 105457.

Takahashi, K., T. Hirose and K. Kawaguchi. 1999. The importance of intertidal Peracarid crustaceans as prey for fish in the surf-zone of a sandy beach in Otsuchi Bay, Northeastern Japan. Fish. Sci. 65: 856–864.

Van Tomme, J., S. Degraer and M. Vincx. 2014. Role of predation on sandy beaches: Predation pressure and prey selectivity estimated by laboratory experiments. J. Exp. Mar. Bio. Ecol. 451: 115–121.

Viola, S.M., D.M. Hubbard, J.E. Dugan and N.K. Schooler. 2014. Burrowing inhibition by fine textured beach fill: Implications for recovery of beach ecosystems. Estuar. Coast. Shelf Sci. 150: 142–148.

Witmer, A.D. and D.L. Roelke. 2014. Human interference prevents recovery of infaunal beach communities from hurricane disturbance. Ocean Coast. Manag. 87: 52–60.

Wolcott, T.G. 1978. Ecological role of ghost crabs, *Ocypode quadrata* (Fabricius) on an ocean sandy beach: Scavengers or predador? J. Exp. Mar. Bio. Ecol. 31: 67–82.

Chapter 2

The Biology and Ecology of Sandy Beach Surf Zones

Jose R. Marin Jarrin[1]*, Alan L. Shanks[2] and Jessica A. Miller[3]

[1]Department of Fisheries Biology, Telonicher Marine Lab, Humboldt State University, 1 Harpst St. Arcata, California, USA (jose.marinjarrin@humboldt.edu).

[2]Oregon Institute of Marine Biology, University of Oregon, 63466 Boat Basin Rd., Charleston, Oregon, USA (ashanks@uoregon.edu).

[3]Department of Fisheries and Wildlife, Coastal Oregon Marine Experiment Station, Hatfield Marine Science Center, Oregon State University, Newport, Oregon, USA (jessica.miller@oregonstate.edu).

INTRODUCTION

Surf zones encompass the region where waves break alongshore, from the area where the break originates offshore to the limit of wave impact on the beach. These zones range from truly reflective, vertical walls, to highly dissipative, very gradual bottom slope. Most surf zones are not one of these 'ideals' and when one describes a surf zone as reflective or dissipative one means this as shorthand for more reflective and more dissipative.

Curiously, the study of surf zones by physical oceanographers and biologists has focused on a limited range of surf zone types. Work has focused on more dissipative surf zones associated with sandy beaches. More reflective surf zones

*Corresponding author: jose.marinjarrin@humboldt.edu

associated with sandy beaches have received far less attention and all types of surf zones associated with rocky shores are essentially unstudied. One suspects that this focus on a small range of surf zone types is a consequence of history, the difficulty of working in reflective surf zones, and research trends within fields of research.

It appears that the study of surf zones by physical oceanographers has been affected by two historical trends. During World War II there were numerous amphibious landings. The first were seriously hampered by a lack of understanding of the morphology of surf zones. This led to the immediate spin up of a research program to study surf zones to support the war effort. Most amphibious landings appear to have taken place on dissipative beaches where bottom slopes are gradual, hence, the research tended to focus on these types of surf zones. The U.S. Office of Naval Research has continued this research bias. Engineering appears to one as the second motivator. Waves mold sandy shores by moving the sand. The moving sand can block harbor entrances and sand can be removed from a beach exposing the inhabited shoreline to erosion. In the U.S., the Army Corp of Engineers actively studies surf zones with what appears to be the primary goal of understanding sand transport. This work tends to be done in dissipative surf zones. For example, an Army Corp of Engineers station devoted to surf zone research is located at Duck, North Carolina where the surf zone is dissipative. On the west coast of North America, research groups focused on surf zone oceanography are located at Scripps Institution of Oceanography in San Diego and the Naval Postgraduate School in Monterey where the more actively studied beaches are dissipative.

Reflective surf zones are difficult, often dangerous places to do field work. The beach slope is steep and the sand is poorly sorted making for tenuous footing. Wading into the surf zone, one abruptly encounters the sharp drop off into the surf; one moves from shallow water to deep in a step or two. The surf zone is narrow often with waves breaking within meters of shore seemingly right in the face of the researcher. When the tide drops, the exposed intertidal zone is often only a few meters wide. The benthic community is invariably sparse as the sand is constantly in motion. Speaking from experience, these are difficult, at times dangerous, and often unpleasant places to work. It is very difficult to maintain oceanographic instrumentation within these surf zones and very difficult to sample the plankton within them. It is understandable that researchers, both physical oceanographers and biologists, have tended to focus their research on dissipative surf zones.

Intertidal ecologists working on rocky shores have largely ignored the surf zone. Surf zones are, of course, present at all rocky shores and, in fact, at high tide the surf zone covers the intertidal zone. Historically the primary interest of rocky shore ecologists with respect to surf zones is the effect of wave exposure (Denny 1988), how do the high current speeds and turbulence of breaking waves affect the survival of organisms in the intertidal zone. This is clearly a perfectly reasonable research focus, if limited. The surf zone is also the body of water separating the shore from the inner shelf and coastal ocean. Subsidies to a rocky intertidal zone, phytoplankton food and larval settlers, must cross from the coastal ocean through the surf zone before they reach the rocky shore (Shanks et al. 2010, Salant and Shanks 2018). The larvae of many species released in the intertidal zone go through their development in the coastal ocean so they must cross the surf zone before they get out to sea and to recruit back into the intertidal zone they must again cross the surf zone. Recent research indicates that the hydrodynamics of the surf zone, whether it is dissipative or reflective, profoundly affects the delivery of subsidies to rocky intertidal zones and sandy shores as well (Shanks et al. 2010, Morgan et al. 2017, Shanks et al. 2017a).

A key paradigm within the community of sandy beach ecologists is that surf zones are semi-closed systems (McLachlan 1980); the waters within the surf zone are more or less isolated from the coastal ocean thus allowing for the development of unique communities of organisms within the surf zone. One of the co-authors of this chapter (ALS) has been pursuing a research program investigating the effect of surf zone hydrodynamics, dissipative to reflective surf zones, on the delivery of subsidies to the intertidal zone of both rocky and sandy shores. What one has found is that, at least as far as water is concerned, surf zones are entirely open systems; coastal waters move through both reflective and dissipative surf zones (Brown et al. 2015, Shanks et al. 2015b). The residence time of water within the surf zones that one has studied is short, hours at the most. The pelagic community within surf zones is, however, different in important ways from the pelagic community on the inner shelf just seaward of the surf zone be it reflective or dissipative. The biological differences are, however, not due to unique populations of organisms developing within an isolated body of water, but rather appear to be due to the interaction of the 'behavior' of the organisms with the hydrodynamics of the surf zone.

PHYSICAL, PHYTO- AND ZOOPLANKTON OBSERVATIONS OF REFLECTIVE AND DISSIPATIVE SURF ZONES

Barnacle populations on rocky shores and rocks within sandy beaches vary tremendously. Densities can be very low or the rocks can be blanketed with a continuous sheet of densely packed individuals with new recruits settled on their shells. Shanks et al. (2010) tested the hypothesis that these differences in populations varied with the types of surf zone, reflective to dissipative, associated with the sample site. Intertidal rocks within sandy beaches were sampled. Populations of barnacles and juvenile limpets were one or more orders of magnitude higher on rocks in beaches where the surf zone was dissipative than on rocks in beaches where the surf zone was reflective. The results of this experiment led to a new hypothesis; dissipative surf zones are open systems allowing barnacle recruits, cyprids, to be delivered to the rocks whereas reflective surf zones are semi-closed systems with the surf zone acting as a barrier to the shoreward migration of larvae. This is a modification of the surf zone ecologist paradigm that surf zones are semi-closed systems.

To test this hypothesis we extensively sampled a reflective and dissipative surf zone associated with sandy beaches. The research was aimed at explaining the distribution of rocky shore taxa, but we sampled sandy beach surf zones for logistical reasons; surf zone physical oceanographers are adept at placing and maintaining instruments in sandy beach environments, but have little experience working on rocky shores. Because one of the primary drivers of the type of surf zone hydrodynamics at a site is the slope of the bottom, not the composition of the bottom, the hydrodynamics and associated biology of the pelagic community within the surf zones at these sandy shores should be generally applicable.

We sampled two surf zones, a steep more reflective surf zone and a flatter more dissipative surf zone (Fig. 2.1). At each site, the physical oceanographers set out an extensive array of instruments, deployed surface drifters and made a number of dye releases within the surf zones. The observations extended for about a month at each site. Concurrent with the oceanographic sampling, biologists sampled the phyto-zooplankton within, and just seaward of the surf zones. If the new hypothesis was correct, then we expected to see slow exchange of water within the reflective surf zone with that offshore coupled with a different community of phyto- and zooplankton within the surf zone than offshore while the pelagic community at the dissipative surf zone would be similar to that seaward.

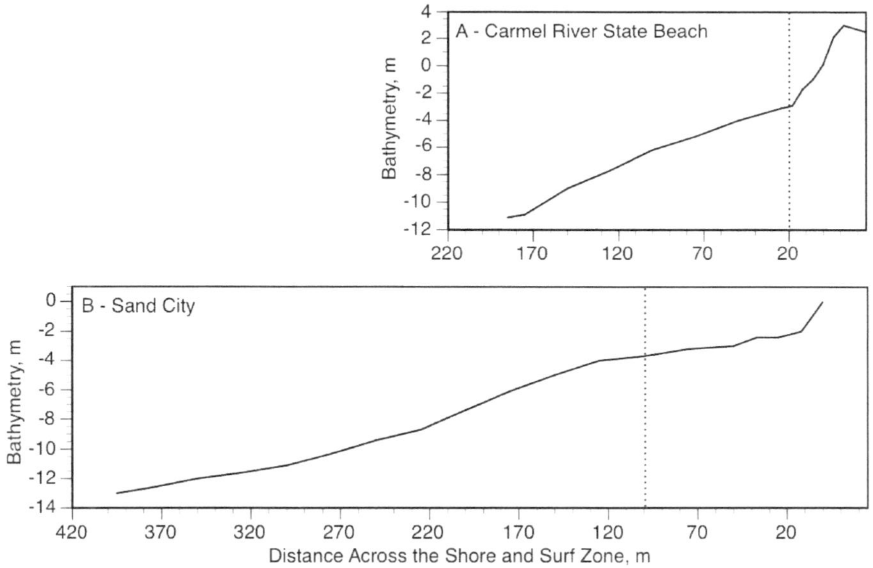

Figure 2.1 Bottom slope at the more reflective (A) and dissipative (B) surf zones. The vertical dotted lines indicate the location of the breaker line.

Reflective Surf Zone

The reflective surf zone sampled, Carmel River State Beach in California (USA) (36.53789°N; 121.928886°W, CRSB) is associated with a crescent shaped beach (Fig. 2.2A). Detailed descriptions of the results from this study can be found in Shanks et al. (2015a) and a summary is presented here. The beach slope was quite steep (Fig. 2.1), the surf zone was narrow ranging in width from a few meters to maybe 10 m during larger wave events, the sand was coarse, and at low tide the intertidal zone was several meters wide.

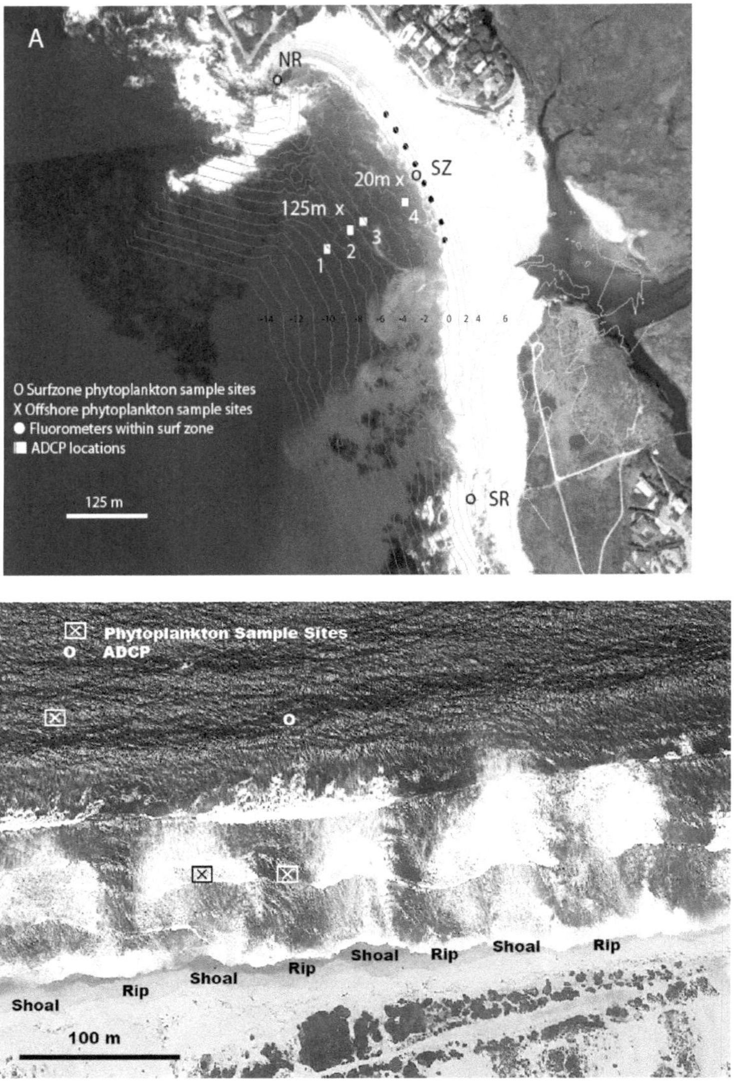

Figure 2.2 Aerial images of Carmel River State Beach (A) and Sandy City (B), California. (A) Phytoplankton was sampled in the surf zone at the rocky intertidal to the north (NR) and south (SR) of the beach and in the beach surf zone (SZ). Zooplankton were also sampled with the surf zone at SZ. Offshore phyto- and zooplankton samples were collected 125 and 20 m from the shore. During dye releases, the movement of the dye within the surf zone was followed with fluorometers set in the surf zone (dots). Acoustic Doppler Current Profiler were positioned across the inner shelf (white squares). Contour lines are meters. Image modified from a Google Earth photograph. (B) The surf zone is characterized by rip currents with deep channels separated by shallow shoals. Rip channels were spaced ~125 m apart. Phyto- and zooplankton samples were collected in the surf zone at the bathymetric rip current and shoals labeled with an X. Offshore phytoplankton samples were collected just outside the surf zone at the Offshore Site also indicated with an X. Image modified from a Google Earth photograph.

There were two sources of water entering the surf zone. (1) There was an alongshore current within the surf zone with water entering the current at the north end of the beach near where one could sample phytoplankton within the surf zone (NR, Fig. 2.2A). This site was just beyond the placement of the oceanographic instrumentation. The alongshore current exited at the south end of the beach (cloudy water in Fig. 2.2A). (2) Water entered the surf zone at the surface with the turbulence of breaking waves and exited the surf zone via the alongshore current and everywhere along the surf zone as an undertow. Figure 2.3A presents a cross section of the surf zone highlighting the cross-surf zone flow. Onshore flow into the surf zone occurred as a thin layer at the breaker line and throughout the remainder of the water column flow was offshore, the undertow. Seaward of the surf zone there was onshore flow at the bottom caused by benthic streaming (discussed below). Dye released within the surf zone was transported alongshore and rapidly offshore by the undertow; within several tens of minutes, the dense dye release was no longer visible. The reflective surf zone was not a semi-closed system rather the water within the surf zone was rapidly and continuously exchanged with that seaward.

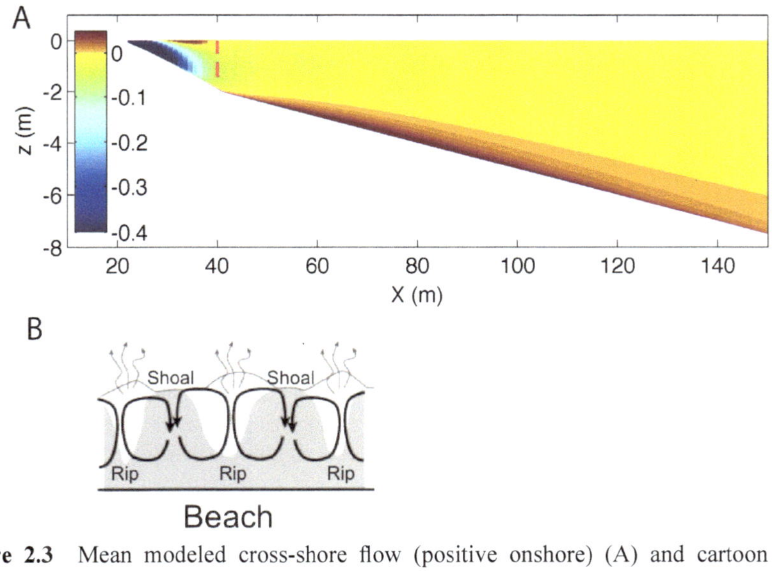

Figure 2.3 Mean modeled cross-shore flow (positive onshore) (A) and cartoon of rip current eddy flow (B) at a beach similar to CRSB using a one-dimensional profile model with Stokes drift (Reniers et al. 2004). The cross-shore flow is driven by normally incident waves only. The vertical red dashed line indicates the other edge of the surf zone where waves start breaking. The undertow generated by the breaking waves within the surf zone is clearly visible (blue cloud). The near-bed streaming (orange layer at the bottom) extends across the model domain. The streaming layer is thicker offshore and thins as the outer edge of the surf zone is approached. The flow within the surf zone is pulsed, onshore with the passage of each breaking wave with offshore undertow flow between waves.

Given that the surf zone was open with relatively rapid exchange of water one expected surf zone populations of phyto- and zooplankton to be identical to that seaward; however this was not the case. Concentrations of phyto- and zooplankton were one or more orders of magnitude lower than seaward (Shanks et al. 2016, Morgan et al. 2017, Shanks et al. 2017b). Phytoplankton concentrations just 20 m seaward of the breaker line were far higher (Fig. 2.4A) than within any of the three surf zone sample sites (i.e., SZ, NR, and SR, Fig. 2.4A).

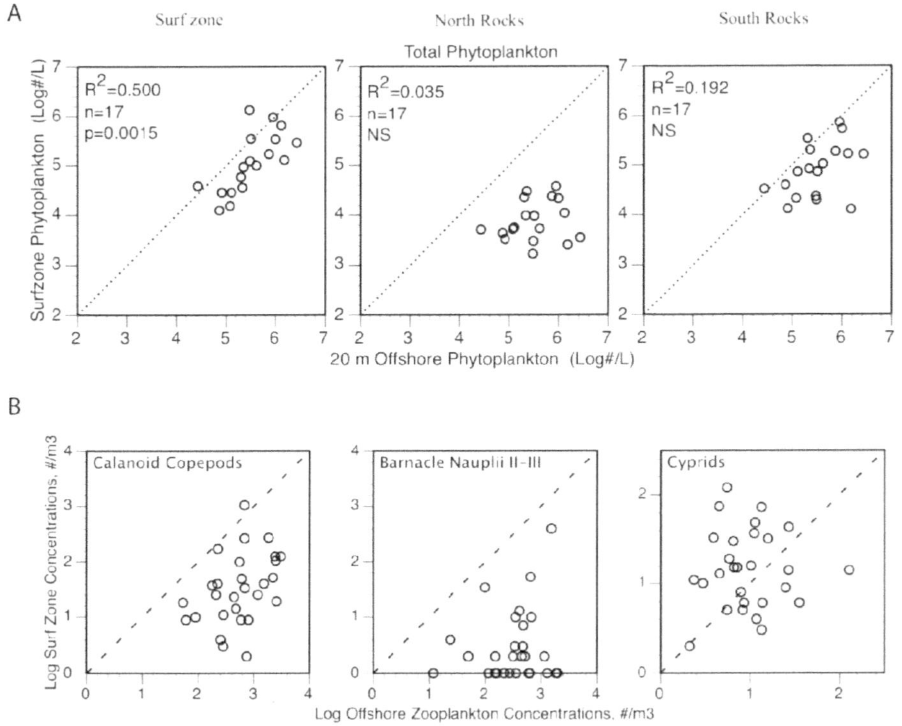

Figure 2.4 Log_{10} concentrations of phytoplankton (A) and zooplankton (B). Phytoplankton was sampled 20 m offshore plotted with concentrations with the beach Surf Zone (SZ) and the surf zone at the north (NR) and south (SR) rocky shore sample sites. The dotted line indicates a one-to-one relationship between the concentrations, which is what one expected given how rapidly the water within the surf zone was exchanged with that offshore. Note that no surf zone diatom taxa were present in the samples. Zooplankters sampled 125 m from shore and with the beach Surf Zone (SZ). Calanoid copepods are representative holoplankters, Barnacle nauplii II-III are pre-competent meroplankters, and barnacle cyprids are competent meroplankters. The dashed line indicates a one-to-one relationship between the concentrations, which is what one expected given how rapidly the water within the surf zone was exchanged with that offshore. Note that the concentrations of the representative holoplankter and pre-competent meroplankter were much lower in the surf zone than offshore while cyprid concentrations were at times higher than offshore.

We were unable to collect zooplankton samples so close to the shore (e.g., within 20 m), but concentrations of zooplankton collected 125 m offshore were also for most taxa far higher than within the surf zone (Fig. 2.4B). Concentrations of holoplankters and pre-competent meroplankters were nearly always one or more orders of magnitude lower within the surf zone than in the offshore samples (Shanks et al. 2016, Morgan et al. 2017, Shanks et al. 2017b). Detritus and competent meroplankters were at times more concentrated in the surf zone than offshore (Shanks et al. 2015a). Higher concentrations tended to occur during periods of smaller waves (Shanks et al. 2015a). A possible explanation for this latter observation is discussed below in the section on benthic streaming.

The water in the surf zone was readily exchanged yet the concentrations of passive to nearly passive particles, i.e., phyto- and zooplankton, were far lower within the surf zone than just seaward. How could this be? The only explanation is that the concentrations of plankton within the waters entering the surf zone, i.e., the surface water pushed into the surf zone by breaking waves (Fig. 2.3A), were very low. We were only able to partially test this hypothesis. Vertical profiles of Chl *a* were made using a CTD a few meters seaward of the breaker line. Consistent with the hypothesis, Chl *a* concentrations within the upper meter of the water column were only 20% of that within the remainder of the water column (Shanks et al. 2017c). We were unable to collect similar data for zooplankton, however, there are a number of experimental studies that have demonstrated that zooplankton will drop, swim or sink, downward upon encountering strong turbulence (Fuchs and Gregory 2016), which is what they would encounter as they approach the breaker line. By dropping down into the water column they would enter the undertow and be pushed away from the surf zone.

In summary, water within this reflective surf zone was readily exchanged with that offshore, the surf zone was not a semi-closed system, but was open. Despite being an open system, the phyto- and zooplankton concentrations within the surf zone were for all phytoplankton and most zooplankton taxa far lower than offshore. The low concentrations appear to be the result of an interaction between the phyto- and zooplankton, their 'behavior', and the surf zone hydrodynamics; they appear to have been avoiding the turbulent surface waters entering the surf zone.

Dissipative Surf Zone

The second surf zone was located at Sand City, Monterey, California, USA (36.615760°N 121.85485°W, Fig. 2.2B). This site is close to the U.S. Naval Postgraduate School, which has supported physical oceanographic studies of this surf zone as well as others for decades (MacMahan et al. 2010, MacMahan et al. 2006). The surf zone and adjacent inner shelf were extensively instrumented and a number of surface drifter and dye studies were carried out. Results from this physical oceanographic study were published by Brown et al. (2015). Daily biological sampling of phyto- and zooplankton within the surf zone and on the inner shelf just seaward of the breaker line took place concurrent with the physical oceanographic work. The biological results can be found in Morgan et al. (2017) and Shanks et al. (2017b). Here the results, conclusions and hypotheses from these studies are summarized.

The slope of the beach and adjacent surf zone is fairly gradual (Fig. 2.1). Waves tended to approach the surf zone at right angles setting up regular rip current flow systems (Fig. 2.2B). Deeper channels were situated under the rip currents with much shallower shoals between the rips. Water exits the surf zone in the rip, waves push water into the surf zone over the shoals, this water flows alongshore into the rip channel and the process is repeated generating an eddy to either side of the rip channel (Fig. 2.3B). Eddies were situated roughly in the middle of the surf zone and between the rip current channel and the shoal. Dye released over the shoals acted like the water, it was carried alongshore into the rip current and then rapidly seaward beyond the breaker line. The residence time of the dye, and presumably the water, was brief on the order of a few tens of minutes (Brown et al. 2015). Talbot and Bate (1987a, b) working in South Africa observed similar residence times of water within the dissipative surf zone they studied. GPS surface drifters released with the dye acted quite differently. Like the dye, the drifters moved alongshore and into the adjacent rip currents where they were carried through the breaker line. Here their behavior changed. The dye spread out and mixed into the water column on the inner shelf, but the drifters moved alongshore a short distance and were then pushed back into the surf zone over the shoals by the waves entering the surf zone from the inner shelf. They were again transported alongshore, into the rip and the process repeated itself. Ultimately, the surface drifters were carried into rip current eddies where they became trapped.

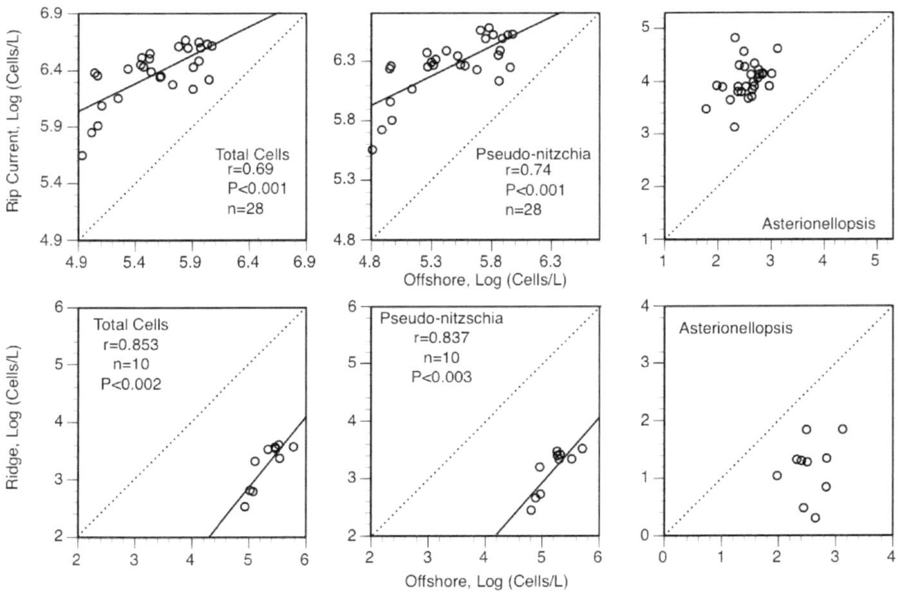

Figure 2.5 Log_{10} of the concentrations of phytoplankton collected offshore plotted against the log of the concentrations in the rip current (upper set of graphs) and over the shoal adjacent to the rip current (lower set of graphs). The dotted lines indicate one-to-one relationships between the variables. The solid lines and statistical results indicate significant correlations between the variables.

The water within the surf zone was not isolated or even semi-isolated from the adjacent shelf water. Water instead passed rapidly through the surf zone and was replaced by water from the inner shelf pushed into the surf zone by the waves. If, however, one looked just at the surface drifters, their behavior gives the impression that the water is retained, but what is actually happening is the rip current eddy system flushed water through the surf zone generating eddies which have the capacity to capture floating objects. Water and 'things' imbedded in the water (e.g., dye) pass through the surf zone while 'floating things', 'things' that do not follow the water either because they are buoyant or swim upward, can be retained by the flow system.

Phyto- and zooplankton were sampled daily at high tide within the rip current, in the water over the shoal adjacent to the sampled rip current, and just seaward of the breaker line (Shanks et al. 2017b). Phytoplankton concentrations within the rip currents were far higher than in the waters seaward of the surf zone while over the shoals the reverse was true (Fig. 2.5). Most of the phytoplankton within the surf zone were typical coastal taxa. *Asterionellopsis,* a surf zone diatom specialist, was present, but accounted for only a small percentage of the phytoplankton present in the surf zone, most were typical coastal phytoplankton such as *Pseudo-nitzchia.* Like phytoplankton, zooplankton concentrations were higher in the rip current samples and lower in the shoal than offshore (Fig. 2.6) (Morgan et al. 2017) although the differences were not as large as in the case of phytoplankton.

Diatom taxa that are surf zone specialists can maintain dense populations within the dissipative surf zone despite the rapid exchange of surf zone water with that offshore (Talbot et al. 1990). Surf zone diatoms produce mucus to which bubbles adhere. The diatoms trapped in foam float at the surface and, like the surface drifters released in the Sand City surf zone, they became trapped in the rip current eddy. Surf zone diatom taxa were not present at CRSB but were in the Sand City surf zone; however, they made up only a small portion of the surf zone phytoplankton community, nearly all of which were typical coastal diatom taxa. Bubbles rising through a water column can capture particles, which become trapped in the surface film of the bubble (Schlichting 1972). On a number of occasions, we sampled foam and, in all cases, coastal diatom taxa were present in the foam. At times the concentration within the foam was similar to the average concentration in the underlying water, in a number of other samples their concentration was orders of magnitude higher (Shanks et al. 2017b). We hypothesize that coastal diatom taxa can become trapped in naturally occurring foam and, like typical surf zone diatom specialists, they become concentrated in the surf zone eddy. In this way, high concentrations within the surf zone can build up over time.

The concentration of phytoplankton was far lower in the samples from over the shoals. We are less certain how this is happening. One possibility is that due to the trapping of diatoms in surf zone foam and their subsequent build up in the rip current eddy there is a redistribution of the phytoplankton into the eddies. The other possibility is that, as at the reflective surf zone at CRSB, the water entering the surf zone over the shoals is surface water pushed in by the breaking waves and these surface waters contain low concentrations of phytoplankton. These hypotheses, which are not mutually exclusive, were not generated until after the fieldwork was complete so we did not collect data to test them.

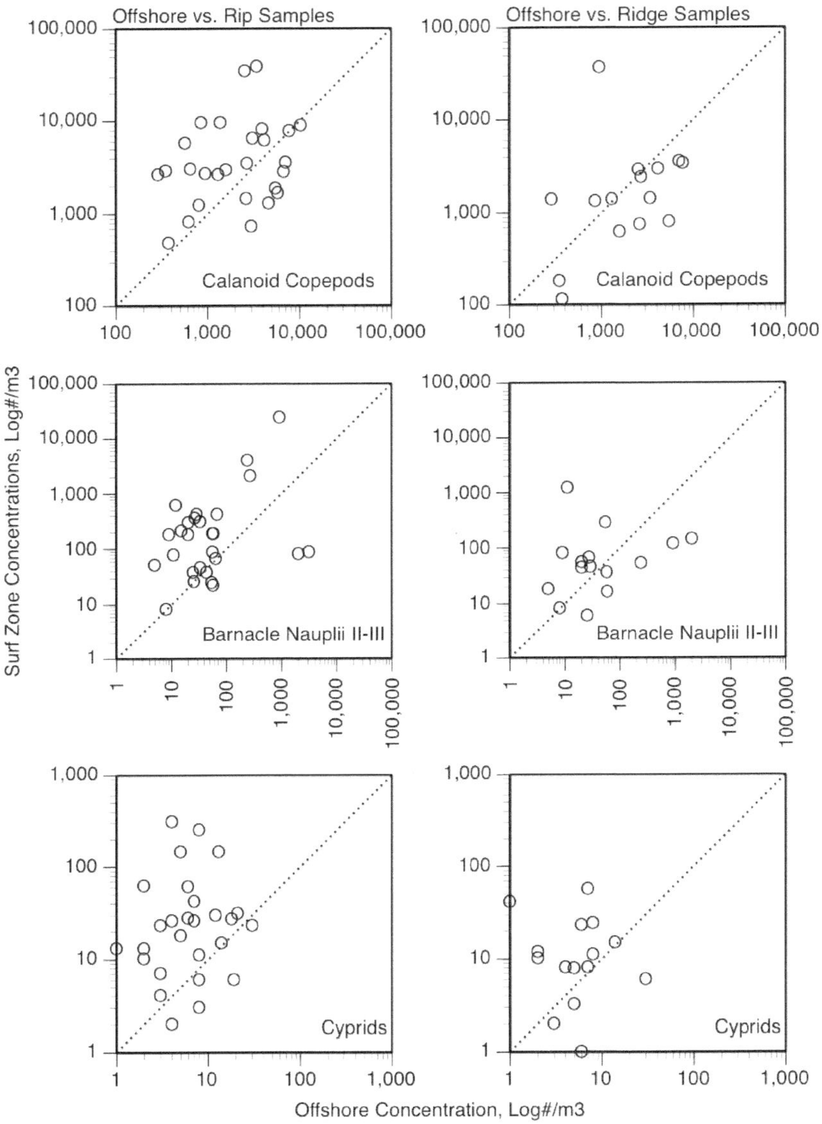

Figure 2.6 Log_{10} concentrations of representative zooplankton in the nearshore samples plotted with concentrations in the rip samples (left hand figures) and the samples from over the shoals (right hand figures). Calanoid copepods are representative holoplankters, Barnacle nauplii II-III are pre-competent meroplankters, and barnacle cyprids are competent meroplankters. The dotted line indicates a one-to-one relationship between the concentrations.

The intensive sampling took place at one reflective and one dissipative surf zone; we are clearly pseudo replicating for surf zone type. It is equally clear that repeating the month-long intensive sampling at multiple reflective and dissipative shores was beyond the capacity of the grant. To address the pseudo-replication problem we made extensive but limited sampling of a number of surf zones.

Concentrations of zooplankton within reflective surf zones were generally far lower than in the water just offshore while the reverse was true at more dissipative surf zones. In our initial investigation of the effect of surf zone hydrodynamics on the delivery of subsidies to the shore we measured the abundance of new barnacle recruits on rocks within beaches at reflective and dissipative surf zones; new recruit density was orders of magnitude higher at dissipative sites than reflective (Shanks et al. 2010). We largely expanded on this sampling. At 40 sites from San Diego, California to the Olympic Peninsula, Washington, a roughly 2,000 km transect, we sampled barnacle populations on rocks within beaches and rocky shores (Shanks et al. 2017a). We measured daily settlement and weekly recruitment at a subset of these sites. Where surf zones were reflective, barnacle population density, recruit density, daily settlement and weekly recruitment were all far lower than at dissipative sites; the results completely consistent with our previous work.

If our observations of phytoplankton within a reflective and dissipative surf zone are generally true then one should find low concentrations of phytoplankton within reflective surf zones and high in dissipative. To attempt to control for alongshore differences in the concentration of phytoplankton in the coastal ocean, we sampled closely spaced sites (8 km total spacing) around Cape Arago, Oregon. The results were completely consistent with the intensive sampling (Shanks et al. 2017c). The experimental observations were run a second time with the same results (Salant and Shanks 2018). In the second experiment, the effect of the variable phytoplankton concentrations on the reproductive output of mussels and barnacles was investigated. At dissipative surf zones, phytoplankton concentrations were far higher and this translated into a far higher reproductive output by both intertidal filter feeders. The conclusions from these studies are completely consistent with the results from our intensive sampling suggesting that, while the intensive sampling was pseudo replicated, the conclusions are general.

BENTHIC STREAMING

At the reflective beach we sampled intensively (CRSB), concentrations of zooplankton within the surf zone were on most days much lower than just offshore. However, on some days, the reverse was true for some types of zooplankton and detritus (mostly pieces of benthic macro algae). We think that shoreward transport by benthic streaming may account for these observations (Shanks et al. 2015a). Benthic streaming, at least the biological consequences of streaming, is a poorly studied phenomenon. Oceanographic studies are uncommon and are primarily models; no physical measurements of streaming currents have been made. Hence, what follows is a bit speculative and is included here in the hopes of motivating research.

As waves approach the shore they begin to 'feel' the bottom at half the wave length of the wave, that is the orbital motion generated by the waves begins to interact with the bottom (Sverdrup et al. 1942). In deep water, the motion beneath waves is orbital with the diameter of the orbits decreasing with depth. The water moves in circles with a slight movement in the direction of wave propagation, the

Stokes drift. When the orbital motion is constrained by the bottom, the circular orbits become elliptical and, in shallow enough water, the orbits become the back and forth motion of the surge. Like the Stokes drift, the back and forth motion is not equal, the shoreward motion is longer than the seaward. This generates a shoreward current on the bottom, which can be seen in Fig. 2.3. This is benthic streaming or bed load transport. The shoreward current should be set up as waves begin to feel the bottom. Under a wave with a 100 m wavelength, benthic streaming may commence at water depths around 50 m. Depending on the bottom slope, benthic streaming could commence 100s of m to km offshore.

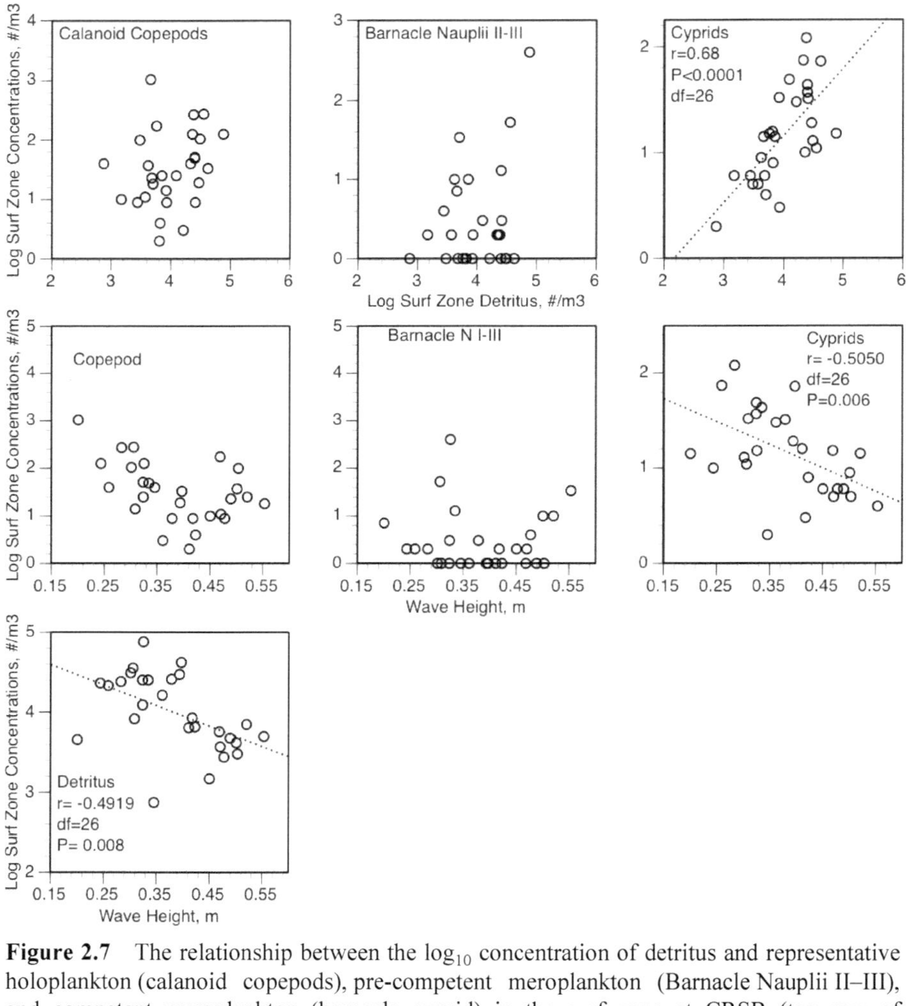

Figure 2.7 The relationship between the $\log_{10}$ concentration of detritus and representative holoplankton (calanoid copepods), pre-competent meroplankton (Barnacle Nauplii II–III), and competent meroplankton (barnacle cyprid) in the surf zone at CRSB (top row of figures). Wave height plotted with the $\log_{10}$ surf zone concentrations of detritus (bottom figure), holoplankton (calanoid copepods), pre-competent meroplankton (Barnacle Nauplii II–III), and competent meroplankton (barnacle cyprid). The dotted line and statistics are from correlation analyses.

At CRSB we observed that on days with small waves detritus, organisms associated with detritus (isopods and amphipods), and competent stage invertebrate larvae were more highly concentrated within the surf zone than offshore (Fig. 2.7) (Shanks et al. 2015a). We hypothesized that they were entering the surf zone via benthic streaming. All were either small weak swimming plankton or near neutrally buoyant material (the detritus). When waves were larger, benthic streaming was present but turbulence was high enough that these elements could not remain within the benthic boundary layer and the layer of benthic streaming. When waves were small, benthic streaming was still occurring, it likely started closer to shore, but the turbulence was low enough that these elements could accumulate in the benthic boundary layer and be carried shoreward by benthic streaming.

Often in the late summer and fall at the Sand City beach there is an influx of *Ulva* detritus. Sand City has a more dissipative surf zone. Piles of *Ulva* accumulate on the beach and there is an abundance in the surf zone. These accumulation events appear to only occur when waves are small. This has not been studied so what follows is speculation, again, added in the hopes of motivating research. In late summer/fall, *Ulva* is senescing and shedding blades; they are producing detritus. When waves are small and turbulence is low, the blades sink into the benthic boundary layer where they can be transported shoreward by benthic streaming. Ultimately, the blades can be transported into the surf zone. This is the same hypothesis as we used to explain the accumulation of detritus and some zooplankters in the CRSB surf zone during small wave events.

In Chile, researchers observed that juvenile mussel recruitment peaks during periods of larger waves, which they also attribute to benthic streaming (Navarrete et al. 2015). In this system, mussels settle and metamorphose offshore and then make a secondary migration as juveniles into the intertidal zone. At this stage in their development, the mussels are non-swimming and they grow a sturdy shell, they are 'heavy' objects. Chilean researchers hypothesize that the shoreward transport is occurring via benthic streaming or bed load transport 'bouncing' the juvenile mussels across the bottom, what geologists would describe as saltation. Another possibility is that the mussels use byssus thread drifting to exploit the shoreward transport of the streaming.

As benthic streaming is set up simply from the interaction of wave orbital motion with the bottom, it should be present on all shores and have the potential to transport material into both reflective and dissipative surf zones; it should be a ubiquitous process with the potential to transport detritus and organisms from 100s of m to km offshore to the shore. The study of the biological consequences of this phenomenon is in its infancy.

DEMERSAL ZOOPLANKTON AND FISHES IN SURF ZONES

Surf zones are also inhabited by demersal zooplankton that are often referred to as suprabenthos, hyperbenthos, epibenthic crustaceans, small swimming surf zone fauna or benthic invertebrates (Mees and Jones 1997, Munilla et al. 1998, McLachlan and Brown 2006, Marin Jarrin and Shanks 2011). This group is primarily composed of peracarid crustaceans such as mysids and amphipods (1–5 mm in length), which spend part of their lives close to or on the seafloor (McLachlan and Brown 2006). Fishes also inhabit surf zones, particularly during their early life stages, such as flatfishes, surfperches, smelts and other forage fish, sculpins, drums or croakers, mojarras, sharks and rays (Ross 1983, Clark et al. 1996, Borland et al. 2017, Olds et al. 2017). Most studies that have sampled these faunas have used zooplankton nets, hyperbenthic sledges, epibenthic trawls, beach seines and, more recently, Baited Remote Underwater Video (BRUV), that were deployed by hand, jet ski, zodiac, and scuba divers (Munilla et al. 1998, Strydom 2007, Marin Jarrin and Miller 2016, Borland et al. 2017). Many authors from around the world including Australia, South Africa, Japan, Belgium, England, Ireland, Spain, Ecuador, Brazil and the US, have studied these faunas mostly at flat dissipative surf zones but also in steep beaches with more reflective surf zones (Mees and Jones 1997, Marin Jarrin et al. 2015, Olds et al. 2017).

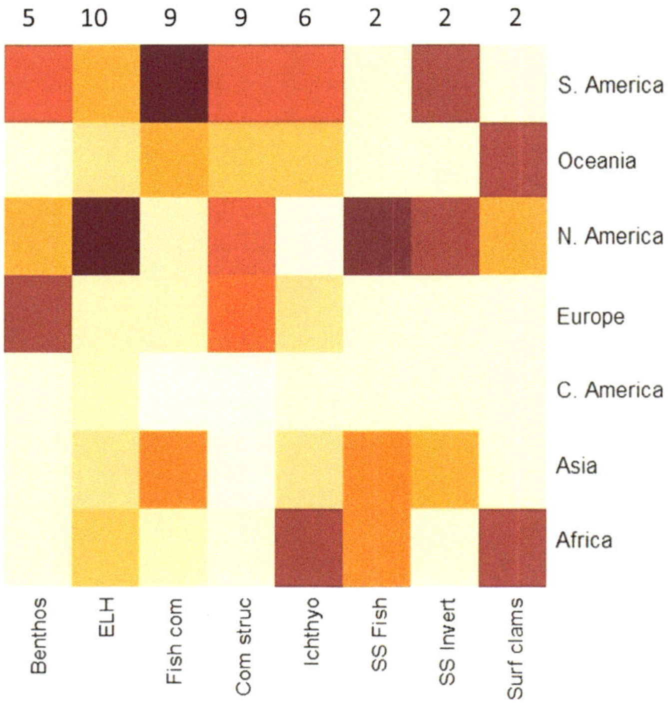

Figure 2.8 Heat map for 124 studies on surf zone fauna. The region of the study is noted along with the dominant faunal group examined. 'Benthos' = all types of benthos; 'ELH' = early life history and recruitment studies; 'Fish comm' = fish community structure; 'Comm struc' = community structure, invertebrates or fish and invertebrates; 'Ichthyo' = ichthyoplankton; 'SS Fish' = single species fish studies; 'SS invert' = single species invertebrate studies; and 'Surf clams'. Color intensity is scaled by columns (faunal group studied).

A review of publications on surf zone fauna identified 124 studies focused on single species or communities (Fig. 2.8). While studies were completed around the globe, most studies originated in South America (n = 31) whereas most studies focused on early life history and recruitment occurred in North America (n = 10). The most common research focus was description or comparison of community structure (n = 55), followed by studies on ichthyoplankton, early life history and recruitment and benthic organisms.

Demersal Zooplankton and Fishes in Reflective Surf Zones

Very little biological work has been conducted in reflective beaches due to safety issues and difficulties deploying field gear. However, the few studies that have been conducted have found similar taxonomic groups but significantly lower abundances, species richness and diversity to those present in nearby dissipative beaches (Munilla et al. 1998, Nakane et al. 2013, Oliveira and Pessanha 2014). At the same time the lower concentrations of zooplankton, these patterns appear to be due to the hydrodynamic conditions and large sand grain size that reduce the amount of particulate organic matter/detritus in these beaches. Organic matter is an important diet article for harpacticoid copepods, and both organic matter and harpacticoids are eaten by peracarid crustaceans, that are in turn the main prey item of many surf fishes (Takahashi and Kawaguchi 1998, Yu et al. 2003, Vilas et al. 2008, Nakane et al. 2011, Marin Jarrin and Miller 2013, Nakane et al. 2013). Supporting this hypothesis and similar to what has been observed for smaller zooplankton, higher abundances of peracarids have been observed seaward (5–10 m in depth) than inside the surf zone (1 m) at a beach in Cataluña, Spain (Munilla et al. 1998, San Vicente and Munilla 2000). Lower abundances of fishes do not seem to be due to predation by larger piscivorous fishes as a tethering study in Japan found similar number of predators and mortality rates in reflective beaches when compared to adjacent intermediate and dissipative beaches (Nakane et al. 2009). Another study in the same Cataluña area also found that the size of fish was larger in the reflective than in the dissipative surf zone suggesting these steep beaches may not be as important for juveniles (Nakane et al. 2013).

Demersal Zooplankton in Dissipative Surf Zones

Demersal zooplankton are highly abundant in flat sandy beach surf zones, where they can average 10 to >10,000 individuals per 100 m^{-2} (Dominguez-Granda et al. 2004, Marin Jarrin et al. 2015). The fauna is mostly composed of peracarid crustaceans, but can also include decapods and pycnogonids (Dominguez-Granda et al. 2004). Within these surf zones, demersal zooplankton often vertically migrate from the sediment, or right above it, during the day to the water column at night to feed, mate, molt and carry out other activities (Colman and Segrove 1955, Tully and Ceidigh 1987, De Ruyck et al. 1991, Marin Jarrin and Shanks 2011). Similarly, some mysid species will horizontally migrate shoreward into the surf zone at night to feed on the abundant phytoplankton in the surf zone (Webb et al. 1987, Webb and Wooldridge 1990), potentially in the rip currents where plankton tends to concentrate in higher abundances (Shanks et al. 2017b). This migratory behavior

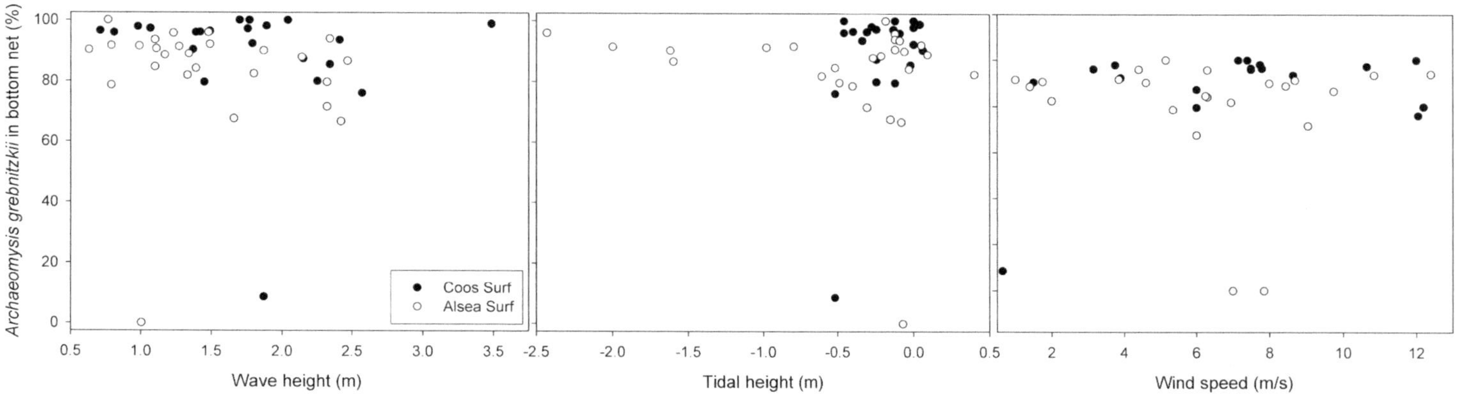

Figure 2.9 Percentage of *Archaeomysis grebnitzkii* present in the bottom half of the water column in relation to wave and tidal heights, and wind speeds (Marin Jarrin and Miller, unpublished data).

in an exposed intermediate to dissipative surf zone in Oregon, USA, where mean tidal variation and wave height during the summer is 2 months and 1–2 months, respectively were investigated by us. Organisms were sampled with a hyperbenthic sledge during the day and night. Significantly higher densities at night of the mysid *Archaeomysis grebnitzkii*, the most abundant species, and the whole community when compared to the day (Mean ± S.D.: 14520 ± 4687 and 15255 ± 4297 vs. 127 ± 127 and 351 ± 228 individuals 100 m^{-2}) were found. Our results confirmed that even in these dynamic surf zones, demersal zooplankters vertically migrate daily (Marin Jarrin and Shanks 2011).

Furthermore, follow up sampling during the summers of 2007–2010 during the day-time lower low tide at Coos and Alsea Surf, Oregon, USA, found that this fauna maintained their position lower in the water column regardless of the tidal heights, wind speed and wave height they encountered (Marin Jarrin and Miller, unpublished data, Fig. 2.9). For this study, samples were taken at 0.5 m depth with a hyperbenthic sledge that had two nets (mouth: 0.25 m high × 0.70 m wide), one on top of the other, to sample the whole water column.

Demersal zooplankton have also been found to vary at other temporal scales. Some burrowing species complete tidal migrations; individuals migrate upwards into the water column at high tide to mate (Hager and Croker 1980) or downwards at low tide to maintain a position in the intertidal/subtidal boundary at its preferred sand grain size. Non-burrowing species may also migrate horizontally with tides in order to stay within the surf zone (Lock et al. 1999, Beyst et al. 2002b).

Lunar cycles also appear to affect some species of demersal zooplankters, exhibiting a higher abundance and species richness during spring tides. The adults of several Excirolanid isopods can be more active during spring tides, mating during the higher spring high tides (De Ruyck et al. 1991). These isopods will remain 'stranded' , burrowed in sediment during the smaller neap tides, waiting for the higher spring high tides to reach the height at which they are located (Jones and Hobbins 1985). Besides larger tidal variation, spring tides also exhibit larger waves that appear to push, potentially via benthic streaming, a different community of larger zooplankton and macrophytes with clinging invertebrates into the surf zone (Marin Jarrin and Shanks 2011).

Seasonally, the demersal zooplankton community varies with higher abundances, species richness and diversity during the spring/summer in temperate regions and the warmer/wet season in the tropics due to higher primary productivity produced by higher water temperatures, nutrient availability and longer day length (Beyst et al. 2001, Ruiz 2002, Bernardo-Madrid et al. 2013, Marin Jarrin et al. 2017). These differences are also due to immigration of certain species during the productive seasons to forage in the surf zone and, in temperate regions, some strong swimming species, such as mysids, migrating offshore during the winter when larger waves make inhabiting the surf zone more stressful (Beyst et al. 2001). Interestingly, seasonal variation also influences the zonation of multiple species of amphipods that move up and down the shore to feed on fish eggs. During the summer, these species will inhabit different tidal zones to reduce inter-specific competition for prey (DeBlois and Leggett 1993, Yu et al. 2002).

Very little work has been done on inter-annual variability (Calles et al. 2002). We sampled the two Oregon surf zones mentioned above during the summers of 2007–2010 and 2008–2010, respectively, and found significant variability among years that appears to have been related to oceanographic conditions in the spring. Specifically, the date of physical spring transition or when conditions switch from mostly downwelling to upwelling, and upwelling intensity during the spring appear to influence the development of this community during the productive summer season (Fig. 2.10).

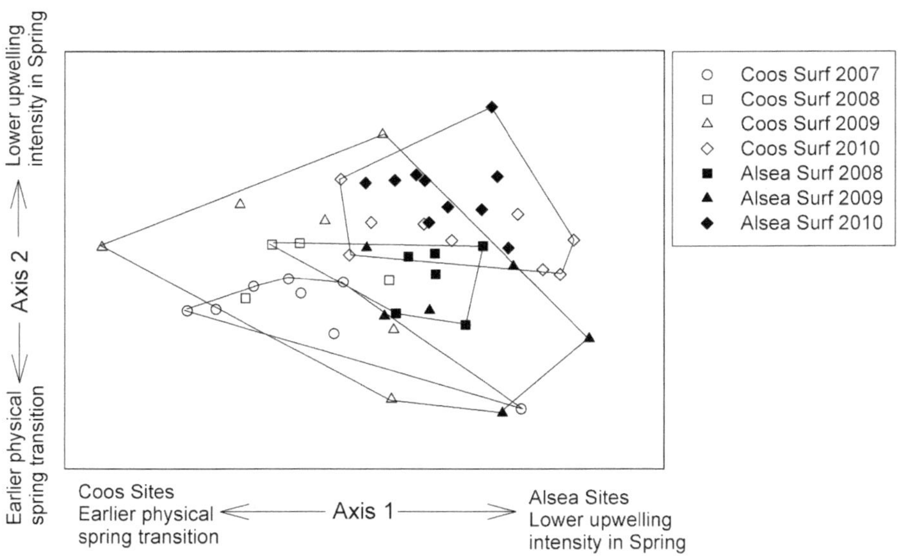

Figure 2.10 Non-Metric Multidimensional Scaling (nMDS) plot of demersal zooplankton community collected with a hyperbenthic sledge in the surf zone of Coos and Alsea Surf during the day-time lower low tide in summer of 2007–2010. Samples show separation among surf zones and years, which were correlated to differences in the day of physical spring transition and upwelling intensity during the spring season. nMDS was conducted following the slow and thorough mode in PCOrd. Prior to analysis, data were $\log_{10}$-transformed with a generalized procedure and similarity matrix created with the Bray-Curtis index. Stress = 12.6, $p < 0.004$, instability < 0.0001 (Marin Jarrin and Miller, unpublished data).

Demersal zooplankton can also vary spatially within and among surf zones. Within a site, the community may vary due to differences in exposure, slope, depth and sand grain size due to their influence on biological zonation on sandy beaches (San Vicente and Sorbe 1999, Dominguez-Granda et al. 2004). There is also some evidence that demersal zooplankton communities can be influenced by troughs or runnels, which develop on many beaches around the world. These physical features are formed during smaller wave seasons, such as the summer, when sand is accreted offshore and transported onshore in the form of sand bars. As bars approach the shore line they produce troughs that exhibit different sand grain size, wave conditions and circulation, and can be inhabited by different demersal zooplankton communities than the surrounding areas within the surf zone (Short and Wright 1983, Komar 1998, Alexander and Holman 2004, Ruggiero et al. 2005, Marin Jarrin and Miller 2016). Among surf zones, invertebrate species present lower in the water column can also vary due to adjacent habitats because of the movement of species from and to other environments such as estuaries, rocky intertidal and the nearshore ocean (Tully and Ceidigh 1987, Webb et al. 1987, Webb and Wooldridge 1990, Marin Jarrin et al. 2017) (Fig. 2.10).

Fishes in Dissipative Surf Zones

Fishes (and ichthyoplankton) are probably the best studied group within dissipative surf zones and were the focus of nearly 50% of the papers included in our chapter (Fig. 2.8) (Ross 1983, Allen and Pondella II 2006, McLachlan and Brown 2006, Olds et al. 2017). Olds et al. (2017) found >150 studies from around the world on fish in surf zones, most of which were descriptive. These authors found that surf zones are inhabited by a diverse fish fauna (171 families use surf zones) that tends to be dominated by few species (~10 species make up ~95% of individuals), particularly during their early life stages, and is highly variable in space and time. The most common families in surf zones are Engraulidae (forage fish), Syngnathidae (pipefish and seahorses), Osmeridae (smelt), Embiotocidae (surfperch), Paralichthydae and Pleuronectidae (flatfishes). Faunal variability appears to be largely driven by prey availability, water conditions and presence of macrophytes that can provide cover and clinging prey (Lasiak and McLachlan 1987, Clark et al. 1996, Beyst et al. 2002a, Marin Jarrin and Shanks 2011). Prey availability and appropriate water conditions are largely driven by seasonality, with many young fish migrating into surf zones to feed during the summer. These migrations can be from adjacent habitats such as rocky intertidal, estuaries and nearshore waters (Oliveira and Pessanha 2014, Borland et al. 2017, Marin Jarrin et al. 2017, Olds et al. 2017). Migration of many species occurs during the cover of night but some move during the day to reduce inter-specific competition (Beyst et al. 2002b, de Araujo Silva et al. 2004, Marin Jarrin and Shanks 2011, Rishworth et al. 2015).

In addition to providing foraging habitat for adults, surf zones also serve as nurseries. Due to their shallowness and soft bottoms, surf zones can provide a warm water habitat where a large number of individuals can forage on the abundant small, size appropriate prey field and grow at rates similar to or higher than in other habitats where the species occurs (Suda et al. 2002, Gillanders et al. 2003, Watt-Pringle and Strydom 2003, Marin Jarrin and Miller 2013, Rishworth et al. 2015). Surf zones are also considered a refuge from predation for juveniles due to the high and low occurrence of young and piscivorous fish, respectively (i.e., 'shallow water refuge hypothesis') (Suda et al. 2002, Inoue et al. 2008, Ryer et al. 2010, Marin Jarrin and Miller 2013). However, several tethering experiments and studies using Baited Remote Underwater Videos found predation in some surf zones is higher than expected (~5 – 20% daily survival rate), and suggest some earlier studies may have been biased by seining selectivity towards smaller/ younger fish (Nakane et al. 2009, Vargas-Fonseca et al. 2016, Olds et al. 2017).

CLIMATE CHANGE

Several studies have explored the potential impacts of climate change on sandy beaches and their surf zones worldwide (Jones et al. 2007, Schlacher et al. 2008, Defeo et al. 2009, Ruggiero et al. 2010, Oliver et al. 2018). These authors have predicted that increasing water temperatures, wave height, sea-level rise, ocean acidification, more frequent and stronger marine heat waves and ENSO events and, in general, changes in ocean conditions, will have direct and indirect effects on surf zone fauna by impacting the sandy beach physical environment and allowing the immigration of non-native species.

In Oregon, when juvenile, subyearling Chinook salmon (*Onchorhynchus tshawytscha*) first enter the marine environment, they inhabit estuaries, their main nursery, and, in lower numbers, surf zones adjacent to estuaries, their alternative nursery habitat (Marin Jarrin et al. 2009, Marin Jarrin 2012, Marin Jarrin and Miller 2013). Due to the high and variable rates of mortality, this early marine residence is a critical period for salmonids (Healey 1991, Pearcy 1992, Quinn 2005). To study the impact of increasing coastal water temperatures, juvenile Chinook salmon summer growth rates in Oregon were modeled by us, and it was determined if there was a scenario that may force these juveniles to leave estuaries early, and potentially use surf zones, at a higher rate. The Wisconsin bioenergetics model and its Chinook salmon parameters to estimated current and future (2050) summer growth rates (Stewart and Ibarra 1991, Hanson 1997). The model was parameterized with *in-situ* data on diet composition, prey energetic content (Joules g^{-1}), juvenile mass (g) and growth rates (mm d^{-1}), and modeled consumption rates (g d^{-1}). Water temperature (°C) was measured *in-situ* at Coos and Alsea Bay and Coos and Alsea Surf in 2010 and modeled to estimate 2008 and 2050 values. Temperature models were developed using fresh, estuarine, surf zone and Sea Surface Temperature (SST), estuarine tidal range and coastal wave height. Then, fresh water temperature (+1.5 and +5.8°C) was modified (Mantua et al. 2009, Mote and Salathe 2009), SST (+1.2, 0 and –1.2°C) (Mote and Mantua 2002, Di Lorenzo et al. 2005, García-Reyes and Largier 2010), and wave height (+0.75 m) (Ruggiero et al. 2010) to estimate future temperatures under six scenarios.

The model suggested that using current consumption rates and prey energetic density, increases in temperature (mean ± SD: 2.3 ± 1.4 and 2.9 ± 1.6°C, respectively, Fig. 2.11) led to decreased growth rates in estuaries and surf zones (3.0 ± 2.8 and 5.3 ± 4.1%, respectively, Fig 2.12). To compensate for these increases in estuarine water temperature, juveniles could increase their consumption rates or prey energy density (0.14 g g^{-1} day^{-1} and 5270 Joules g^{-1}) by 3.5 ± 2.8 and 4.1 ± 3.2%, respectively. These increases could be possible as these were below the maximum values observed (88% of maximum consumption and 5360 Joules g^{-1}). However, even if juveniles increased their consumption rates and prey energetic content, estuarine growth rates would be below observed values on an average of 3 to 9% of summer days, depending on the scenario, suggesting more salmonids may use surf zones in the near future. This modeling exercise highlights how habitat characteristics will change under future climate change, likely influencing habitat selection and migration patterns.

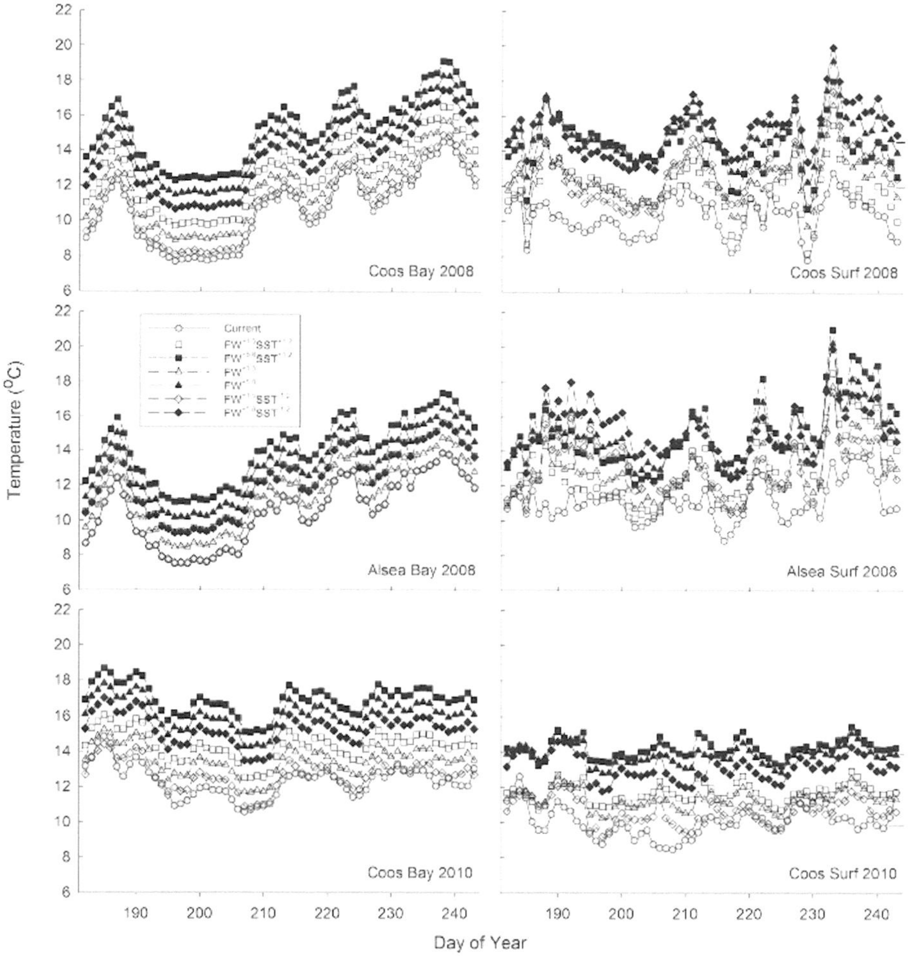

Figure 2.11 Current and future (2050) summer water temperatures (°C) in two estuaries and two surf zones on the Oregon coast. Temperature in future scenarios was estimated using previously developed habitat temperature models (Marin Jarrin 2012) and modifying Freshwater temperature (FW, +1.5 and +5.8°C), Sea Surface Temperature (SST, +1.2, 0 and −1.2°C) and Wave Height (WH +0.75 m).

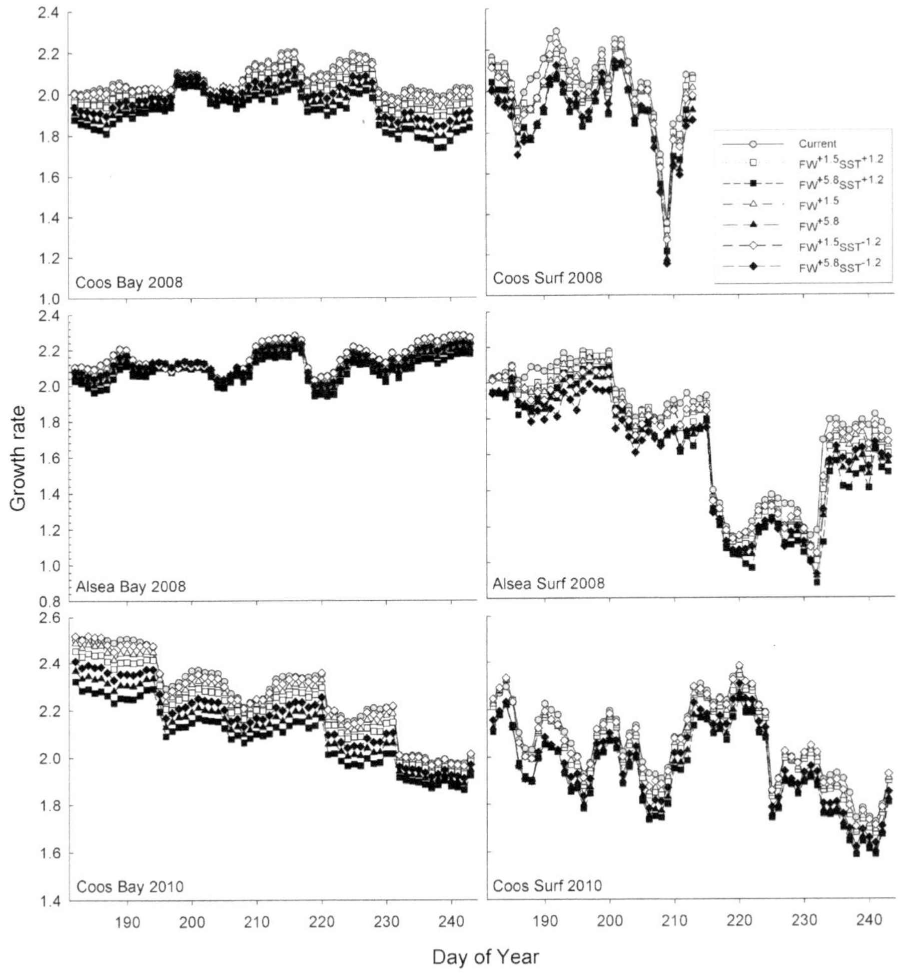

Figure 2.12 Modeled current and future (+50 years) growth rates (% body weight day^{-1}) of juvenile Chinook salmon at two estuaries and two surf zones on the Oregon coast using current and predicted future water temperatures under six scenarios (Fig. 2.11).

Conclusion

Sandy beach surf zone organisms are poorly studied but diverse and can be residents or immigrant species from the open ocean and adjacent coastal habitats such as estuaries and rocky reefs. These biological communities include phytoplankton, zooplankton, demersal zooplankton and fish. These communities vary with the beach type (dissipative–reflective), among beaches from different latitudes or oceanographic regions, and within a beach due to local hydrodynamic factors (differences in beach exposure, current velocity, beach slope) and seasonal sand movement (sand bars and troughs). Temporal variation can also occur on multiple scales, from the duration of a tidal cycle to years in length. Surf zones are impacted by climate change, with increasing water temperatures, wave height, sea-level rise, ocean acidification, and more frequent and stronger marine heat waves and ENSO events having a particular strong impact on the habitat as a whole.

REFERENCES

Alexander, P.S. and R.A. Holman. 2004. Quantification of nearshore morphology based on video imaging. Mar. Geol. 208: 101–111.

Allen, L.G. and D.J. Pondella II. 2006. Surf zone, coastal pelagic zone and harbors. pp. 149–166. *In:* M.H. Horn, L.G. Allen and R. Lea [eds]. The Ecology of Marine Fishes, California and Adjacent Waters. University of California Press, Berkeley and Los Angeles, California, USA.

Bernardo-Madrid, R., J.M. Martínez-Vázquez, J.M. Viéitez and J. Junoy. 2013. Two year study of swash zone suprabenthos of two Galician beaches (NW Spain). Main results from the XVII Iberian Symposium of Marine Biology Studies 83(0): 152–162.

Beyst, B., D. Buysse, A. Dewicke and J. Mees. 2001. Surf zone Hyperbenthos of Belgian sandy beaches: Seasonal patterns. Estuar. Coast. Shelf Sci. 53: 877–895.

Beyst, B., K. Hostens and J. Mees. 2002a. Factors influencing the spatial variation in fish and macrocrustacean communities in the surf zone of sandy beaches in Belgium. J. Mari. Biol. Ass. U.K. 82: 181–187.

Beyst, B., J. Vanaverbeke, M. Vincx and J. Mees. 2002b. Tidal and diurnal periodicity in macrocrustaceans and demersal fish of an exposed sandy beach, with special emphasis on juvenile plaice *Pleuronectes platessa*. Mar. Ecol. Prog. Ser. 225: 263–274.

Borland, H., T. Schlacher, B. Gilby, R. Connolly, N. Yabsley and A. Olds. 2017. Habitat type and beach exposure shape fish assemblages in the surf zones of ocean beaches. Mar. Ecol. Prog. Ser. 570: 203–211.

Brown, J.A., J.H. MacMahan, A.J.H.M. Reniers and E.B. Thornton. 2015. Field observations of surf zone-inner shelf exchange on a rip-channeled beach. J. Phys. Oceanogr. 45: 2339–2355.

Calles, A.K., L. Domínguez, S. Guartatanga, V. Ruiz, K. González, C.R. de Grunauer, et al. 2002. Interannual variability of the Meiobenthos and Hyperbenthos communities from two Ecuadorian sandy beaches (1999–2001). Investig. Mar. 30(1): 135–137.

Clark, B.M., B.A. Bennett and S.J. Lamberth. 1996. Factors affecting spatial variability in seine net catches of fish in the surf zone of False Bay, South Africa. Mar. Ecol. Prog. Ser. 131: 17–34.

Colman, J.S. and F. Segrove. 1955. The tidal plankton over Stoupe Beck Sands, Robin Hood's Bay (Yorkshire, North Riding). J. Anim. Ecol. 24(2): 445–462.

de Araujo Silva, M., F.G. Araujo, M.C. Costa de Azevedo and J. Neto de Sousa Santos. 2004. The nursery function of sandy beaches in a Brazilian tropical bay for 0-group anchovies (Teleostei: Engraulidae): diel, seasonal and spatial patterns. J. Mar. Biol. Ass. U.K. 84: 1229–1232.

De Ruyck, A.M.C., A. McLachlan and T.E. Donn. 1991. The activity of three intertidal sand beach isopods (Flabellifera: Cirolanidae). J. Exp. Mar. Biol. Ecol. 146(2): 163–180.

DeBlois, E.M. and W.C. Leggett. 1993. Importance of biotic and abiotic regulators of abundance of the intertidal amphipod *Calliopius laeviusculus*. Mar. Biol. 115: 75–83.

Defeo, O., A. McLachlan, D.S. Schoeman, T.A. Schlacher, J. Dugan, A. Jones, et al. 2009. Threats to sandy beach ecosystems: A review. Estuar. Coast. Shelf Sci. 81: 1–12.

Denny, M.W. 1988. Biology and Mechanics of the Wave-Swept Environment. Princeton University Press, Princeton.

Di Lorenzo, E., A.J. Miller, N. Schneider and J.C. McWilliams. 2005. The warming of the California Current System: Dynamics and ecosystem implications. J. Phys. Oceanogr. 35: 336–362.

Dominguez-Granda, L., N. Fockedey, M. De Mey, B. Beyst, M. del Pilar Cornejo, J. Calderon, et al. 2004. Spatial patterns of the surf zone hyperbenthic fauna of Valdivia Bay (Ecuador). Hydrobiologia 529(1–3): 205–224.

Fuchs, H.L. and P. Gregory. 2016. Seascape-level variation in turbulence- and wave-generated hydrodynamic signals experienced by plankton. Prog. Oceanogr. 141: 109–129.

García-Reyes, M. and J. Largier. 2010. Observations of increased wind-driven coastal upwelling off central California. J. Geophys. Res. 115: C04011.

Gillanders, B.M., K.W. Able, J.A. Brown, D.B. Eggleston and P.F. Sheridan. 2003. Evidence of connectivity between juvenile and adult habitats for mobile marine fauna: an important component of nurseries. Mar. Ecol. Prog. Ser. 247: 281–295.

Hager, R.P. and R.A. Croker. 1980. The sand-burrowing amphipod *Amphiporeia virginiana* Shoemaker 1933 in the tidal plankton. Can. J. Zool. 58: 860–864.

Hanson, P.C. 1997. Fish Bioenergetics 3.0. Board of Regents, University of Wisconsin System.

Healey, H.C. 1991. Life history of Chinook salmon (*Oncorhynchus tshawytscha*). pp. 311–395. *In*: C. Groot and L. Margolis [eds]. Pacific Salmon Life Histories. UBC Press, Vancouver, Canada.

Inoue, T., Y. Suda and M. Sano. 2008. Surf zone fishes in an exposed sandy beach at Sanrimatsubara, Japan: Does fish assemblage structure differ among microhabitats? Estuar. Coast. Shelf Sci. 77(1): 1–11.

Jones, A., W. Gladstone and N. Hacking. 2007. Australian sandy-beach ecosystems and climate change: ecology and management. Aust. Zool. 34(2): 190–202.

Jones, D.A. and C.S.C. Hobbins. 1985. The role of biological rhythms in some sand beach cirolanid Isopoda. J. Exp. Mar. Biol. Ecol. 93(1): 47–59.

Komar, P.D. 1998. The Pacific Northwest Coast: Living with the Shores of Oregon and Washington. Duke University Press Books, Durham.

Lasiak, T.A. and A. McLachlan. 1987. Opportunistic utilization of mysid shoals by surf-zone teleosts. Mar. Ecol. Prog. Ser. 37(1): 7.

Lock, K., B. Beyst and J. Mees. 1999. Circadiel patterns in the tidal plankton of a sandy beach in Zeebrugge (Belgium). Belg. J. Zool. 129(2): 339–352.

MacMahan, J.H., E.B. Thornton and A.J.H.M Reniers. 2006. Rip current review. Coast. Eng. 53: 191–208.

MacMahan, J.H., J.W. Brown, J.A. Brown, E.B. Thornton, A.J.H.M. Reniers, T.P. Stanton, et al. 2010. Mean Lagrangian flow behavior on an open coast rip-channeled beach: A new perspective. Mar. Geol. 268: 1–15.

Mantua, N., I. Tohver and A. Hamlet. 2009. Impacts of climate change on key aspects of freshwater salmon habitat in Washington State. Chapter 6 *in* The Washington Climate Change Impacts Assessment: Evaluating Washington's Future in a Changing Climate, Climate Impacts Group, University of Washington, Seattle, Washington.

Marin Jarrin, J.R., A.L. Shanks and M.A. Banks. 2009. Confirmation of the presence and useof sandy beach surf-zones by juvenile Chinook salmon. Environ. Biol. Fishes 85(2): 119–125.

Marin Jarrin, J.R. and A.L. Shanks. 2011. Spatio-temporal dynamics of the surf-zone faunal assemblages at a Southern Oregon sandy beach. Mar. Ecol. 32: 232–242.

Marin Jarrin, J.R. 2012. Sandy Beach Surf Zones: What is their Role in the Early Life History of Juvenile Chinook Salmon (*Oncorhynchus tshawytscha*)? Ph.D. Dissertation, Oregon State University, Corvallis, Oregon, USA.

Marin Jarrin, J.R. and J.A. Miller. 2013. Sandy beach surf zones: An alternative nursery habitat for 0-age Chinook salmon. Estuar. Coast. Shelf Sci. 135: 220–230.

Marin Jarrin, J.R., S.L.M. Quezada, L.E. Dominguez-Granda, S.M.G. Argudo and M. del P.C.R. de Grunauer. 2015. Spatio-temporal variability of the surf-zone fauna of two Ecuadorian sandy beaches. Mar. Freshw. Res. 67(5): 566–577.

Marin Jarrin, J.R. and J.A. Miller. 2016. Spatial variability of the surf zone fish and macroinvertebrate community within dissipative sandy beaches in Oregon, USA. Mar. Ecol. 37(5): 1027–1035.

Marin Jarrin, J.R., J. Vanaverbeke, N. Fockedey, M. del P.C.R. de Grunauer and L. Dominguez-Granda. 2017. Surf zone fauna of Ecuadorian sandy beaches: spatial and temporal patterns. J. Sea Res. 120: 41–49.

McLachlan, A. 1980. Exposed sandy beaches as semi-closed ecosystems. Mar. Environ. Res. 4, 59–63.

McLachlan, A., and A. C. Brown. 2006. The Ecology of Sandy Shores. 2nd Ed. Elsevier Academic Press, Burlington, MA.

Mees, J., and M. B. Jones. 1997. The Hyperbenthos. Oceanography and Marine Biology: An Annual Review 35: 221–255.

Morgan, S.G., A.L. Shanks, A.G. Fujimura, A.J.H.M. Reniers, J. MacMahan, C.D. Griesemer, et al. 2017. Surf zones regulate larval supply and zooplankton subsidies to nearshore communities. Limnol. Oceanogr. 62: 2811–2828.

Mote, P.W. and N.J. Mantua. 2002. Coastal upwelling in a warmer future. Geophys. Res. Lett. 29(23): 2138.

Mote, P. and E. Salathe. 2009. Future Climate in the Pacific Northwest. Chapter 1 *in* The Washington Climate Change Impacts Assessment: Evaluating Washington's Future in a Changing Climate, Climate Impacts Group, University of Washington, Seattle, Washington.

Munilla, T., M.J. Corrales and C. San Vicente. 1998. Suprabenthic assemblages from Catalan beaches: zoological groups. Orsis 13: 67–78.

Nakane, Y., Y. Suda and Y. Hayakawa. 2009. Predation pressure for a juvenile fish on an exposed sandy beach 3 comparison among beach types using tethering. La mer 46: 109–115.

Nakane, Y., Y. Suda and M. Sano. 2011. Food habits of fishes on an exposed sandy beach at Fukiagehama, South-West Kyushu Island, Japan. Helgol. Mar. Res. 65(2): 123.

Nakane, Y., Y. Suda and M. Sano. 2013. Responses of fish assemblage structures to sandy beach types in Kyushu Island, southern Japan. Mar. Biol. 160(7): 1563–1581.

Navarrete, S.A., J.L. Largier, G. Vera, F.J. Tapia, M. Parrague, E. Ramos, et al. 2015. Tumbling under the surf: wave-modulated settlement of intertidal mussels and the continuous settlement-relocation model. Mar. Ecol. Prog. Ser. 520: 101–102.

Olds, A.D., E. Vargas-Fonseca, R.M. Connolly, B.L. Gilby, C.M. Huijbers, G.A. Hyndes, et al. 2017. The ecology of fish in the surf zones of ocean beaches: A global review. Fish Fish. 19(1): 78–89.

Oliveira, R.E.M.C.C. and A.L.M. Pessanha. 2014. Fish assemblages along a morphodynamic continuum on three tropical beaches. Neotrop. Ichthyol. 12(1): 165–175.

Oliver, E.C.J., M.G. Donat, M.T. Burrows, P.J. Moore, D.A. Smale, L.V. Alexander, et al. 2018. Longer and more frequent marine heatwaves over the past century. Nat. Commun. 9(1): 1–12.

Pearcy, W.G. 1992. Ocean Ecology of North Pacific Salmonids. Washington Sea Grant Program, Seattle, WA.

Quinn, T.P. 2005. The Behavior and Ecology of Pacific Salmon and Trout. University of Washington Press, Bethesda, MD.

Reniers, A.J.H.M., J.A. Roelvink and E.B. Thornton. 2004. Morphodynamic modeling of an embayed beach under wave group forcing. J. Geophys. Res. 109: C01030.

Rishworth, G.M., N.A. Strydom and W.M. Potts. 2015. The nursery role of a sheltered surf-zone in warm temperate southern Africa. African Zoology 50(1): 11–16.

Ross, S.T. 1983. A review of surf zone Icthyofaunas in the Gulf of Mexico. Proceedings of the Northern Gulf of Mexico Estuaries and Barrier Islands Research Conference: 25–34.

Ruggiero, P., G.M. Kaminsky, G. Gelfenbaum and B. Voigt. 2005. Seasonal to interannual morphodynamics along a high-energy dissipative littoral cell. J. Coast. Res. 21(3): 553–578.

Ruggiero, P., P.D. Komar and J.C. Allan. 2010. Increasing wave heights and extreme value projections: The wave climate of the U.S. Pacific Northwest. Coastal Engineering 57: 539–552.

Ruiz, V. 2002. Estudio preliminar de la variacion anual del Hiperbentos intermareal de una playa arenosa de la provincia del Guayas (CENAIM-San Pedro). B.Sc. Thesis, Escuela Superior Politécnica del Litoral, Guayaquil, Ecuador.

Ryer, C.H., B.J. Laurel and A.W. Stoner. 2010. Testing the shallow water refuge hypothesis in flatfish nurseries. Mar. Ecol. Prog. Ser. 415: 275–282.

Salant, C.D. and A.L. Shanks. 2018. Surfzone hydrodynamics alter phytoplankton subsidies affecting reproductive output and growth of *Mytilus californianus* and *Balanus glandula*. Ecology 99: 1878–1889.

San Vicente, C. and J.C. Sorbe. 1999. Spatio-temporal structure of the suprabenthic community from Creixell beach (western Mediterranean). Acta Oecol. 20(4): 377–389.

San Vicente, C. and T. Munilla. 2000. Misidáceos suprabentónicos de las playas catalanas (Mediterráneo nordoccidental). Orsis 15: 45–55.

Schlacher, T.A., D.S. Schoeman, J. Dugan, M. Lastra, A. Jones, F. Scapini, et al. 2008. Sandy beach ecosystems: key features, sampling issues, management challenges and climate change impacts. Mar. Ecol. 29: 70–90.

Schlichting, H.E.J. 1972. Seafoam, algae and protozoa. J. Elisha Mitchell Sci. Soc. J. 88: 186–187.

Shanks, A.L., S.G. Morgan, J. MacMahan and A.J.H.M. Reniers. 2010. Surf zone physical and morphological regime as determinants of temporal and spatial variation in larval recruitment. J. Exp. Mar. Biol. Ecol. 392: 140–150.

Shanks, A.L., J. MacMahan, S.G. Morgan, A.J.H.M. Reiniers, M. Jarvis, J. Brown, et al. 2015a. Transport of larvae and detritus across the surf zone of a steep reflective pocket beach. Mar. Ecol. Prog. Ser. 528: 71–86.

Shanks, A.L., S.G. Morgan, J. MacMahan, A.J.H.M. Reniers, M. Jarvis, J. Brown, et al. 2015b. Transport of larvae and detritus across the surf zone of a steep reflective pocket beach. Mar. Ecol. Prog. Ser. 528: 71–86.

Shanks, A.L., S.G. Morgan, J. MacMahan, A.J.H.M. Reiniers, R. Kudela, M. Jarvis, et al. 2016. Variation in the abundance of Pseudo-nitzschia and domoic acid with surf zone type. Harmful Algae 55: 172–178.

Shanks, A.L., S.G. Morgan, J. MacMahan and A.J.H.M. Reniers. 2017a. Alongshore variation in barnacle populations is determined by surfzone hydrodynamics. Ecological Monographs 87: 508–532.

Shanks, A.L., S.G. Morgan, J. MacMahan, A.J.H.M. Reniers, M. Jarvis, J. Brown, et al. 2017b. Persistent differences in horizontal gradients in phytoplankton concentration maintained by surfzone hydrodynamics. Estuaries Coast. 41: 158–176.

Shanks, A.L., P. Sheeley and L. Johnson. 2017c. Phytoplankton subsidies to the intertidal zone are strongly affected by surfzone hydrodynamics. Mar. Ecol. 38: e12441.

Short, A. D. and L. D. Wright. 1983. Physical variability of sandy beaches. Sandy beaches as Ecosystems: 133–144.

Stewart, D. J. and M. Ibarra. 1991. Predation and production by Salmonine fishes in Lake Michigan, 1978–88. Can. J. Fish. Aqua. Sci. 48: 909–922.

Strydom N.A. 2007. Jetski-based plankton towing as a new method of sampling larval fishes in shallow marine habitats. Environ. Biol. Fish. 78: 299–306.

Suda, Y., T. Inoue and H. Uchida. 2002. Fish communities in the surf zone of a protected sandy beach at Doigahama, Yamaguchi Prefecture, Japan. Estuar. Coast. Shelf Sci. 55(1): 81–96.

Sverdrup, H.U., M.W. Johnson and R.H. Fleming. 1942. The Oceans Their Physics, Chemistry, and General Biology. Prentice-Hall, Inc., Englewood Cliffs.

Takahashi, K. and K. Kawaguchi. 1998. Diet and feeding rhythm of the sand-burrowing mysids *Archaeomysis kokuboi* and *A. japonica* in Otsuchi Bay, northeastern Japan. Mar. Ecol. Prog. Ser. 162: 191–199.

Talbot, M.M.B. and G.C. Bate. 1987a. Rip current characteristics and their role in the exchange of water and surf diatoms between the surf zone and nearshore. Estuar. Coast. Shelf Sci. 25: 707–720.

Talbot, M.M.B. and G.C. Bate. 1987b. The spatial dynamics of surf diatom patches in a medium energy, cuspate beach. Bot. Mar. 30: 459–465.

Talbot, M.M.B., G.C. Bate and E.E. Campbell. 1990. A review of the ecology of surf-zone diatoms, with special reference to *Anaulus australis*. Oceanography Marine Biology Annual Review 28: 155–175.

Tully, O. and P.O. Ceidigh. 1987. Investigations of the plankton of the West coast of Ireland-VIII. The neustonic phase and vertical migratory behaviour of benthic Peracaridea in Galway Bay. Proceedings of the Royal Irish Academy 87b: 43–64.

Vargas-Fonseca, E., A.D. Olds, B.L. Gilby, R.M. Connolly, D.S. Schoeman, C.M. Huijbers, et al. 2016. Combined effects of urbanization and connectivity on iconic coastal fishes. Divers. Distrib. 22(12): 1328–1341.

Vilas, C., P. Drake and N. Fockedey. 2008. Feeding preferences of estuarine mysids *Neomysis integer* and *Rhopalophthalmus tartessicus* in a temperate estuary (Guadalquivir Estuary, SW Spain). Estuar. Coast. Shelf Sci. 77(3): 345–356.

Watt-Pringle, P. and N.A. Strydom. 2003. Habitat use by larval fishes in a temperate South African surf zone. Estuar. Coast. Shelf Sci. 58: 765–774.

Webb, P., R. Perissinotto and T.H. Wooldridge. 1987. Feeding of *Mesopodopsis slabberi* (Crustacea, Mysidacea) on naturally occurring phytoplankton. Mar. Ecol. Prog. Ser. 38: 115–123.

Webb, P. and T.H. Wooldridge. 1990. Diel horizontal migration of *Mesopodopsis slabberi* (Crustacea: Mysidacea) in Algoa Bay, South Africa. Mar. Ecol. Prog. Ser. 62: 73–77.

Yu, O.H., H.Y. Soh and H.-L. Suh. 2002. Seasonal zonation patterns of benthic amphipods in a sandy shore surf zone of Korea. J. Crust. Biol. 22(2): 459–466.

Yu, O.H., H. Suh and Y. Shirayama. 2003. Feeding ecology of three amphipod species *Synchelidium lenorostralum*, *S. trioostegitum* and *Gitanopsis japonica* in the surf zone of a sandy shore. Mar. Ecol. Prog. Ser. 258: 189–199.

Chapter 3

Human Impacts over Sandy Beaches

María Victoria Laitano*, Nicolás Mariano Chiaradia and Jesús Darío Nuñez

Marine and Coastal Research Institute (IIMyC), National University of Mar del Plata-National Council for Scientific and Technical Research (UNMdP-CONICET), 7600, Mar del Plata, Argentina (vlaitano@mdp.edu.ar, nicolaschiaradia@conicet.gov.ar, jdnunez@mdp.edu.ar).

INTRODUCTION

Coastal zones have always captivated human beings for different reasons such as recreational, cultural, economic or due to other ecosystem services they provide. Nowadays, coastal zones are the most densely populated regions in the world and projections indicate important increases in this pattern. Even within a low world population growth scenario, a rise of more than 50% (from the year 2000) is predicted for 2030, which would lead to 880 million people living in coastal zones and more than a billion people by 2060 (Neumann et al. 2015). Furthermore, among the world coastlines sandy beaches occupy the largest area, and are also preferred by tourists among other coastal landscapes.

Human footprints on nature as a result of its interaction with the different components of natural environments, are not usually positive. As a general rule, modern humans transform the environment according to their needs and thus, sandy beaches are not exceptions. Many human activities performed close or over sandy beaches generate impacts which commit the ecosystem services they provide. These activities alter the environment at different time and space scales,

*Corresponding author: vlaitano@mdp.edu.ar

from weeks to centuries and from less than one to thousands of kilometers (Defeo et al. 2009) and are projected to escalate in the near future.

In this chapter it is first presented how scientists assess the human impacts over sandy beaches and then an overview of the main human activities which produce alterations in these ecosystems and evidences of such impacts are provided.

ASSESSMENT OF HUMAN IMPACTS ON SANDY BEACHES

Scientists have evaluated the impact of human activities over sandy beaches for decades, this topic being strongly represented in the general sandy beach research literature (Nel et al. 2014). However, as with most disciplines, this task is not homogeneous around the world (Fig. 3.1). Sandy beaches from countries like the United States, Brazil, England, Portugal, Spain, France, Italy, China and Australia account for the most studied regarding human impacts. This heterogeneity can hamper the assessment of general patterns or responses, given the multiple morpho dynamics and other features that characterize different sandy beaches. However, the exponential growth through time of sandy beach studies in different topics (Nel et al. 2014, Costa et al. 2020), could give rise to a more balanced world map.

Human impacts over sandy beaches can be assessed from different perspectives, the physical or chemical effects, the biological or ecological ones or a combination of both. The assessment of human impacts over sandy beaches from an ecological perspective faces multiple challenges, and one of them are the methodological approaches. In general, there are few studies focusing on sampling methodologies to this ecosystem (Schlacher et al. 2008b). Many studies have evaluated the impact of one human activity at a time, for instance comparing fauna from a beach with a seawall with that of a beach without a seawall. This approach has been questioned because it is likely that several other human activities would be present in the impacted beach causing additional effects. To cope with this issue, many authors have developed experimental designs, either through the exclusion of certain areas of the studied beach or through Before-After-Control-Impact (BACI) approaches. Also, some studies compare beaches with and without (or less) human intervention (e.g., urbanized) that comprise multiple activities. Nonetheless, it is yet difficult to find truly control beaches and to separate the natural variability from the human-induced (McLachlan and Defeo 2017). Despite the different interrogations around these approaches to assess human effects on sandy beaches, studies using them demonstrate some kind of impact and are accepted in the international scientific literature, sometimes with some caution in their conclusions.

In this context, González et al. (2014) proposed an urbanization index to characterize the beaches to be studied, which includes measures or estimations of different urban indicators: proximity to urban centers, building on the sand, beach cleaning, solid waste on the sand, vehicle traffic on the sand, quality of the night sky and frequency of visitors. Through this index, beaches can be classified according with their level of urbanization (i.e., high, intermediate and low) and it has been applied by many authors (e.g., Machado et al. 2016, Orlando et al. 2020). Furthermore, McLachlan et al. (2013) proposed a Conservation Index and a Recreation Index, which assess the condition of the studied beaches and the aptness and convenience of different uses of them. Factors related to the physical environment, as well as ecological factors and socio-economic issues are included to develop them. Although these indexes were mainly developed for management purposes, they have been also used by some ecologists to characterize beaches in human impacts studies (e.g., Cardoso et al. 2016). The utilization of these indexes provides more accurate results since they make the reference sites more reliable and because several human interventions are being taken into account at the same time. However, to differentiate anthropogenic from natural variability as well as to estimate the effects of pulse and press disturbances, large-scale and long-term studies are specifically needed, as has been highlighted by many expert sandy beach ecologists (e.g., Schoeman et al. 2000, McLachlan and Defeo 2017, Fanini et al. 2020, Orlando et al. 2020).

Figure 3.1 World map showing the distribution of research articles regarding human impacts on sandy beaches. The map was performed with a shiny app of Bibliometrix software (version 1.0, https://www.bibliometrix.org/Biblioshiny.html)*. Different shades of gray indicate different productivity rates: black=high productivity; light gray=no articles.

*Figure 3.1 was obtained applying a bibliometric review method (See Aria and Cuccurullo 2017) using SCOPUS database. SCOPUS is the largest abstract and citation database of peer reviewed literature (Mongeon and Paul-Hus 2016). The subject area was limited to studies in Life Sciences and Agricultural and Biological Sciences due to the topical foci of this chapter. Articles listed in the SCOPUS database were reviewed in alignment with our objective to extract all research articles related to Human Impacts Over Sandy Beaches. The timeframe for our document search was limited from 2000 to 2019. The extraction of the dataset was conducted at the middle of the year 2020. The keywords and free words were 'Sandy beaches' AND 'Human Impacts' OR "Sandy" OR 'Beach' OR 'Anthropogenic' on topic field. The search resulted in 2,513 documents published in the period from 2000–2020. Applying the following screening criteria (i.e,. excluding studies on climatic change, only dunes, only surf zone, assessment of sustainable strategies) reduced the original dataset to 639 documents.

Studies evaluating biological or ecological effects of human activities assess the structure of different communities (e.g., meio or macrofauna) or certain metrics of a single species (e.g., abundance, distribution, ecological biomarkers). As a result, many species or taxonomic groups have been proposed as efficient bioindicators of human impact on sandy beaches, such as ghost crabs (Barros 2001), Talitrid amphipods and Tylid isopods (Gonçalves et al. 2013), mole crabs and sandhoppers (Cardoso et al. 2016), among many others. Schlachler et al. (2014) made an exhaustive evaluation of different variables used to assess the impact of human interventions on sandy beaches and dunes systems, contrasting them according to the degree of reflection of such impacts and ecological indicator performance criteria. They found the following broadly applicable metrics in sandy beach impacts studies: the attributes of bird populations and assemblages; the breeding/reproductive performance in a broad sense, particularly for birds and turtles (also suitable for invertebrates and plants); population features and distribution of vertebrates mainly inhabiting dunes and the supralittoral beach zone; and composite measurements of the abundance/cover/biomass of plants, invertebrates or vertebrates at both, population or community level. However, the particular context of each location to be studied, including local fauna composition, landscape features, level of environmental degradation, among other issues, will influence this metric selection (Schlacher et al. 2014). On the other hand, Costa et al. (2020) through a meta-analysis of researches using macroinvertebrates as bioindicators, identified useful indicators for specific disturbances: ghost crabs for vehicle traffic and nourishment, Talitrids for trampling and mechanical cleaning and clams for harvesting. However, the authors underline the potential of some organisms as the tiger beetles (larvae and adults) for being good bioindicators but this has been poorly studied until now. The study and knowledge of the bio-ecology of key macrofauna species in order to differentiate the effects of human disturbances from natural variations would give rise to a powerful tool to evaluate sandy beaches ecological quality conditions (Gonçalves et al. 2013). Moreover, bio-ecology and population dynamics models of key macrofauna species can also allow scientists and managers to make predictions on the impacts of human disturbances over sandy beaches (Gonçalves and Marques 2017).

HUMANS IMPACTING SANDY BEACHES

Coastal regions are home to large human populations, which leads to great coastal development and urbanization, directly impacting sandy beaches (Fig. 3.2). Many human activities, as coastal defense constructions, nourishment and grooming are carried out next or over sandy beaches altering different features of these ecosystems. Sandy beaches also do not escape from the unavoidable pollution footprint of humans' interaction with the environment. On the other hand, sandy beaches host living and non-living resources susceptible to human exploitation rarely included in management programs. Many of these activities, and consequently their impact over sandy beaches, are enhanced by tourism. Ocean and coastal tourism involves the movement and arrival of thousands of million people to coastal areas

(Gössling et al. 2018). Thus, tourism is the main economic activity of many coastal regions which are consequently prepared to receive it (Hall 2001). In this context, sandy beaches are conditioned in different ways to generate quality and comfortable places to their users (Fig 3.2). In general, these human activities generate changes on sandy beach physical properties. Since sandy beach fauna, and in particular benthic communities, depend largely on these physical features (McLachlan et al. 1993, McLachlan 1996, Defeo and McLachlan 2005), impacts over beach inhabitants are therefore expected and in several cases demonstrated.

Figure 3.2 Coastal urbanization (up) and sandy beach modification for touristic purposes (below). Photos by Pablo González (up) and Nicolás Chiaradia (below).

As mentioned earlier, many authors have considered urbanization as a whole when assessing human impacts over sandy beach residents. Thus, urbanization has been associated with abundance reductions of certain species (e.g., *Emerita brasiliensis*, *Talitrus saltator*) when the entire macrofaunal community is analyzed, which indicates different susceptibility among species probably related with life history traits or their distribution along the beach (Veloso et al. 2006, Bessa et al. 2014). Urbanized sandy beach macrofauna would also be more vulnerable to natural events, such as storm waves (Machado et al. 2016) and natural environmental gradients, as that of salinity due to the presence of a large estuary (Orlando et al. 2020), compared to non-urbanized beach communities. Adverse effects over macrofauna can lead to reductions of the energy transfer efficiency through trophic levels altering the trophic functioning of these ecosystems (Costa et al. 2017).

Studies evaluating the effects of the urbanization of sandy beaches on a single or two species also yield interesting results in this regard. Significant reductions on the abundances of the crustacean *Emerita brasiliensis* and *Atlantorchestoidea brasiliensis* (Cardoso et al. 2016), the darkling beetles *Phaleria maculata* (González et al. 2014), the tiger beetles *Cylindera nivea* (Costa and Zalmon 2019) and ghost crabs *Ocypode cordimana* (Barros 2001) have been reported as a consequence of beach urbanization. When the affected species has a strong density-dependence with other species (e.g., due to space competition), like the clams *Amarilladesma mactroides* and *Donax hanleyanus*, the abundance reduction of one of them can result in higher abundances of the less vulnerable one (Laitano et al. 2019). Adverse effects of urbanization over sandy beach inhabitants also include issues other than abundance. The amphipod *Talitrus saltator* living on urbanized beaches shows morphometric asymmetries (a common biomarker in this species of developmental disruptions caused by stressors) likely related to habitat destruction and modification, reduced food resources and refuges or exposition to various substances as consequence of coastal pollution (Barca-Bravo et al. 2008). Ghost crabs alter their distribution by restricting it to the edges on urbanized beaches and change their burrow morphology making them deeper, steeper and smaller (Gül and Griffen 2018). Further, disturbances of beach urbanization induce smaller and shallower claws on these crabs, which could directly affect their ecology and physiology, likely altering sexual selection, foraging efficiency, success in agonistic behaviors and mating success (Gül and Griffen 2020).

On the other hand, a great amount of evidence of human impacts over sandy beach bioindicators can be found from studies assessing human activities separately, which will be presented below.

Coastal Defenses

The presence of infrastructures over coastal environments is an unavoidable part of the growing population and tourism therein, and since they are hazardous zones many types of defenses have been consequently constructed to protect constructions (Fig. 3.3). With the aim of neutralizing coastal erosion, sea level rise and flooding as threats not only to human constructions but also to people themselves, as well as to gain appropriate conditions to maritime operations, shorelines have been modified through hardening and armoring (Table 3.1). The most common deployed coastal defenses include groins, seawalls, breakwaters, sills and sand fences (Nordstrom 2014) made of concrete, granite or sandstone rock, marl, wood, and vinyl sheeting (Gittman et al. 2016), although there are many others. One instantaneous impact of these structures is the beach loss, as a portion of it is covered by the structure, although the magnitude will depend on the type of defense and the size of the beach (Griggs 2005). Hard structures also disrupt the natural sediment dynamic and consequently the sediment supply to sandy beaches either by sand drift (Rangel-Buitrago et al. 2018) or by cliffs or dunes (Griggs and Patsch 2019) generating new erosion hotspots. Hard structures along the coastline also induce passive erosion: the loss of the beach in front of the defense because the natural retreat has been interrupted and the landward migration of adjacent beaches (Griggs and Patsch 2019). All these processes may cause habitat fragmentation and loss for sandy beach resident organisms impacting at population and community levels. Furthermore, the introduction of hard artificial substrates in areas where they do not naturally occur would disrupt the dispersion of larvae due to water flow interference and favor the spread of exotic species (Bulleri and Chapman 2010).

Figure 3.3 Seawall (left) and groin (right). Photos by Guillermina Laitano.

The ecological effects of coastal defenses in sandy beaches or open coast shores have been less studied than in other soft sediment environments (Dugan et al. 2018). The principal impact of the introduction of engineering structures for coastal protection to sandy beach biota is related to habitat fragmentation and loss. Related negative effects on diversity and abundance of seabirds (Dugan and Hubbard 2006), intertidal crustaceans (Rodil et al. 2016), survival of intertidal spawning fish embryos (Rice 2006) and ghost crabs susceptibility to storms (Lucrezi et al. 2010) due to seawalls, on supralittoral arthropodofauna community structure (Fanini et al. 2009) due to breakwaters, on macrofaunal diversity, abundance and species composition due to a small groyne (Walker et al. 2008), have been already reported. Broader scale studies spanning many kilometers of coast with different types of armoring (e.g., seawalls, ripraps, bulkheads) describe reductions in beach wrack and associated invertebrates (Sobocinski et al. 2010, Dethier et al. 2016). Furthermore, armoring is the human impact with the greatest potential effects on sea turtles nesting ground according to the perception of 21 experts in the field (Nelson Sella et al. 2019). In fact, Behera et al. (2016) reported that, as a consequence of nesting beach reduction due to the installation of cement tetrapod, olive ridley sea turtles were forced to nest into the artificial structure which resulted in 209 injured and 24 dead individuals over a 3 year period.

Table 3.1 The extent in numbers of some coastal defenses and beach nourishment events

Impact	Location/date	Numbers	Source
Coastal defenses	Northern Ireland coast	32% of an 650 km coastline fronted by man-made structures	Cooper et al. (2020)
	California State, USA	13.9% (239.3 km) of coastline with seawalls and revetments	Griggs and Patsch (2019)
	Tarawa Atoll, Kiribati	29% of 26 km of coastline (mostly seawalls)	Duvat (2013)
	Gulf of Manfredonia, Italy	300 coastal defenses in 22 km of coastline	Infante et al. (2012)
Beach nourishment	Germany (North and Baltic Sea) From 2000 to 2017	1.9 million m^3 of sand	Staudt et al. (2020)
	Varadero, Cuba Since 1980	2.5 million m^3 of sand	Pranzini et al. (2016)
	Mar del Plata, Argentina (three beaches nourished simultaneously) 1998	1.67 million m^3 (Bristol beach), 0.66 million m^3 (Playa Grande beach) and 0.15 million m^3 (Varese beach) of sand	Marcomini and Lopez (2006)
	California State, USA From 1927 to 2020	700 million m^3 of sand	ASBPA (2020)

These impacts would vary among the different types of defense structures (Gittman et al. 2016), with their location in the shore profile (Martin et al. 2005, Dugan et al. 2018) and their size, shape, orientation and material used (Nordstrom 2014). In addition, the different beach zones would show different responses to engineering structures; for instance, the uppermost zones (supralittoral and high intertidal) would be the most physically and biologically affected by seawalls (Dugan et al. 2008). However, lower intertidal level organisms can experience lower recovery if the seawall is removed and the beach nourished as a restoration process (Toft et al. 2014). As the ecological effects are mainly a consequence of beach morpho dynamics and sediment dynamic modification by engineering structures, their age would also modulate such effects. While the studies assessing coastal defenses after some or many years of installation clearly show significant negative effects in supralittoral and intertidal sandy beach biota, contrasting results have been found at the moment of installation or a few months later. The macroinfaunal community indexes of a south central Chile sandy beach did not indicate changes between before and after the construction of a seawall, as neither more than one year after (Jaramillo et al. 2002). Nourisson et al. (2018) studied the supralittoral arthropod community structure of beaches in the western Mediterranean coast of Italy during the construction of a seawall and many groins (some in front of the seawall) and a few months after. They found changes in the community structure only in the most impacted site (closest to the seawall), although they expected to find changes sometime later in other sites due to an expected new alongshore sediment transport regimen.

Nourishment

Another way by which coastal managers protect sandy beaches from erosion processes is beach nourishment that is, fill the beach with sediments extracted from another place in the sea (Table 3.1). This sediment used to nourish beaches is chosen from offshore sites according to its characteristics (deposit size, grain size and color) looking for compatibility, although sometimes, the economic factor (costs of shipping) are prioritized (Staudt et al. 2020). Often, filling sediments are extracted from harbors through a dredging operation with navigation logistics as its aim. The material is deposited on the beach (beach or shore nourishment), on the sublittoral (shoreface nourishment) or on dunes (dune nourishment) (Staudt et al. 2020). This activity has been generally considered as a soft engineering approach and environmental-friendly, even more so when compared to coastal hardening and armoring (Schoonees et al. 2019). However, many changes in the environment are generated by beach filling procedures, which may impact on the resident organisms of both, the extraction (Rosov et al. 2016) and the recipient sites. At the recipient environment, changes in sand mean grain size, color and beach slope which generate modifications on beach morphodynamics are consequences of beach nourishment when the borrowed material is not appropriately chosen (Pranzini et al. 2018). An unusual and great swash zone turbidity has also been associated with beach nourishment and explained by changes in mineralogical composition and higher particle wear of the post-nourishment sand (Chiva et al. 2018). Finally, when sediments come from ports dredging, contaminants are transferred to the beach and can remain there for many years (Laitano et al. 2015).

The effects of nourishment procedures over sandy beach inhabitants are mainly focused, but not exclusively, on intertidal fauna. Negative effects on beach inhabitants can be immediate causing mortality by burial, crushing, suffocation or, in the case of filter-feeding organisms by clogged gills and palps. Thus, a decrease in abundance and density of intertidal macroinvertebrates (Peterson et al. 2006, Schlacher et al. 2012, Manning et al. 2014, Wooldridge et al. 2016, van Egmond et al. 2018) as well as in community indexes (Wooldridge et al. 2016) and meiofaunal copepods and polychaetes (Menn et al. 2003) has been already documented. Consequently, shorebirds show foraging activity reductions in filled beaches likely due to the decrease in prey density and abundance and physical disruption of bill probing and prey capture (Peterson et al. 2006, 2014). Sea turtles are also disturbed by this activity, as the loggerhead, green, and leatherback ones, which exhibit lower nesting and hatching success due to beach filling (Cisneros et al. 2017). Contrastingly, more than one study reported an enhanced density of an opportunistic polychaete species, *Scolelepis squamata*, following beach nourishment events, probably related with its adaptation to finer sediments (Leewis et al. 2012, Manning et al. 2014). However, Leewis et al. (2012) suggested a potential change in community structure given this increment, which was not measured but would have impacts at a higher level of biological organization.

Since beach replenishment activity is in many cases an isolated event in time, a great deal of focus has been put in beach fauna recovery assessment. Fast recovery of beach resident organisms ranging between weeks to 12 months has been demonstrated (Burlas et al. 2001, Menn et al. 2003, Jones et al. 2008, Leewis et al. 2012). However, those recovery times correspond to nourishment operations where natural sediment matching was reached, either at the moment or a few months later and thus, sand features (grain size, mineralogy, organic matter content, etc.) are not largely disturbed by the event. When deposited sediment traits differ significantly from the natural ones at the receiving beach, recovery times are different. The abundance of five of the six dominant infaunal invertebrates off Topsail Island, North Carolina (USA), has been still negatively affected after one year of beach nourishment with finer sediments (Manning et al. 2014). Similar results were obtained when deposited sediments are coarser than the natural ones (Peterson et al. 2006). Indeed, a long monitoring of 3–4 years after a beach replenishment with coarser sand, revealed that some macroinvertebrates, like haustoriid amphipods and *Donax* spp did not recover their natural densities after such a period (Peterson et al. 2014). Ghost crab abundances and the foraging activity of seabirds was not normalized after 4 years from the event. Besides the sand matching between deposited and natural sand, sediment drift regimen of the different coastal areas would affect the recovery times: high along-shore sediment transport would dilute impact over sediments as well as facilitate the recolonization of the beach by invertebrates through their plankton larvae (Peterson et al. 2006). Furthermore, different taxa display varying degrees of negative effects due to beach filling, and therefore, different recovery times, which has been associated with species life history traits and other attributes. Species which display fast reproduction and high mobility (Menn et al. 2003), the ability to burrow in a range of sediment grain sizes (Peterson et al. 2014) and/or planktonic dispersing stage (Schlacher et al. 2012, Wooldridge et al. 2016) will recover faster than those with contrasting characteristics. Diverse recolonization times of disturbed beaches would give rise to alterations in the community structure which could last much more time than individual recoveries.

Recreation

Maintaining the sandy beach width against erosion is not only a necessity to keep urban infrastructure safe, but also to provide quality beaches for tourism, an economically important activity of most coastal regions. Recreational activities, like sunbathing, beach sports (walking, running, volleyball, etc.), off-road vehicle driving, among others, which are developed by the local coastal population and extensively multiplied by mass tourism, generate similar impacts to sandy beach habitats. Hundreds of people may be walking on the sand at the same time during warm seasons (Reyes-Martínez et al. 2015), whilst high traffic volumes of off-road vehicles can reach as much as an average of 727 vehicles crossing the beach per day (Schlacher et al. 2008a) (Fig. 3.4). Human trampling on the beach induce sediment compaction and crushing of resident organisms (Reyes-Martínez et al. 2015), whilst off-road vehicle driving over the beach has been associated with compaction, rutting and displacement of the sand matrix (Davies et al. 2016). Both trampling and off-road or all-terrain vehicle driving over dunes also compromise their stability and erosion (Davenport and Davenport 2006) and consequently, the sediment dynamics of the associated beach is changed.

Figure 3.4 Tracks left by off-road vehicles in the sand (up) and massive tourism at Mar del Plata, Argentina (below). Photos by Nicolás Chiaradia (up) and Luciano Gargiulo/Qué digital (below).

Evidence on the impact of recreational activities over sandy beach inhabitants are mostly related to trampling and Off-Road Vehicles (ORVs) traffic. Negative effects of trampling on ghost crabs abundance (Barros 2001), orientation of the sandhopper *Talitrus saltator* (Scapini et al. 2005) and intertidal macroinfauna community (Veloso et al. 2006) have been suggested. However, in such studies, as stated by their authors, trampling is not isolated from other human-induced habitat modification which can be simultaneously acting (as armoring, grooming, etc.). The assessment of trampling impacts is maybe the most questioned and discussed in this sense among other human activities or modifications (Schlacher and Thompson 2012, Costa et al. 2019b, Zielinski et al. 2019). Thus, studies evaluating the impact of trampling through experimental designs, which account for a direct effect of trampling, increased in recent years forming a considerable body of research which demonstrates measurable negative effects of this activity. Decreased abundance, species richness and diversity and the resulting change in community structure have been reported for intertidal macrobenthos exposed to trampling (Schlacher and Thompson 2012, Reyes-Martínez et al. 2015, Machado et al. 2017). Differences in the response of species within the community are mainly explained by the possession or lack of body protection structures (Machado et al. 2017) or by different burrowing depths (Moffett et al. 1998, Schlacher and Thompson 2012). Upper shore invertebrate assemblages can also experience declines in abundance and species richness due to trampling, with relevant potential consequences to their predators (e.g., shorebirds) (Schlacher et al. 2016). Furthermore, abundances of particular species like the amphipod *Talitrus saltator* (Ugolini et al. 2008) and those of the genus *Ocypode* (ghost crabs; Lucrezi et al. 2009, Costa et al. 2019b) have been successfully used as ecological response variables to demonstrate negative effects of trampling. Many of these studies report that there is a relationship between the intensity of trampling and the degree of adverse effects it causes.

The use of ORVs over sandy beaches is associated with reduced species richness and diversity of intertidal macrobenthic assemblages (Schlacher et al. 2008a, Davies et al. 2016) and mortality (Schlacher et al. 2008c) and impaired burrowing performance of the surf clam *Donax deltoides* (Sheppard et al. 2009). Ghost crabs have been extensively selected to study the impacts of ORVs traffic on sandy beaches. A decrease in burrows density due to ORVs use in beaches is well documented (Moss and McPhee 2006, Schlacher and Lucrezi 2010, Lucrezi et al. 2014). Burrow sizes can also be reduced by half due to ORVs driving on beaches, which indicate smaller sized individuals and thus, a change in population structure could be assumed (Lucrezi et al. 2014). In addition to these types of impact, other related behavioral effects can be highlighted. While under natural conditions crabs are mainly located at the upper and middle zones of the beach, under ORVs disturbance they are forced to concentrate on the lower beach (Lucrezi and Schlacher 2010, Schlacher and Lucrezi 2010, Lucrezi et al. 2014). Therefore, ghost crabs habitat is narrowed and the possibility of competition for food and space among smaller and larger crabs would be incremented (Lucrezi et al. 2014). Furthermore, vehicle traffic lead to changes in the shape of burrows, these being deeper, probably to avoid burrow collapse and simpler, because due to burrow collapse crabs have to construct them more frequently (Lucrezi and Schlacher 2010). It is worth noting that even at low traffic rates and imposed limitations to ORVs use, impacts of this activity are still evident (Lucrezi et al. 2014, Davies et al. 2016), highlighting the relevant ecological consequences it can generate.

Grooming

Cleanliness is a feature especially required by tourists on sandy beaches. Great amounts of litter reach beaches through stranding of material that drifts into the ocean, as well as through land drainages, fishery or by beach users themselves (Martinez-Ribes et al. 2007, Vlachogianni et al. 2018). Litter devalues the landscape quality (Rangel-Buitrago et al. 2017), but also involves potential injuries to beach users (Campbell et al. 2016). Important potential loss of beachgoers and consequently loss of income from tourism due to a significant presence of litter on sandy beaches has also been reported (Krelling et al. 2017). Additionally, stranded natural debris mainly composed by seaweed (wrack) can accumulate several meters of material undesirable for tourists. This wrack may limit the access of beachgoers to the sea, besides accumulating pathogens and, particularly during warm days, wrack undergoes decomposition generating odor and attracting flies. For all the above reasons, and particularly with the aim to offer better scenic beauty to tourists, local governments allocate large funds each year for beach cleaning (Krelling et al. 2017). Beach cleaning is carried out by mechanical machinery (generally heavy) that rakes and sieves the sand, or by hand. Considering litter, this may be a positive action since physical contaminants are being removed from the beach (later the impact of contamination is explained in detail); however, grooming also impacts the nature of the sandy beach through different ways. Although until now, particularly the effects of the heavy machines on beach organisms has not been yet determined (Zielinski et al. 2019), it could be assumed that the sediment compaction and sieving would imply stress and other negative consequences for them. On the other hand, wrack removal takes out a relevant source of food and habitat for many macrofauna species (MacMillan and Quijón 2012, Michaud et al. 2019) and consequently, a food source for shorebirds (Hubbard and Dugan 2003). Furthermore, macroalgal wrack plays a role in nutrient supply to the surf zone (Dugan et al. 2011) and sand dynamics (Dugan and Hubbard 2010) and thus, its extraction could interfere with these processes.

Since wrack subsidies are an important component of many sandy beaches as providing food sources and microhabitat refuges (Gonçalves and Marques 2011), the effects of its removal by beach grooming over associated beach fauna has been already addressed. Lower species richness, biodiversity and biomass of macrofauna (Dugan et al. 2003, Gilburn 2012) and a reduction in abundance of ghost crab burrows (Stelling-Wood et al. 2016), sandhoppers (Fanini et al. 2005) and bacteria production (Malm et al. 2004) were reported as consequences of wrack removal. A smaller abundance of shorebirds related to less standing crop of macrophyte wrack and a lower abundance of wrack-associated macrofauna has also been found in groomed beaches (Dugan et al. 2003). The negative effects of stranded material removal found in the species richness of the fauna dependent on it, is associated with a decrease in wrack depth (Griffin et al. 2018). However, adverse effects can be observed at the same wrack depth in control and impacted sites, showing some additional effect of grooming than just the decrease of wrack depth (Griffin et al. 2018). Within wrack-associated taxa different responses according to their different habits could be expected. Species which reproduce in the wrack (and therefore their larvae depend on wrack) and those which rely on old deposited wrack, are more sensitive than those which just feed on fresh wrack (Gilburn 2012). Similarly, species which distribute beyond wrack and feed on bare sand (e.g., Staphylinid *Bledius bonariensis*) will be less vulnerable to cleaning procedures as well as those which burrow in the sand and therefore do not need wrack as refuge (Vieira et al. 2016). On the other hand, little is known about the impact of grooming in not-wrack-associated fauna, which could be affected by heavy machines and sand sieve. In beaches where a small amount of wrack reaches the shore no differences in the abundance of bacteria, meiofaunal and macrofaunal assemblages nor in ocypodid crabs burrows abundance were found between cleaned and uncleaned sites within the same beach (Morton et al. 2015). Malm et al. (2004) also found no significant difference in not-wrack-associated infauna nor epifauna biodiversity between groomed and ungroomed beaches. Although less sensitive than macroinvertebrates associated with wrack, Schooler et al. (2019) determined lower species richness, abundance and biomass of macroinvertebrates communities of lower intertidal levels in groomed beaches. However, this study assessed beaches impacted not only by grooming but also by filling, which modifies grain size, a key factor to burrowing invertebrates usually found at lower intertidal levels. It is clear that more evidence would be necessary on the response in regions where the stranded wrack input is not substantial.

Grooming regimen varies according to beach features, particularly to the number of beach users, and the frequency of grooming would have a direct relation with its impact. Stelling-Wood et al. (2016) studied the abundance of ghost crab burrows in beaches with different cleaning regimes: no mechanical cleaning, infrequently cleaned (3 times or less per week) and frequently cleaned (5–7 times per week). Although they found the lowest burrows abundance in the frequently cleaned beaches, the infrequently cleaned ones supported the highest burrows abundance. Morton et al. (2015) also suggested that the infrequent grooming (1–2 times per week) at the beaches of their study could explain the results of no effects that they reported. This relationship of frequency-effects could be explained from the time that the affected fauna needs to recovery from the impact. In this sense, some experiments have been done in order to estimate time recovery of strandline meiofauna to a one-off cleaning event (Gheskiere et al. 2006). Meiofauna recovered approximately 36 hours later, in the next second high tide, probably due to passive fauna reposition from deeper layers. In another case, after 10 days of an experimental wrack removal, supralittoral arthropods regained their densities after 3, 6 and 16 days after, varying according to the species (Vieira et al. 2016). Willmott and Smith (2003) studied amphipod species from a beach two years after the cessation of cleaning and found recovered populations besides an extra 1m of sand dune stabilized by vegetation. A recovery of strandline macroinvertebrate community diversity has also been reported during the offseason when cleaning stopped, although without reaching the diversity of ungroomed beaches (Griffin et al. 2018). Both, the different sensitivity to beach cleaning events and the different recovery times of species would lead to changes in community structure.

Pollution

Litter is composed mainly by the visible part of plastic residues produced by humans. Microplastics, the fraction of such debris that are less than five millimeters in length, are produced from the wear of larger plastic pieces or manufactured as part of health and cosmetic products (Fendall and Sewell 2009). While larger plastic constitute a threat to bigger animals due to consumption or entanglement (Li et al. 2016), microplastics would also have an impact on smaller ones, particularly filter-feeders (Moore et al. 2001) and wrack consumers (Carrasco et al. 2019) (Table 3.2). As mentioned earlier, many of these plastics reach sandy beaches by land-based sources (e.g., stormwater and wastewater effluents), although also by beach users and fishing activities (Fig. 3.5). Such land drainages carry many other pollutants, originating in industrial processes or domestic activities, which ultimately get to sandy beaches. Agriculture chemicals (Ma et al. 2001), heavy metals (Fernández-Severini et al. 2019), pharmaceutical and personal care products (Beretta et al. 2014) and pathogen microorganisms (Halliday and Gast 2011) are just some examples. The exposition to such contaminants is well known to trigger a wide variety of sublethal effects in organisms. Another land-based source of pollution to sandy beaches and closely related to urbanization is artificial lighting at night (Table 3.2). Since many behavioral patterns and physiological and ecological processes of organisms rely on daily, seasonal and lunar cycles, it is evident that lighting at night by anthropogenic sources alters such natural activities (Gaston et al. 2017). Further sources of pollution to sandy beaches come from coastal human activities. Besides those related to the beach use itself, which have been already mentioned, shipping and maritime facilities as ports, marinas and shipyards can be the origin of beach pollution by hydrocarbons, butyltins, heavy metals, among other contaminants. Although maritime facilities are generally installed in enclosed areas, the high dynamics of coastal zones allow pollutants to reach the intertidal and shallow subtidal sediments of beaches (Paz-Villarraga et al. 2015). On the other hand, the main pollution produced by shipping on sandy beaches is related to oil spills (Bejarano and Michel 2016).

The main impact of plastic debris on sandy beach organisms is related with its consumption. Large plastic litter or macroplastics in sandy beaches affect big animals. Plastic ingestion has been extensively documented in shorebirds and turtles (Li et al. 2016), and its consumption produce entanglement (Ryan 2018), gastro-intestinal tract obstruction (Roman et al. 2019), lower total seasonal reproductive output (Marn et al. 2020), higher probabilities of death (Wilcox et al. 2018, Roman et al. 2019), among many other effects. Beyond the effects due to plastic ingestion, beach litter may hinder nest site selection of marine turtles and impede hatchlings to rapidly reach the sea (Nelms et al. 2016). Compared with these impacts, toxicity due to chemicals adsorbed to plastic on large animals is expected to be less dangerous, although not despicable (Wilcox et al. 2016).

Table 3.2 The extent in numbers of plastic pollution in some countries and artificial light pollution by continents

Kind of pollution	Location	Numbers	Source
Plastics	Portugal (five beaches)	Macro and microplastics: 185.1 items/m^2 72% of microplastics	Martins and Sobral (2011)
	Poland (12 beaches)	Microplastics: between 76 and 295 items/kg dry sediment	Urban-Malinga et al. (2020)
	Russia (13 beaches)	Microplastics: between 1.3 and 36.3 items/kg dry sediment	Esiukova (2017)
	México (21 beaches)	Microplastics: between 16 and 312 items/kg dry sediment	Piñon-Colin et al. (2018)
	Brazil (16 beaches)	Macroplastics: between 0.19 and 31.5 items/m^2	Fernandino et al. (2016)
	Chile (six beaches)	Macroplastics: between 1.1 and 4 items/m^2	Gómez et al. (2020)
Artificial lighting	Europe	54.3% of coastline affected	Davies et al. (2014)
	Asia (excluding Russia)	34.2% of coastline affected	
	Africa	22.1% of coastline affected	
	South America	15.5% of coastline affected	
	North America	11.8% of coastline affected	
	Oceania	7.9% of coastline affected	
	Russia	6.1% of coastline affected	

The consumption of microplastics by sandy beach inhabitants have been already demonstrated through the analysis of organisms sampled in the environment (Iannilli et al. 2018, Carrasco et al. 2019, Costa et al. 2019a, Horn et al. 2019). Nonetheless, their accumulation and tissue translocation seems to depend on plastic particle size (Tosetto et al. 2016, Bruck and Ford 2018). Since the effects of microplastic pollution is a relatively recent research topic, particularly in sandy beaches (Fig. 3.6), the general effects of this consumption are still being evaluated. No effects of experimentally microplastic ingestion were reported for growth rate (Carrasco et al. 2019), survival (Ugolini et al. 2013) nor consumption rates, weight and molting (Bruck and Ford 2018) of different amphipod species. However, all these authors highlight the possibility to find significant effects carrying out experiments for longer term periods. Some studies demonstrated negative consequences of exposure to microplastics. Feeding activity of lugworm *Arenicola marina* decreases when animals are placed in sand contaminated with microplastics (Besseling et al. 2013, Browne et al. 2013, Green et al. 2016) while metabolic rates are enhanced, the bioturbation diminished and the biomass of microalgae on the sediment surface reduced, probably due to the above or by direct impact of microplastics (Green et al. 2016). Other evidence shows that microplastics added to food decreased the consumption rate of the amphipod *Orchestoidea tuberculata* (Carrasco et al. 2019) as well as the survival and jump height (28% lower) of the sandhopper *Platorchestia smithi* (Tosetto et al. 2016). Crabs *Carcinus maenas* fed with diet containing polypropylene rope microfibers also showed a reduced scope for growth, probably related to a decreased consumption rate (Watts et al. 2015). It is worth noting that the negative effects of microplastics have been reported to vary with the dose and type of plastic (e.g., Green et al. 2016).

Figure 3.5 Storm drain discharging on the beach (left) and plastics retained in the sand (right).

Besides the physical impact of microplastics, another threat related to them appears from its capacity to adsorb and carry Persistent Organic Pollutants (POPs, as PCBs, PAHs and DDT) and their additive chemicals (such as brominated flame retardants and phthalates), both known for being highly toxic compounds. However, different results are shown in this sense regarding sandy beach inhabitants. It was demonstrated that microplastics enhanced the bioaccumulation of PCBs from contaminated sediments in the lugworm *Arenicola marina* (Besseling et al. 2013). On the other hand, Browne et al. (2013) studied the same species and although they found a transfer of pollutants (nonylphenol and phenanthrene) and additives (Triclosan and PBDE-47) from microplastic to the worm tissues, bioaccumulation of these pollutants was higher from contaminated sand than from microplastics. Nonetheless, pollutants transferred by microplastics significantly altered the immune function of worms, their sediment engineering ability and increased their mortality. Finally, Scopetani et al. (2018) demonstrated that clean microplastics can act as scavengers of POPs in contaminated amphipods; thus, and considering other pollutant sources like water or sediments, they question the real importance of microplastic as transfers of pollutants to biota. Since evidence on this is made up by experimental trails, in which the type of plastic used, the duration of the experiments and the tested species among other issues are highly variable, many questions about the effects of microplastics on sandy beach organisms are still unanswered and hence more research is needed on this.

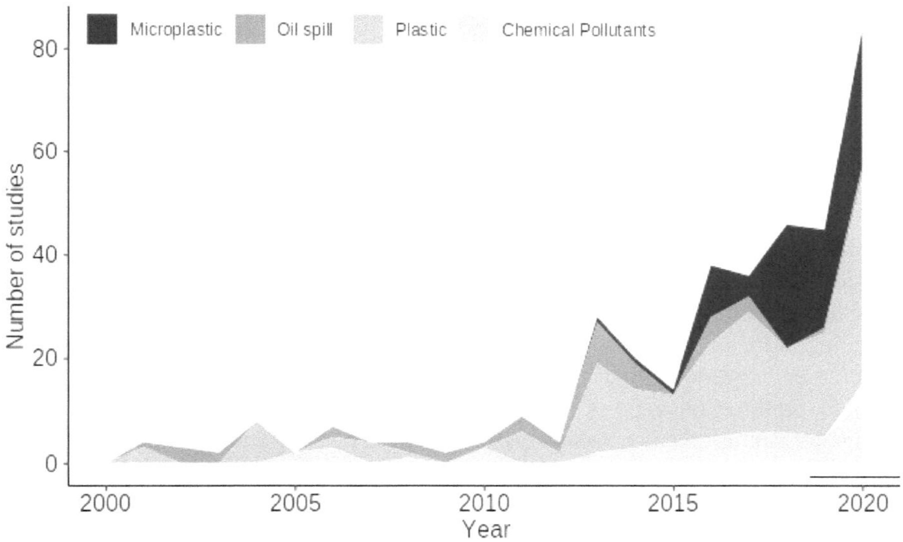

Figure 3.6 Trends in different types of pollution research over time. The graph was obtained by the same methodology as Fig. 3.1. Patterns were constructed from selected words in keywords and titles of the final selected articles.

The effects of exposure to sewage and other wastewater effluents in sandy beach inhabitants is less studied than other types of pollution (Fig. 3.6), probably because these facilities are generally located at rocky shores or have an ocean outfall. However, there is some evidence of negative effects when discharges encounter sandy beaches. Clams (*Donax sp.*) effectively bioaccumulates heavy metals released by domestic and industrial wastewaters (Haynes et al. 1997, Idardare et al. 2011). Although the toxicity of heavy metals in these organisms has been experimentally tested (Ong and Din 2001, Neuberger-Cywiak et al. 2003), assessments *in situ* have not been done yet. During an exceptional domestic sewage effluent exposure due to the rupture of an effluent pipeline, Boere et al. (2011) found high ammonia concentrations and lower sized individuals of the mole crab *Emerita brasiliensis* indicating a higher proportion of juveniles or males since reproductive females are bigger; the burying latency of *E. brasiliensis* was not affected. On the other hand, the congener *Emerita asiatica* showed to be very sensitive to discharges from an atomic power station (Suresh et al. 1995). Power plant effluents discharge hot water (thermal pollution) that was previously used in refrigeration systems. As much as 100% of mole crabs mortality has been found when exposed to this kind of pollution, being attributable only to temperature among other altered environmental conditions (Suresh et al. 1995). Hussain et al. (2010) found lower abundances of the wedge clam *Donax cuneatus* close to an atomic power station discharge, but they could not relate this pattern exclusively to thermal pollution since other interacting factors were present. Finally, freshwater courses that flow into the sea, either natural or artificial, carry many land-based contaminants (e.g., herbicides), which ultimately reach sandy beach residents. Moreover, a synergy between pollutants and low salinity in these scenarios was suggested (Sauco et al. 2010). A lower salinity would generate stress in marine organisms under which, detoxification mechanisms would be more difficult to develop.

A much more studied chemical pollution source for sandy beaches are oil spills (Fig. 3.6), maybe due to their dimensions. Bejarano and Michel (2016) made an extensive review of oil spill effects on sandy beach invertebrate communities, collecting evidence from laboratory and field studies. Both major and small oil spills produce firstly a drastic reduction in diversity, abundance and density of meio and macroinvertebrate communities at the intertidal environment. Besides the chemical toxicity of hydrocarbons, oil spills impact on sandy beach invertebrates coming from some physical disturbances generated by fouling which results in decreased ingestion, filtering and feeding rates. In the long term, these effects would produce impacts at the ecological level, although this is not confirmed from literature. Oil also changes several properties of sediments as temperature, permeability, compaction, water content, resulting in habitat degradation to sandy beach residents. Recoveries from oil spills of the intertidal fauna may last from weeks as much as 10 years, depending on the size of the spill, the cleanup activities and the physical and biological features of beaches. Moreover, there is a species-specific time of recovery where the sensitivity to the impact and life history traits plays a significant role. Similar results are found in literature posterior to the mentioned review by Bejarano and Michel (2016) (Kang et al. 2016, Brannock et al. 2017, Park et al. 2020). Regarding sea turtle eggs and hatchlings, physical and chemical adverse effects of oil spills are not entirely known and literature gives contrasting results (Wallace et al. 2020). On the other hand, further beach cleanup activities may crush eggs, provoke hatchlings disorientation, alter female adults nesting and hinder the access to the beach or the sea (Wallace et al. 2020). This issue has been highlighted by Michel et al. (2017) who stated that intensive beach treatments imply other impacts such as trampling, off-road vehicle traffic and grooming, but in great magnitudes and thus, negative effects are clearly expected, although this has been poorly documented.

Relatively recent attention has been paid to artificial light pollution and its effects on wildlife. In sandy beach environments sea turtles are the most studied organisms to assess such an impact. It has been demonstrated that sea turtles avoid beaches with intense artificial light pollution decreasing nest densities on such locations (Mazor et al. 2013, Hu et al. 2018, Windle et al. 2018) probably because nesting activity is made at night and light inhibits this behavior. Night artificial lighting causes abandoning nesting activity by adult female turtles, higher nesting times and incorrect orientation or even disorientation to return to the sea (Silva et al. 2017). Hatchlings orientation to reach the ocean once they emerge from the nest is also disrupted by light pollution causing longer crawl duration (Rivas et al. 2015). This would make hatchlings more vulnerable to predation. Even more so if predators are stimulated by night lighting as demonstrated by Silva et al. (2017) with ghost crabs. In the experiments carried out by these authors, the presence of light increased the abundance of crabs and affected their behavior, showing strong and aggressive activity in the presence of food. Compared to sea turtles, less is known about the effects of artificial light pollution on intertidal invertebrates. Low light pollution as that generated by small beach parties would not be a threat to mobile arthropod fauna (Fanini et al. 2014). On the other hand, combining field and laboratory studies, Luarte et al. (2016) and Duarte et al. (2019) demonstrated adverse effects of light pollution at night in two sandy beach invertebrate species. The amphipod *Orchestoidea tuberculata* showed reduced locomotor activity levels and lower consumption and growth rates when exposed to artificial light at night (Luarte et al. 2016). Furthermore, light pollution reduced the distribution and abundance of the intertidal isopod *Tylos spinulosus* which would be related to the effects of night lighting on their circadian rhythm of locomotor activity: the emersion from their burrows at night to migrate down-shore (Duarte et al. 2019). Garratt et al. (2019) found changes in community composition, species richness and biomass of sandy beach macroinvertebrates exposed to night lighting. However, while some of the non-rare species presented decreased abundance in illuminated beaches, others showed an increased one, indicating different species-specific responses to artificial light pollution. In addition to the scarcity of studies, since many physiological and behavioral processes are closely related or governed by circadian rhythms in animals, the underlying mechanisms of night lighting effects are still not understood.

Mining

Besides recreation, there are other human activities related to resource exploitation which cause an impact on sandy beaches, such as mining and fisheries or extraction. Sand is, together with gravel, the most exploited solid material worldwide and since it is a common-pool resource, the regulation of its extraction is difficult which generally leads to overexploitation (Torres et al. 2017). During mining operations, sand is extracted from dunes, intertidal and subtidal environments and used for construction purposes (Masalu 2002), to obtain heavy minerals (Ghosh and Prasad 2006) or diamonds (McLachlan 1996). Heavy machines are employed, which traverse on the beach impacting directly on beach fauna and the organisms extracted along with the sand are removed. Sediment grain size is modified whilst erosion scarp is accentuated (Jonah et al. 2015). Sediment depletion caused by mining leads to beach and dunes erosion and changes on beach morpho dynamics (Borges et al. 2002). Heavy mineral mining generally dump waste water and sediments contaminated with heavy metals onto the beaches (Peter et al. 2017). Although in many countries sand mining has been forbidden or strictly regulated, illegal sand extraction can still be found as important local industries, particularly in developing nations (Masalu 2002, Gavriletea 2017).

Although there is quite a large amount of evidence on mining effects in the physical conditions of sandy beaches (e.g., Borges et al. 2002, Thornton et al. 2006, Peduzzi 2014), information about the effects of this activity on beach inhabitants is relatively scarce. A cooper mine close to the coast of North Chile discharged Cu-contaminated tailings to beaches causing a reduction in meiofauna densities and biodiversity (Lee and Correa 2005). Sometimes, tailings from mining operations are constituted only by sand (without contaminants) which is discharged on beaches leading to changes in sediment grain size and beach morpho dynamics. Tailings disposal of a diamond mine in Elizabeth Bay, Namibia, decreased abundance, biomass and species richness of intertidal macrofauna (McLachlan 1996). These effects are related to the physical changes generated to the beach although smothering is not discarded as an additional impact. A long term study (1994–2012) at the same sites reported that after 3 years of mining cessation, and thus sediments redistributed uniformly along the beach, intertidal macrofauna reached relatively natural abundances (Pulfrich and Branch 2014). An independent 20 years work demonstrated a 4-fold decline in shorebirds density due to the same diamond mine and suggested that it was associated with a decrease in the abundance of shorebird preys (Simmons 2005). Sand extraction by mining operations can also lead to coarser sediment grain size which has been linked to decreased ghost crabs densities and burrows diameter (Jonah et al. 2015). In this case, coarser sand could be avoided by crabs due to less resistance of burrows or less water holding capacity of sediments.

Fisheries and Recreational Extractions

Sandy beach fisheries are generally developed at small scales and have relevant socioeconomic connotations, mainly in developing countries (Defeo 2011). According to their objectives, three main types of sandy beach fisheries can be distinguished: subsistence, when the resource is exploited for food; recreational, when the resource is exploited for bait or for food but with no dependence on it; and commercial, when the resource is exploited for sale (McLachlan and Defeo 2017). The species usually harvested by sandy beach fisheries are clams (McLachlan et al. 1996), crabs (Kyle et al. 1997), polychaetes (Cole et al. 2018) and fishes (Cabral et al. 2003). Like other fishery operations, resource overexploitation, habitat destruction and mortality due to by-catch are some of the impacts related to this activity. In particular, the strong recreational component of sandy beach fisheries and the easy access to the resource make this exploitation notably difficult to manage (Defeo 2003) (Fig. 3.7).

Figure 3.7 Recreational extraction of clams (up) and its subsequent sediment modification (below). Photos by Nicolás Chiaradia.

Although polychaete harvesting occurs mainly at muddy environments where they located in high abundances, there are some examples at sandy beaches, particularly in Australia (Cole et al. 2018). Collection of polychaetes for bait in sandy beaches can lead to changes in population size structure (due to size selection) and further contribute to a whole population decline, although a relatively fast recovery can be expected to this group of organisms (Olive 1993). However, it has been stated recently that the amount of catches of bait fisheries are overlooked in the management context and its impact on natural environments may be underestimated (Watson et al. 2017). Another example of fishing for bait or food is that reported by Kyle et al. (1997) on ghost crabs and mole crabs at a marine reserve in South Africa. In such area, the harvest catches of ghost crabs (4800 km^{-1} $year^{-1}$) are relatively low considering the standing stock of the species (4400–15200 km^{-1}). Mole crabs harvested sizes are above the first maturity size and their catches per unit effort are stable in time. Therefore, Kyle et al. (1997) suggested that these intertidal fisheries can be considered sustainable, but they recommend not to introduce fishing gears, as pit falls or nets, because they could lead to higher extraction above the sustainable degree. The case is different with clams, which are harvested for both food and commercial purposes and are maybe the most studied sandy beach intertidal fishery regarding its impact. Harvesting size selection in sandy beach of clams leads to a depletion of adult individuals and consequently to recruitment reduction, which will further result in an impact to the whole population structure (Defeo 1996, Castilla and Defeo 2001). In fact, Defeo (2003) through an extensive analysis of sandy beach invertebrate fisheries stated that sandy beach stocks are fully exploited or even overexploited or collapsed. Collateral adverse effects (e.g., decreased density) on non-target species (Defeo and de Alava 1995, Defeo 1996) or size class (Brazeiro and Defeo 1999) are generated probably by physical damage or habitat modification. Other indirect effects of sandy beach fisheries has been reported for a fish weir, an artisanal fixed trap deployed on the low intertidal zone to catch fishes, which leads to relevant changes in the macroinfaunal intertidal community probably due to the registered alterations in sediment grain size and water and organic matter content (dos Santos and Aviz 2020).

The difference in study intensity between the different impacts could be noted here. Some have been historically understudied and remain the same, while others are on the rise. It is also worth pointing out that in most cases of the above description of each activity, always at least another is mentioned, which highlight the imbricated structure of human impacts over sandy beaches and their close relationship. Therefore, it is difficult to study the impact of these activities separately, but not impossible. Both studies of these impacts separately and those studying them together provide valuable information for understanding the extent of human intervention on sandy beaches and its effects on their residents.

FINAL CONSIDERATIONS

In this chapter the main human activities that impact sandy beaches have been described. A significant amount of evidence of the adverse effects generated by such impacts has been presented. In general, there is consensus that human activities and therefore their impacts, are expected to grow in an accelerated manner in the next decades. Furthermore, this increase would be exacerbated by the global climate change. In this context, some refinements in the scientific methodologies to assess impacts, the research on newly accepted sources of impacts (as artificial light and microplastic pollution) and long-term studies, are of paramount importance.

The collection of evidence on human impacts over sandy beaches is meaningless if it just to be stored in scientific literature. Although many authors are dedicated to the generation of management tools in order to provide them to stakeholders, most sandy beaches in the world are still uncontrollably exploited. In other cases, the decisions and measures taken are solutions in the short term and thus, the problem is finally unresolved. A main issue in this sense is the balance among economic, social and ecological (including all ecosystem services) value of sandy beaches: the last one is poorly perceived and assumed by both the decision managers and the society in general. Therefore, in addition to continuing to provide evidence and tools to decision makers, it is extremely relevant to make society aware of the importance of these ecosystems through scientific dissemination.

Acknowledgments

We would like to thank to *Consejo Nacional de Investigaciones Científicas y Técnicas* (CONICET) and *Universidad Nacional de Mar del Plata* (UNMdP) for their support in our researches.

REFERENCES

Aria, M. and C. Cuccurullo. 2017. Bibliometrix: An R tool for comprehensive science mapping analysis. J. Informetr. 11: 959–975.

ASBPA (American Shore and Beach Preservation Association). 2020. National Beach Nourishment Database. https://gim2.aptim.com/ASBPANationwideRenourishment.

Barca-Bravo, S., M.J. Servia, F. Cobo and M.A. Gonzalez. 2008. The effect of human use of sandy beaches on developmental stability of *Talitrus saltator* (Montagu, 1808) (Crustacea, Amphipoda). A study on fluctuating asymmetry. Mar. Ecol. 29: 91–98.

Barros, F. 2001. Ghost crabs as a tool for rapid assessment of human impacts on exposed sandy beaches. Biol. Conserv. 97: 399–404.

Behera, S., B. Tripathy, K. Sivakumar and B.C. Choudhury. 2016. Beach dynamics and impact of armouring on olive ridley sea turtle (*Lepidochelys olivacea*) nesting at Gahirmatha rookery of Odisha coast, India. Indian J. Geo-Mar. Sci. 45(2): 233–238.

Bejarano, A.C. and J. Michel. 2016. Oil spills and their impacts on sand beach invertebrate communities: A literature review. Environ. Poll. 218: 709–722.

Beretta, M., V. Britto, T.M. Tavares, S.M. da Silva and A.L. Pletsch. 2014. Occurrence of pharmaceutical and personal care products (PPCPs) in marine sediments in the Todos os Santos Bay and the north coast of Salvador, Bahia, Brazil. J. Soils Sed. 14: 1278–1286.

Bessa, F., S.C. Gonçalves, J.N. Franco, J.N. André, P.P. Cunha and J.C. Marques. 2014. Temporal changes in macrofauna as response indicator to potential human pressures on sandy beaches. Ecol. Ind. 41: 49–57.

Besseling, E., A. Wegner, E.M. Foekema, M.J. Van Den Heuvel-Greve and A.A. Koelmans. 2013. Effects of microplastic on fitness and PCB bioaccumulation by the lugworm *Arenicola marina* (L.). Environ. Sci. Technol. 47: 593–600.

Boere, V., E.R. Cansi, A.B.B. Alvarenga and I.O. Silva. 2011. The burying behavior of the mole crabs before and after an accident with urban sewage efluents in Bombinhas Beach, Santa Catarina, Brazil. Rev. Ambient. Água. 6: 70–76.

Borges, P., C. Andrade and M.C. Freitas. 2002. Dune, bluff and beach erosion due to exhaustive sand mining—the case of Santa Barbara Beach, São Miguel (Azores, Portugal). J. Coast. Res. 36: 89–95.

Brannock, P.M., J. Sharma, H.M. Bik, W.K. Thomas and K.M. Halanych. 2017. Spatial and temporal variation of intertidal nematodes in the northern Gulf of Mexico after the Deepwater Horizon oil spill. Mar. Environ. Res. 130: 200–212.

Brazeiro, A. and O. Defeo. 1999. Effects of harvesting and density dependence on the demography of sandy beach populations: the yellow clam *Mesodesma mactroides* of Uruguay. Mar. Ecol. Prog. Ser. 182: 127–135.

Browne, M.A., S.J. Niven, T.S. Galloway, S.T. Rowland and R.C. Thompson. 2013. Microplastic moves pollutants and additives to worms, reducing functions linked to health and biodiversity. Curr. Biol. 23: 2388–2392.

Bruck, S. and A.T. Ford. 2018. Chronic ingestion of polystyrene microparticles in low doses has no effect on food consumption and growth to the intertidal amphipod *Echinogammarus marinus*? Environ. Poll. 233: 1125–1130.

Bulleri, F. and M.G. Chapman. 2010. The introduction of coastal infrastructure as a driver of change in marine environments. J. Appl. Ecol. 47: 26–35.

Burlas, M., G. Ray and D. Clarke. 2001. The New York District's Biological Monitoring Program for the Atlantic Coast of New Jersey, Asbury Park to Manasquan Sections Beach Erosion Control Project: Draft Pre-construction Baseline Studies Report. US Army Corps of Engineers. Washington, D.C., USA.

Cabral, H., J. Duque and M.J. Costa. 2003. Discards of the beach seine fishery in the central coast of Portugal. Fish. Res. 63: 63–71.

Campbell, M.L., C. Slavin, A. Grage and A. Kinslow. 2016. Human health impacts from litter on beaches and associated perceptions: a case study of 'clean' Tasmanian beaches. Ocean Coast. Manag. 126: 22–30.

Cardoso, R.S., C.A.M. Barboza, V.B. Skinner and T.M.B. Cabrini. 2016. Crustaceans as ecological indicators of metropolitan sandy beaches health. Ecol. Ind. 62: 154–162.

Carrasco, A., K. Pulgar, D. Quintanilla-Ahumada, D. Perez-Venegas, P.A. Quijón and C. Duarte. 2019. The influence of microplastics pollution on the feeding behavior of a prominent sandy beach amphipod, *Orchestoidea tuberculata* (Nicolet, 1849). Mar. Poll. Bull. 145: 23–27.

Castilla, J.C. and O. Defeo. 2001. Latin American benthic shellfisheries: emphasis on co-management and experimental practices. Rev. Fish Biol. Fish. 11: 1–30.

Chiva, L., J.I. Pagán, I. López I, A.J. Tenza-Abril, L. Aragonés and I. Sánchez. 2018. The effects of sediment used in beach nourishment: study case El Portet de Moraira beach. Sci. Total Environ. 628: 64–73.

Cisneros, J.A., T.R Briggs and K. Martin. 2017. Placed sediment characteristics compared to sea turtle nesting and hatching patterns: a case study from Palm Beach County, FL. Shore Beach. 85(2): 35–40.

Cole, V.J., R.C. Chick and P.A. Hutchings. 2018. A review of global fisheries for polychaete worms as a resource for recreational fishers: diversity, sustainability and research needs. Rev. Fish Biol. Fish. 28: 543–565.

Cooper, J.A.G., M.C. O'Connor and S. McIvor. 2020. Coastal defences versus coastal ecosystems: a regional appraisal. Mar. Policy. 102332.

Costa, L.L., D.C. Tavares, M.C. Suciu, D.F. Rangel and I.R. Zalmon. 2017. Human-induced changes in the trophic functioning of sandy beaches. Ecol. Ind. 82: 304–315.

Costa, L.L. and I.R. Zalmon. 2019. Sensitivity of macroinvertebrates to human impacts on sandy beaches: a case study with tiger beetles (Insecta, Cicindelidae). Estuar. Coast. Shelf Sci. 220: 142–151.

Costa, L.L., V.F. Arueira, M.F. da Costa, A.P. Di Beneditto and I.R. Zalmon. 2019a. Can the Atlantic ghost crab be a potential biomonitor of microplastic pollution of sandy beaches sediment? Mar. Poll. Bull. 145: 5–13.

Costa, L.L., J.F. Madureira and I.R. Zalmon. 2019b. Changes in the behaviour of *Ocypode quadrata* (Fabricius, 1787) after experimental trampling. J. Mar. Biol. Assoc. UK. 99: 1135–1140.

Costa, L.L., I.R. Zalmon, L. Fanini and O. Defeo. 2020. Macroinvertebrates as indicators of human disturbances on sandy beaches: a global review. Ecol. Ind. 118: 106764.

Davenport, J. and J.L. Davenport. 2006. The impact of tourism and personal leisure transport on coastal environments: a review. Estuar. Coast. Shelf Sci. 67: 280–292.

Davies, T.W., J.P. Duffy, J. Bennie and K.J. Gaston. 2014. The nature, extent, and ecological implications of marine light pollution. Front. Ecol. Environ. 12(6): 347–355.

Davies, R., P.C. Speldewinde and B.A. Stewart. 2016. Low level off-road vehicle (ORV) traffic negatively impacts macroinvertebrate assemblages at sandy beaches in south-western Australia. Sci. Rep. 6: 1–8.

Defeo, O. and A. de Alava. 1995. Effects of human activities on long-term trends in sandy beach populations: the wedge clam *Donax hanleyanus* in Uruguay. Mar. Ecol. Progr. Ser. 123: 73–82.

Defeo, O. 1996. Experimental management of an exploited sandy beach bivalve population. Rev. Chil. Hist. Nat. 69: 605–614.

Defeo, O. 2003. Marine invertebrate fisheries in sandy beaches: an overview. J. Coast. Res. 35: 56–65.

Defeo, O. and A. McLachlan. 2005. Patterns, processes and regulatory mechanisms in sandy beach macrofauna: a multi-scale analysis. Mar. Ecol. Prog. Ser. 295: 1–20.

Defeo, O., A. McLachlan, D.S Schoeman, T.A. Schlacher, J. Dugan, A. Jones et al. 2009. Threats to sandy beach ecosystems: a review. Estuar. Coast. Shelf Sci. 81: 1–12.

Defeo, O. 2011. Sandy beach fisheries as complex social-ecological systems: emerging paradigms for research, management and governance. pp. 111–112. *In*: A. Bayed [ed.]. Sandy Beaches and Coastal Zone Management—Proceedings of the Fifth International Symposium on Sandy Beaches, 19th–23rd October 2009, Rabat, Morocco. Travaux de l'Institut Scientifique, Rabat, Morocco.

Dethier, M.N., W.W. Raymond, A.N. McBride, J.D. Toft, J.R. Cordell, A.S. Ogston, et al. 2016. Multiscale impacts of armoring on Salish Sea shorelines: evidence for cumulative and threshold effects. Estuar. Coast. Shelf Sci. 175: 106–117.

dos Santos, T.M.T. and D. Aviz. 2020. Effects of a fish weir on the structure of the macrobenthic community of a tropical sandy beach on the Amazon coast. J. Mar. Biol. Assoc. UK. 100: 211–219.

Duarte, C., D. Quintanilla-Ahumada, C. Anguita, P.H. Manríquez, S. Widdicombe, J. Pulgar, et al. 2019. Artificial light pollution at night (ALAN) disrupts the distribution and circadian rhythm of a sandy beach isopod. Environ. Poll. 248: 565–573.

Dugan, J.E., D.M. Hubbard, M.D. McCrary and M.O. Pierson. 2003. The response of macrofauna communities and shorebirds to macrophyte wrack subsidies on exposed sandy beaches of southern California. Estuar. Coast. Shelf Sci. 58: 25–40.

Dugan, J.E. and D.M. Hubbard. 2006. Ecological responses to coastal armoring on exposed sandy beaches. Shore Beach. 74: 10–16.

Dugan, J.E., D.M. Hubbard, I.F. Rodil, D.L. Revell and S. Schroeter. 2008. Ecological effects of coastal armoring on sandy beaches. Mar. Ecol. 29: 160–170.

Dugan, J.E., and D.M. Hubbard. 2010. Loss of coastal strand habitat in Southern California: The role of beach grooming. Estuaries Coast. 33: 67–77.

Dugan, J.E., D.M. Hubbard, H.M. Page and J.P. Schimel. 2011. Marine macrophyte wrack inputs and dissolved nutrients in beach sands. Estuaries Coast. 34: 839–850.

Dugan, J.E., K.A. Emery, M. Alber, C.R. Alexander, J.E. Byers, A.M. Gehman, et al. 2018. Generalizing ecological effects of shoreline armoring across soft sediment environments. Estuaries Coast. 41: 180–196.

Duvat, V. 2013. Coastal protection structures in Tarawa Atoll, Republic of Kiribati. Sustain. Sci. 8: 363–379.

Esiukova, E. 2017. Plastic pollution on the Baltic beaches of Kaliningrad region, Russia. Mar. Poll. Bull. 114(2): 1072–1080.

Fanini, L., C.M. Cantarino and F. Scapini. 2005. Relationships between the dynamics of two Talitrus saltator populations and the impacts of activities linked to tourism. Oceanologia 47(1): 93–112.

Fanini, L., G.M. Marchetti, F. Scapini and O. Defeo. 2009. Effects of beach nourishment and groynes building on population and community descriptors of mobile arthropodofauna. Ecol. Ind. 9: 167–178.

Fanini, L., G. Zampicinini and E. Pafilis. 2014. Beach parties: a case study on recreational human use of the beach and its effects on mobile arthropod fauna. Ethol. Ecol. Evol. 26: 69–79.

Fanini, L., O. Defeo and M. Elliott. 2020. Advances in sandy beach research-Local and global perspectives. Estuar. Coast. Shelf Sci. 234:106646

Fendall, L.S. and M.A. Sewell. 2009. Contributing to marine pollution by washing your face: microplastics in facial cleansers. Mar. Poll. Bull. 58: 1225–1228.

Fernández-Severini, M.D., M.C. Menéndez, N.S. Buzzi, A.L. Delgado, M.C. Piccolo and J.E. Marcovecchio. 2019. Metals in the particulate matter from surf zone waters of a Southwestern Atlantic sandy beach (Monte Hermoso, Argentina). Reg. Stud. Mar. Sci. 29: 100646.

Fernandino, G., C.I. Elliff, S.I. Reimão, T. de Souza Brito and A.C. da Silva Pinto Bittencourt. 2016. Plastic fragments as a major component of marine litter: a case study in Salvador, Bahia, Brazil. J. Int. Coast. Zone. Manag. 16(3): 281–287.

Garratt, M.J., S.R. Jenkins and T.W. Davies. 2019. Mapping the consequences of artificial light at night for intertidal ecosystems. Sci. Total Environ. 691: 760–768.

Gaston, K.J., T.W. Davies, S.L. Nedelec and L.A. Holt. 2017. Impacts of artificial light at night on biological timings. Annu. Rev. Ecol. Evol. Syst. 48: 49–68.

Gavriletea, M.D. 2017. Environmental impacts of sand exploitation. Analysis of sand market. Sustainability 9: 1118.

Gheskiere, T., V. Magda, P. Greet and D. Steven. 2006. Are strandline meiofaunal assemblages affected by a once-only mechanical beach cleaning? Experimental findings. Mar. Environ. Res. 61: 245–264.

Ghosh, A.K. and C.D. Prasad. 2006. Mechanisation of beach placer mining above water table-some preliminary considerations. J. Mines Met. Fuels 54: 64–71.

Gilburn, A.S. 2012. Mechanical grooming and beach award status are associated with low strandline biodiversity in Scotland. Estuar. Coast. Shelf Sci. 107: 81–88.

Gittman, R.K., S.B. Scyphers, C.S. Smith, I.P. Neylan and J.H. Grabowski. 2016. Ecological consequences of shoreline hardening: a Meta-Analysis. BioScience 66: 763–773.

Gómez, V., K. Pozo, D. Nuñez, P. Přibylová, O. Audy, M. Baini, et al. 2020. Marine plastic debris in Central Chile: characterization and abundance of macroplastics and burden of persistent organic pollutants (POPs). Mar. Poll. Bull. 152: 110881.

Gonçalves, S.C. and J.C. Marques. 2011. The effects of season and wrack subsidy on the community functioning of exposed sandy beaches. Estuar. Coast. Shelf Sci. 95: 165–177.

Gonçalves, S.C., P.M. Anastácio and J.C. Marques. 2013. Talitrid and Tylid crustaceans bioecology as a tool to monitor and assess sandy beaches' ecological quality condition. Ecol. Ind. 29: 549–557.

Gonçalves, S.C. and J.C. Marques. 2017. Assessement and management of environmental quality conditions in marine sandy beaches for its sustainable use—Virtues of the population based approach. Ecol. Ind. 74: 140–146.

González, S.A., K. Yáñez-Navea and M. Muñoz. 2014. Effect of coastal urbanization on sandy beach coleoptera *Phaleria maculata* (Kulzer, 1959) in northern Chile. Mar. Poll. Bull. 83: 265–274.

Gössling, S., C.M. Hall and D. Scott. 2018. Coastal and ocean tourism. pp. 773–790. *In*: M. Salomon and T. Markus [eds]. Handbook on Marine Environment Protection. Springer, Berlin.

Green, D.S., B. Boots, J. Sigwart, S. Jiang and C. Rocha. 2016. Effects of conventional and biodegradable microplastics on a marine ecosystem engineer (*Arenicola marina*) and sediment nutrient cycling. Environ. Poll. 208: 426–434.

Griffin, C., N. Day, H. Rosenquist, M. Wellenreuther, N. Bunnefeld and A.S. Gilburn. 2018. Tidal range and recovery from the impacts of mechanical beach grooming. Ocean Coast. Manag. 154: 66–71.

Griggs, G.B. 2005. The impacts of coastal armoring. Shore Beach. 73: 13–22.

Griggs, G. and K. Patsch. 2019. The protection/hardening of California's coast: times are changing. J. Coast. Res. 35: 1051–1061.

Gül, M.R. and B.D. Griffen. 2018. Impacts of human disturbance on ghost crab burrow morphology and distribution on sandy shores. PloS One. 13: e0209977.

Gül, M.R. and B.D. Griffen. 2020. Changes in claw morphology of a bioindicator species across habitats that differ in human disturbance. Hydrobiologia 847: 3025–3037.

Hall, C.M. 2001. Trends in ocean and coastal tourism: the end of the last frontier? Ocean Coast. Manag. 44: 601–618.

Halliday, E. and R.J Gast. 2011. Bacteria in beach sands: an emerging challenge in protecting coastal water quality and bather health. Environ. Sci. Technol. 45: 370–379.

Haynes, D., J. Leeder and P. Rayment. 1997. A comparison of the bivalve species *Donax deltoides* and *Mytilus edulis* as monitors of metal exposure from effluent discharges along the Ninety Mile Beach, Victoria, Australia. Mar. Poll. Bull. 34: 326–331.

Horn, D., M. Miller, S. Anderson and C. Steele. 2019. Microplastics are ubiquitous on California beaches and enter the coastal food web through consumption by Pacific mole crabs. Mar. Poll. Bull. 139: 231–237.

Hu, Z., H. Hu and Y. Huang. 2018. Association between nighttime artificial light pollution and sea turtle nest density along Florida coast: A geospatial study using VIIRS remote sensing data. Environ. Poll. 239: 30–42.

Hubbard, D.M. and J.E. Dugan. 2003. Shorebird use of an exposed sandy beach in southern California. Estuar. Coast. Shelf Sci. 58: 41–54.

Hussain, K.J., A.K. Mohanty, K.K. Satpathy and M.V.R. Prasad. 2010. Abundance pattern of wedge clam *Donax cuneatus* (L.) in different spatial scale in the vicinity of a coastal nuclear power plant. Environ. Monitor. Assess. 163: 185–194.

Iannilli, V., A. Di Gennaro, F. Lecce, M. Sighicelli, M. Falconieri, L. Pietrelli, et al. 2018. Microplastics in *Talitrus saltator* (Crustacea, Amphipoda): new evidence of ingestion from natural contexts. Environ. Sci. Poll. Res. 25: 28725–28729.

Idardare, Z., A. Moukrim, J.F. Chiffoleau and A.A. Alla. 2011. Trace metals in the clam *Donax trunculus* L. from the Bouadisse sandy beach, discharge zone of a plant sewage outfall in Agadir Bay (Morocco). pp. 51–58. *In*: A. Bayed [ed.]. Sandy Beaches and Coastal Zone Management—Proceedings of the Fifth International Symposium on Sandy Beaches, 19th–23rd October 2009, Rabat, Morocco. Travaux de l'Institut Scientifique, Rabat, Morocco.

Infante, M., A. Marsico and L. Pennetta. 2012. Some results of coastal defences monitoring by ground laser scanning technology. Environ. Earth Sci. 67: 2449–2458.

Jaramillo, E., H. Contreras and A. Bollinger. 2002. Beach and faunal response to the construction of a seawall in a sandy beach of south central Chile. J. Coast. Res. 18(3): 523–529.

Jonah, F.E., N.W. Agbo, W. Agbeti, D. Adjei-Boateng and M.J. Shimba. 2015. The ecological effects of beach sand mining in Ghana using ghost crabs (Ocypode species) as biological indicators. Ocean Coast. Manag. 112: 18–24.

Jones, A.R., A. Murray, T.A. Lasiak and R.E. Marsh. 2008. The effects of beach nourishment on the sandy-beach amphipod *Exoediceros fossor*: impact and recovery in Botany Bay, New South Wales, Australia. Mar. Ecol. 29: 28–36.

Kang, T., J.H. Oh, J.-S. Hong and D. Kim. 2016. Effect of the Hebei Spirit oil spill on intertidal meiofaunal communities in Taean, Korea. Mar. Poll. Bull. 113: 444–453.

Krelling, A.P., A.T. Williams and A. Turra. 2017. Differences in perception and reaction of tourist groups to beach marine debris that can influence a loss of tourism revenue in coastal areas. Mar. Pol. 85: 87–99.

Kyle, R., W.D. Robertson and S.L. Birnie. 1997. Subsistence shellfish harvesting in the Maputaland Marine Reserve in northern KwaZulu-Natal, South Africa: sandy beach organisms. Biol. Conserv. 82: 173–182.

Laitano, M.V., Í.B. Castro, P.G. Costa, G. Fillmann and M. Cledón. 2015. Butyltin and PAH contamination of Mar del Plata port (Argentina) sediments and their influence on adjacent coastal regions. Bull. Environ. Cont. Toxicol. 95: 513–520.

Laitano, M.V., N.M. Chiaradia and J.D. Nuñez. 2019. Clam population dynamics as an indicator of beach urbanization impacts. Ecol. Ind. 101: 926–932.

Lee, M.R. and J.A. Correa. 2005. Effects of copper mine tailings disposal on littoral meiofaunal assemblages in the Atacama region of northern Chile. Mar. Environ. Res. 59: 1–18.

Leewis, L., P.M. van Bodegom, J. Rozema and G.M. Janssen. 2012. Does beach nourishment have long-term effects on intertidal macroinvertebrate species abundance? Estuar. Coast. Shelf Sci. 113: 172–181.

Li, W.C., H.F. Tse and L. Fok. 2016. Plastic waste in the marine environment: a review of sources, occurrence and effects. Sci. Total Environ. 566: 333–349.

Luarte, T., C.C. Bonta, E.A. Silva-Rodriguez, P.A. Quijón, C. Miranda, A.A. Farias, et al. 2016. Light pollution reduces activity, food consumption and growth rates in a sandy beach invertebrate. Environ. Poll. 218: 1147–1153.

Lucrezi, S., T.A. Schlacher and W. Robinson. 2009. Human disturbance as a cause of bias in ecological indicators for sandy beaches: Experimental evidence for the effects of human trampling on ghost crabs (*Ocypode* spp.). Ecol. Ind. 9: 913–921.

Lucrezi, S. and T.A. Schlacher. 2010. Impacts of off-road vehicles (ORVs) on burrow architecture of ghost crabs (Genus *Ocypode*) on sandy beaches. Environ. Manag. 45: 1352–1362.

Lucrezi, S., T.A. Schlacher and W. Robinson. 2010. Can storms and shore armouring exert additive effects on sandy-beach habitats and biota? Mar. Freshw. Res. 61: 951–962.

Lucrezi, S., M. Saayman and P. Van der Merwe. 2014. Impact of off-road vehicles (ORVs) on ghost crabs of sandy beaches with traffic restrictions: a case study of Sodwana Bay, South Africa. Environ. Manag. 53: 520–533.

Ma, M., Z. Feng, C. Guan, Y. Ma, H. Xu and H. Li. 2001. DDT, PAH and PCB in sediments from the intertidal zone of the Bohai Sea and the Yellow Sea. Mar. Poll. Bull. 42: 132–136.

Machado, P.M., L.L. Costa, M.C. Suciu, D.C. Tavares and I.R. Zalmon. 2016. Extreme storm wave influence on sandy beach macrofauna with distinct human pressures. Mar. Poll. Bull. 107: 125–135.

Machado, P.M., M.C. Suciu, L.L. Costa, D.C. Tavares and I.R. Zalmon. 2017. Tourism impacts on benthic communities of sandy beaches. Mar. Ecol. 38: e12440.

MacMillan, M.R. and P.A. Quijón. 2012. Wrack patches and their influence on upper-shore macrofaunal abundance in an Atlantic Canada sandy beach system. J. Sea Res. 72: 28–37.

Malm, T., S. Raberg, S. Fell and P. Carlsson. 2004. Effects of beach cast cleaning on beach quality, microbial food web, and littoral macrofaunal biodiversity. Estuar. Coast. Shelf Sci. 60: 339–347.

Manning, L.M., C.H. Peterson and M.J. Bishop. 2014. Dominant macrobenthic populations experience sustained impacts from annual disposal of fine sediments on sandy beaches. Mar. Ecol. Prog. Ser. 508: 1–15.

Marcomini, S.C. and R.A. López. 2006. Evolution of a Beach Nourishment Project at Mar del Plata. J. Coast. Res. 39: 834–837.

Marn, N., M. Jusup, S.A.L.M. Kooijman and T. Klanjscek T. 2020. Quantifying impacts of plastic debris on marine wildlife identifies ecological breakpoints. Ecol. Lett. 23(10): 1479–1487.

Martin, D., F. Bertasi, M.A. Colangelo, M. de Vries, M. Frost, S.J. Hawkins, et al. 2005. Ecological impact of coastal defence structures on sediment and mobile fauna: evaluating and forecasting consequences of unavoidable modifications of native habitats. Coast. Eng. 52: 1027–1051.

Martins, J. and P. Sobral. 2011. Plastic marine debris on the Portuguese coastline: a matter of size? Mar. Poll. Bull. 62: 2649–2653.

Martinez-Ribes, L., G. Basterretxea, M. Palmer and J. Tintoré. 2007. Origin and abundance of beach debris in the Balearic Islands. Sci. Mar. 71: 305–314.

Masalu, D.C. 2002. Coastal erosion and its social and environmental aspects in Tanzania: a case study in illegal sand mining. Coast. Manag. 30: 347–359.

Mazor, T., N. Levin, H.P. Possingham, Y. Levy, D. Rocchini, A.J. Richardson, et al. 2013. Can satellite-based night lights be used for conservation? The case of nesting sea turtles in the Mediterranean. Biol. Conserv. 159: 63–72.

McLachlan, A., E. Jaramillo, T.E. Donn and F. Wessels. 1993. Sandy beach macrofauna communities and their control by the physical environment: a geographical comparison. J. Coast. Res. 15: 27–38.

McLachlan, A. 1996. Physical factors in benthic ecology: effects of changing sand particle size on beach fauna. Mar. Ecol. Prog. Ser. 131: 205–217.

McLachlan, A., J.E. Dugan, O. Defeo, A.D. Ansell, D.M. Hubbard, E. Jaramillo, et al. 1996. Beach clam fisheries. Oceanogr. Mar. Biol. Annu. Rev. 34: 163–232.

McLachlan, A., O. Defeo, E. Jaramillo and A.D. Short. 2013. Sandy beach conservation and recreation: guidelines for optimising management strategies for multi-purpose use. Ocean Coas. Manag. 71: 256–268.

McLachlan, A. and O. Defeo. 2017. The Ecology of Sandy Shores. Academic Press, London, UK.

Menn, I., C. Junghans and K. Reise. 2003. Buried alive: effects of beach nourishment on the infauna of an erosive shore in the North Sea. Senck. marit. 32: 125–145.

Michaud, K.M., K.A. Emery, J.E. Dugan, D.M. Hubbard and R.J. Miller. 2019. Wrack resource use by intertidal consumers on sandy beaches. Estuar. Coast. Shelf Sci. 221: 66–71.

Michel, J., S.R. Fegley, J.A. Dahlin and C. Wood. 2017. Oil spill response-related injuries on sand beaches: when shoreline treatment extends the impacts beyond the oil. Mar. Ecol. Prog. Ser. 576: 203–218.

Moffett, M.D., A. McLachlan, P.E.D. Winter and A.M.C. De Ruyck. 1998. Impact of trampling on sandy beach macrofauna. J. Coast. Cons. 4: 87–90.

Mongeon, P. and A. Paul-Hus. 2016. The journal coverage of Web of Science and Scopus: a comparative analysis. Scientometrics 106: 213–228.

Moore, C.J., S.L. Moore, M.K. Leecaster and S.B. Weisberg. 2001. A comparison of plastic and plankton in the North Pacific central gyre. Mar. Poll. Bull. 42: 1297–1300.

Morton, J.K., E.J. Ward and K.C. de Berg. 2015. Potential small- and large-scale effects of mechanical beach cleaning on biological assemblages of exposed sandy beaches receiving low inputs of beach-cast macroalgae. Estuar. Coast. 38: 2083–2100.

Moss, D. and D.P. McPhee. 2006. The impacts of recreational four-wheel driving on the abundance of the ghost crab (*Ocypode cordimanus*) on a subtropical sandy beach in SE Queensland. Coast. Manag. 34: 133–140.

Nel, R., E.E. Campbell, L. Harris, L. Hauser, D.S. Schoeman, A. McLachlan, et al. 2014. The status of sandy beach science: Past trends, progress, and possible futures. Estuar. Coast. Shelf Sci. 150: 1–10.

Nelms, S.E., E.M. Duncan, A.C. Broderick, T.S. Galloway, M.H. Godfrey, M. Hamann, et al. 2016. Plastic and marine turtles: a review and call for research. ICES J. Mar. Sci. 73: 165–181.

Nelson Sella, K.A., L. Sicius and M.M. Fuentes. 2019. Using expert elicitation to determine the relative impact of coastal modifications on marine turtle nesting grounds. Coast. Manag. 47: 492–506.

Neuberger-Cywiak, L., Y. Achituv and E.M. Garcia. 2003. Effects of Zinc and Cadmium on the burrowing behavior, LC_{50}, and LT_{50} on *Donax trunculus* Linnaeus (Bivalvia-Donacidae). Bull. Environ. Cont. Toxicol. 70: 0713–0722.

Neumann, B., A.T. Vafeidis, J. Zimmermann and R.J. Nicholls. 2015. Future coastal population growth and exposure to sea-level rise and coastal flooding—a global assessment. PloS one 10: e0118571.

Nordstrom, K.F. 2014. Living with shore protection structures: a review. Estuar. Coast. Shelf Sci. 150: 11–23.

Nourisson, D.H., F. Scapini and A. Milstein. 2018. Small-scale changes of an arthropod beach community after hard-engineering interventions on a Mediterranean beach. Reg. Stu. Mar. Sci. 22: 21–30.

Olive, P.J.W. 1993. Management of the exploitation of the lugworm *Arenicola marina* and the ragworm *Nereis virens* (Polychaeta) in conservation areas. Aquat. Conserv.: Mar. Freshw. Ecosyst. 3: 1–24.

Ong, E.S. and Z.B. Din. 2001. Cadmium, copper, and zinc toxicity to the clam, *Donax faba* C., and the blood cockle, *Anadara granosa* L. Bull. Environ. Cont. Toxicol. 66: 86–93.

Orlando, L., L. Ortega and O. Defeo. 2020. Urbanization effects on sandy beach macrofauna along an estuarine gradient. Ecol. Ind. 111: 106036.

Park, H.J., Y.-J. Lee, E. Han, K.-S. Choi, J.H. Kwak, E.J. Choy, et al. 2020. Effect of the Hebei Spirit oil spill on the condition, reproduction, and energy storage cycle of the manila clam *Ruditapes philippinarum* on the West Coast of Korea. Estuar. Coast. 43: 602–614.

Paz-Villarraga, C.A., Í.B. Castro, P. Miloslavich and G. Fillmann. 2015. Venezuelan Caribbean Sea under the threat of TBT. Chemosphere 119: 704–710.

Peduzzi, P. 2014. Sand, rarer than one thinks. Environ. Develop. 11: 208–218.

Peter, T.S., N. Chandrasekar, J.J. Wilson, S. Selvakumar, S. Krishnakumar and N.S. Magesh. 2017. A baseline record of trace elements concentration along the beach placer mining areas of Kanyakumari coast, South India. Mar. Poll. Bull. 119: 416–422.

Peterson, C.H., M.J. Bishop, G.A. Johnson, L.M. D'Anna and L.M. Manning. 2006. Exploiting beach filling as an unaffordable experiment: benthic intertidal impacts propagating upwards to shorebirds. J. Exp. Mar. Biol. Ecol. 338: 205–221.

Peterson, C.H., M.J. Bishop, L.M. D'Anna and G.A. Johnson. 2014. Multi-year persistence of beach habitat degradation from nourishment using coarse shelly sediments. Sci. Total Environ. 487: 481–492.

Piñon-Colin, T.J., R. Rodriguez-Jimenez, M.A. Pastrana-Corral, E. Rogel-Hernandez and F. Toyohiko Wakida. 2018. Microplastics on sandy beaches of the Baja California Peninsula, Mexico. Mar. Poll. Bull. 131: 63–71.

Pranzini, E., G. Anfuso, C. Botero, A. Cabrera, Y. Apin Campos, G. Casas Martinez, et al. 2016. Beach colour at Cuba and management issues. Ocean. Coast. Manag. 126: 51–60.

Pranzini, E., G. Anfuso, I. Cinelli, M. Piccardi and G. Vitale. 2018. Shore protection structures increase and evolution on the Northern Tuscany Coast (Italy): influence of tourism industry. Water 10: 1647.

Pulfrich, A. and G.M. Branch. 2014. Using diamond-mined sediment discharges to test the paradigms of sandy-beach ecology. Estuar. Coast. Shelf Sci. 150: 165–178.

Rangel-Buitrago, N., A. Williams, G. Anfuso, M. Arias and A. Gracia. 2017. Magnitudes, sources, and management of beach litter along the Atlantico department coastline, Caribbean coast of Colombia. Ocean Coast. Manag. 138: 142–157.

Rangel-Buitrago, N., A.T. Williams and G. Anfuso. 2018. Hard protection structures as a principal coastal erosion management strategy along the Caribbean coast of Colombia. A chronicle of pitfalls. Ocean Coast. Manag. 156: 58–75.

Reyes-Martínez, M.J., M.C. Ruíz-Delgado, J.E. Sánchez-Moyano and F.J. García-García. 2015. Response of intertidal sandy-beach macrofauna to human trampling: an urban vs. natural beach system approach. Mar. Environ. Res. 103: 36–45.

Rice, C.A. 2006. Effects of shoreline modification on a Northern Puget Sound beach: microclimate and embryo mortality in surt smelt (*Hypomesus pretiosus*). Estuar. Coast. 29: 63–71.

Rivas, M.L., P.S. Tomillo, J.D. Uribeondo and A. Marco. 2015. Leatherback hatchling sea-finding in response to artificial lighting: interaction between wavelength and moonlight. J. Exp. Mar. Biol. Ecol. 463: 143–149.

Rodil, I.F., E. Jaramillo, E. Acuña, M. Manzano and C. Velasquez. 2016. Long-term responses of sandy beach crustaceans to the effects of coastal armouring after the 2010 Maule earthquake in South Central Chile. J. Sea Res. 108: 10–18.

Roman, L., B.D. Hardesty, M.A. Hindell and C. Wilcox. 2019. A quantitative analysis linking seabird mortality and marine debris ingestion. Sci. Rep. 9: 3202.

Rosov, B., S. Bush, T.R. Briggs and N. Elko. 2016. The state of understanding the impacts of beach nourishment activities on infaunal communities. Shore Beach 84: 51–55.

Ryan, P.G. 2018. Entanglement of birds in plastics and other synthetic materials. Mar. Poll. Bull. 135: 159–164.

Sauco, S., G. Eguren, H. Heinzen and O. Defeo. 2010. Effects of herbicides and freshwater discharge on water chemistry, toxicity and benthos in a Uruguayan sandy beach. Mar. Environ. Res. 70: 300–307.

Scapini, F., L. Chelazzi, I. Colombini, M. Fallaci and L. Fannini. 2005. Orientation of sandhoppers at different points along a dynamic shoreline in southern Tuscany. Mar. Biol. 147: 919–926.

Schlacher, T.A., D. Richardson and I. McLean. 2008a. Impacts of off-road vehicles (ORVs) on macrobenthic assemblages on sandy beaches. Environ. Manag. 41: 878–892.

Schlacher, T.A., D.S Schoeman, J.E. Dugan, M. Lastra, A. Jones, F. Scapini, et al. 2008b. Sandy beach ecosystems: key features, sampling issues, management challenges and climate change impacts. Mar. Ecol. 29: 70–90.

Schlacher, T.A., L.M. Thompson and S.J. Walker. 2008c. Mortalities caused by off-road vehicles (ORVs) to a key member of sandy beach assemblages, the surf clam *Donax deltoides*. Hydrobiologia 610: 345–350.

Schlacher, T.A. and S. Lucrezi. 2010. Compression of home ranges in ghost crabs on sandy beaches impacted by vehicle traffic. Mar. Biol. 157: 2467–2474.

Schlacher, T.A. and L. Thompson. 2012. Beach recreation impacts benthic invertebrates on ocean-exposed sandy shores. Biol. Conserv. 147: 123–132.

Schlacher, T.A., R. Noriega, A. Jones and T. Dye. 2012. The effects of beach nourishment on benthic invertebrates in eastern Australia: impacts and variable recovery. Sci. Total Environ. 435: 411–417.

Schlacher, T.A., D.S. Schoeman, A.R. Jones, J.E. Dugan, D.M. Hubbard, O. Defeo, et al. 2014. Metrics to assess ecological condition, change, and impacts in sandy beach ecosystems. J. Environ. Manag. 144: 322–335.

Schlacher, T.A., L.K. Carracher, N. Porch, R.M. Connolly, A.D. Olds, B.L. Gilby, et al. 2016. The early shorebird will catch fewer invertebrates on trampled sandy beaches. PloS One. 11: e0161905.

Schoeman, D.S., A. McLachlan and J.E. Dugan. 2000. Lessons from a disturbance experiment in the intertidal zone of an exposed sandy beach. Estuar. Coast. Shelf Sci. 50: 869–884.

Schooler, N.K., J.E. Dugan and D.M. Hubbard. 2019. No lines in the sand: impacts of intense mechanized maintenance regimes on sandy beach ecosystems span the intertidal zone on urban coasts. Ecol. Ind. 106: 105457.

Schoonees, T., A.G. Mancheño, B. Scheres, T.J. Bouma, R. Silva, T. Schlurmann, et al. 2019. Hard structures for coastal protection, towards greener designs. Estuar. Coast. 42: 1709–1729.

Scopetani, C., A. Cincinelli, T. Martellini, E. Lombardini, A. Ciofini, A. Fortunati, et al. 2018. Ingested microplastic as a two-way transporter for PBDEs in *Talitrus saltator*. Environ. Res. 167: 411–417.

Sheppard, N., K.A. Pitt and T.A. Schlacher. 2009. Sub-lethal effects of off-road vehicles (ORVs) on surf clams on sandy beaches. J. Exp. Mar. Biol. Ecol. 380: 113–118.

Silva, E., A. Marco, J. da Graça, H. Pérez, E. Abella, J. Patino-Martinez, et al. 2017. Light pollution affects nesting behavior of loggerhead turtles and predation risk of nests and hatchlings. J. Photochem. Photobiol. B 173: 240–249.

Simmons, R.E. 2005. Declining coastal avifauna at a diamond-mining site in Namibia: comparisons and causes. Ostrich 76: 97–103.

Sobocinski, K.L., J.R. Cordell and C.A. Simenstad. 2010. Effects of shoreline modifications on supratidal macroinvertebrate fauna on Puget Sound, Washington beaches. Estuar. Coast. 33: 699–711.

Staudt, F., R. Gijsman, C. Ganal, F. Mielck, J. Wolbring, H.C. Hass, et al. 2020. The sustainability of beach nourishments: a review of nourishment and environmental monitoring practice. Preprint: https://doi.org/10.31223/osf.io/knrvw.

Stelling-Wood, T.P., G.F. Clark and A.G. Poore. 2016. Responses of ghost crabs to habitat modification of urban sandy beaches. Mar. Environ. Res. 116: 32–40.

Suresh, K., M.S. Ahamed, G. Durairaj and K.V.K. Nair. 1995. Environmental physiology of the mole crab *Emerita asiatica*, at a power plant discharge area on the east coast of India. Environ. Poll. 88: 133–136.

Thornton, E.B., A. Sallenger, J.C. Sesto, L. Egley, T. McGee and R. Parsons. 2006. Sand mining impacts on long-term dune erosion in southern Monterey Bay. Mar. Geol. 229: 45–58.

Toft, J.D., J.R. Cordell and E.A. Armbrust. 2014. Shoreline armoring impacts and beach restoration effectiveness vary with elevation. Northwest Sci. 88: 367–375.

Torres, A., J. Brandt, K. Lear and J. Liu. 2017. A looming tragedy of the sand commons. Science 357: 970–971.

Tosetto, L., C. Brown and J.E. Williamson. 2016. Microplastics on beaches: ingestion and behavioural consequences for beachhoppers. Mar. Biol. 163: 199.

Ugolini, A., G. Ungherese, S. Somigli, G. Galanti, D. Baroni, F. Borghini, et al. 2008. The amphipod *Talitrus saltator* as a bioindicator of human trampling on sandy beaches. Mar. Environ. Res. 65: 349–357.

Ugolini, A., G. Ungherese, M. Ciofini, A. Lapucci and M. Camaiti. 2013. Microplastic debris in sandhoppers. Estuar. Coast. Shelf Sci. 129: 19–22.

Urban-Malinga B., M. Zalewski, A. Jakubowska, T. Wodzinowski, M. Malinga, B. Pałys, et al. 2020. Microplastics on sandy beaches of the southern Baltic Sea. Mar. Poll. Bull. 155: 111170.

van Egmond, E.M., P.M. van Bodegom, M.P. Berg, J.W. Wijsman, L. Leewis, J.M. Janssen, et al. 2018. A mega-nourishment creates novel habitat for intertidal macroinvertebrates by enhancing habitat relief of the sandy beach. Estuar. Coast. Shelf Sci. 207: 232–241.

Veloso, V.G., E.S. Silva, C.H. Caetano and R.S. Cardoso. 2006. Comparison between the macroinfauna of urbanized and protected beaches in Rio de Janeiro State, Brazil. Biol. Conserv. 127: 510–515.

Vieira, J.V., M.C. Ruiz-Delgado, M.J. Reyes-Martínez, C.A. Borzone, A. Asenjo, J.E. Sánchez-Moyano, et al. 2016. Assessment the short-term effects of wrack removal on supralittoral arthropods using the M-BACI design on Atlantic sandy beaches of Brazil and Spain. Mar. Environ. Res. 119: 222–237.

Vlachogianni, T., T. Fortibuoni, F. Ronchi, C. Zeri, C. Mazziotti, P. Tutman, et al. 2018. Marine litter on the beaches of the Adriatic and Ionian Seas: an assessment of their abundance, composition and sources. Mar. Poll. Bull. 131: 745–756.

Walker, S.J., T.A. Schlacher and L.M. Thompson. 2008. Habitat modification in a dynamic environment: the influence of a small artificial groyne on macrofaunal assemblages of a sandy beach. Estuar. Coast. Shelf Sci. 79: 24–34.

Wallace, B.P., B.A. Stacy, E. Cuevas, C. Holyoake, P.H. Lara, A.C. Marcondes, et al. 2020. Oil spills and sea turtles: documented effects and considerations for response and assessment efforts. Endanger. Species Res. 41: 17–37.

Watson, G.J., J.M. Murray, M. Schaefer and A. Bonner. 2017. Bait worms: a valuable and important fishery with implications for fisheries and conservation management. Fish Fish. 18: 374–388.

Watts, A.J., M.A. Urbina, S. Corr, C. Lewis and T.S. Galloway. 2015. Ingestion of plastic microfibers by the crab *Carcinus maenas* and its effect on food consumption and energy balance. Environ. Sci. Technol. 49: 14597–14604.

Wilcox, C., N.J. Mallos, G.H. Leonard, A. Rodriguez and B.D. Hardesty. 2016. Using expert elicitation to estimate the impacts of plastic pollution on marine wildlife. Mar. Pol. 65: 107–114.

Wilcox, C., M. Puckridge, Q.A. Schuyler, K. Townsend and B.D. Hardesty. 2018. A quantitative analysis linking sea turtle mortality and plastic debris ingestion. Sci. Rep. 8: 12536.

Willmott, H. and T. Smith. 2003. Effects of mechanical cleaning, and its cessation, on the strandline fauna at sand bay. Somerset Archaeol. Nat. Hist.: Ecol. Somerset 147: 263–273.

Windle, A.E., D.S. Hooley and D.W. Johnston. 2018. Robotic vehicles enable high-resolution light pollution sampling of sea turtle nesting beaches. Front. Mar. Sci. 5: 493.

Wooldridge, T., H.J. Henter and J.R. Kohn. 2016. Effects of beach replenishment on intertidal invertebrates: a 15-month, eight beach study. Estuar. Coast. Shelf Sci. 175: 24–33.

Zielinski, S., C.M. Botero and A. Yanes. 2019. To clean or not to clean? A critical review of beach cleaning methods and impacts. Mar. Poll. Bull. 139: 390–401.

Chapter 4

Urbanization of Coastal Areas: Loss of Coastal Dune Ecosystems in Japan

Hajime Matsushima[1*] and Susana M.F. Ferreira[2]

[1]Research Faculty of Agriculture, Hokkaido University, Japan (matts@res.agr.hokudai.ac.jp).

[2]MARE—Marine and Environmental Sciences Centre, School of Tourism and Maritime Technology, Polytechnic of Leiria, Portugal (susana.ferreira@ipleiria.pt).

INTRODUCTION

Coastal areas are highly populated because of their economic benefits, including revenue from tourism and food production, industrial and urban development, improved transportation links, amongst others. Such population growth in many of the world's deltas, barrier islands and estuaries has led to widespread conversion of natural coastal landscapes to agriculture, aquaculture, silviculture, as well as industrial and residential uses (Valiela 2006). Generally, there is no single definition of 'coast' and 'coastal zone/area', where the latter emphasizes the area or extent of the coastal ecosystems. Many reports refer to it as a within 100 km distance from the sea (CIESIN 2012, IPCC 2014, Kummu et al. 2016, Merkens et al. 2016). It has been estimated that the world's population living in a near-coast zone (within a 100 km distance from the coastline and <100 m above sea level) has increased from 23% in 1990 (1.2 billion people; Small and Nicholls 2003) to 28% in 2010

*Corresponding author: matts@res.agr.hokudai.ac.jp

(1.9 billion people; Kummu et al. 2016). Moreover, population densities in coastal regions are about three times higher than the global average (Small and Nicholls 2003). In relation to a potential sea level rise exposure, the Low-Elevation Coastal Zone (LECZ) has been used as an assessment parameter in recent years, which refers to a specific area and its population up to 10 m above sea level (Vafeidis et al. 2011). The LECZ constitutes only 2% of the world's land area, but contains 10% of the world's population (over 600 million people) and 65% of the world's largest cities, with populations greater than 5 million inhabitants (McGranahan et al. 2007; Kummu et al. 2016).

Japan is an archipelagic country in East Asia, which consists in 6,852 islands (with >100 m perimeter). Although its total land area is 377,800 km^2, the total length of its coastlines rounds 35,000 km. Japan is one of world's countries with a very long coastline (being the 6th) compared to a small land area (ranking in 63rd place) (CIA 2020). Its total population density is 316 $ind.km^{-2}$. As the Japanese land forms a narrow and steep topography, its population has concentrated in small coastal planes. In the case of Japan, most of all land areas (93%) are included in the near-coast zone, so limiting the coastal area to municipalities with coastlines. The population density in these coastal municipalities is almost twice higher (477 $ind.km^{-2}$) than the one inland (243 $ind.km^{-2}$).

Coastal areas have played an important role in Japan since ancient times. They have been important for fisheries, harbor and port activities, transportation hubs, industrial development, beach-going and recreation, etc., (Fig. 4.1a). People also admire the coastal landscape scenery. Most Japanese gardens have coastal landscapes motifs because of their beauty and religious connotation (Fig. 4.1b), portraying the image of a Buddhist paradise. The pond or white sands represent the sea, the rock formations or set of shrubs symbolize the islands where the hermits lived, white pebbles picture the beach and stone lanterns set by the pond correspond to lighthouses. In Japanese gardens, large rugged rocks depict upstream parts of rivers and mountain areas, whereas small round stones represent downstream parts of rivers and coastal beaches. Their presence symbolizes the transformation suffered by large rugged rocks, which become smaller and rounder, as they roll down the river towards the sea. The use of landscape motifs, expressed by the size and shape of rocks and trees, is a technique called *Shuku-kei* (which means 'minimization'). However, such stereotypes have resulted in an instilled false sense of a coastal landscape in the Japanese. For example, there is an idiomatic phrase —*Hakusya-Seisyo* (meaning 'white sand and green pine')—to describe the beauty of a coastal landscape in Japan. This phrase expresses the beauty of the color and structural contrast between the white stretches of sand from the beach and the green from the vertical black pine trees in the adjacent forest. This type of scenery has gradually turned into the 'natural' coastal landscape in Japan. However, it has been engineered by human intervention through time. The pine trees have been planted long ago for coastal protection, which has become a major driving force for afforestation on the coast. In addition, white sand beaches are considered the most appealing ones. For this reason, even if they were originally composed by dark color volcanic ash soil, beach nourishments on eroded coasts have been often conducted with white sand imported from abroad (Saito 2004, Seino 2010).

Regardless, the rapid increase of the Japanese population and high economic growth after World War II (1950s to 1980s) brought a highly intensive development of the coastlines, especially in metropolitan bay areas, as in the cases of Tokyo and Osaka. As a result of this development, natural coastal ecosystems were lost, especially coastal sand dune systems. Even though dunes present a high resilience to natural disturbances (Calafat et al. 2021), their loss occurred not only in metropolitan areas, but also in almost all coastlines in Japan (Iki et al. 2017). Accordingly, this chapter makes an overview of the situation of the Japanese sand dunes as coastal ecosystems, their issues and considerations about future prospects.

Figure 4.1 (a) Old fishing village (Kyoto, Japan). (b) Typical Japanese garden (Sapporo, Japan).

COASTAL SAND DUNE SYSTEMS AND ECOSYSTEM SERVICES

Coastal sand dune systems result from sand transportation by rivers, from mountain areas to the sea, but also by nearshore currents, wave and wind actions, from the beach to inland areas. These ecosystems are sustained by the balance between input and output of sand transportation. Vegetation plays an important role to stabilize the sand and form the dunes. The dune plants can survive severe conditions, such as drought, salt, low nutrient availability, strong wind and sand blast (McLachlan and Brown 2006). As the dune plants catch and stock sand from the beach, they foster the dune higher and higher as they grow. The larger dunes work as natural sand embankments, which protect inland areas from sand transportation, salt spray, strong wind and tidal waves (Fig. 4.2a; McLachlan and Brown 2006).

The sand dunes provide a stable environment landward, allowing inland plants and other living organisms to inhabit there. This sheltered environment also contributes to soil fertility. As nutrients increase in the soil of the dunes, shrubs and trees can grow, replacing herbaceous plants, in a natural ecological succession through time (Hesp 2000). Transitions of vegetation also occur in coastal sand dune systems, in response to spatial gradients of environmental parameters, such as: salt spray, sand transport, organic matter, calcium carbonate or nitrogen (Fig. 4.2a). Therefore, a spatial zonation can be observed along the sand dunes, as a sequence of different vegetation communities found with increasing distance landward

(Fig. 4.2b; Hesp 2000, Walker et al. 2007, Acosta et al. 2007, Marcenò et al. 2018). As a sequence from the shoreline towards inland, the dune vegetation is generally composed by: an annual community of nitrophilous plants on the back beach; a pioneer community of perennial halophilous plants on the low embryo dune and a perennial herbaceous community on the fore dune (both comprising the shifting zone of a dune, also known as yellow or white dune), a perennial community on the semi-fixed dune and pioneer shrubs of the fixed dune (both comprising the stable dune, also known as gray dune), salt marsh/wetlands on the dune slack, followed in the end by forest trees on the mature dune (Hesp 2000, Walker et al. 2007, Acosta et al. 2007, Marcenò et al. 2018).

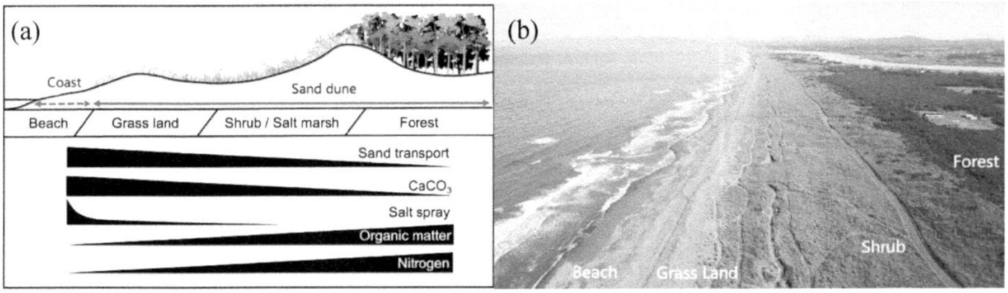

Figure 4.2 (a) Coastal sand dune systems and ecotone (adapted from McLachlan 1991). (b) Zonation of coastal dune plants in Hokkaido (Japan).

As mentioned earlier, dunes are resilient ecosystems. Their most seaward vegetation zone is composed usually by sand trapping species. These plants can cope with high sediment mobility, by collecting airborne sand grains with their leaves and stems, protecting the sand from being blown away by the wind with grass blades, plus stabilizing sand by fast growing extensive and/or branched roots and rhizomes (Wootton et al. 2016). They can grow upward fast, being able to surpass sand burial, and towards the sea, expanding the dune to zones of bare sand (Wootton et al. 2016). Thus are responsible for dune formation and restoration after stochastic and/or extreme weather events (Fig. 4.3), coastal erosion and human intervention (NSW Department of Land and Water Conservation 2001).

Figure 4.3 Resilience of coastal sand dunes in Hokkaido, Japan. Eroded dunes were almost recovered within four years after a storm.

Coastal dunes are dynamic and ever-changing ecosystems in the vicinity of others so diverse, like shallow water bodies (coastal lagoons, estuaries, mangroves, fjords, etc.), sandy beaches, grasslands, shrubs, saltmarshes, forests, etc. (Wootton et al. 2016). Such areas, in which different biological communities, ecosystems and biotic regions contact each other, are called ecotones. They often are steep transitional areas, resulting from spatial gradients of abiotic and biotic environmental parameters (Kark 2013). In an ecotone, each ecosystem is maintained by the mutual relationship with the adjacent one. Thus it is important not only to protect individual ecosystems, but also to maintain its connectivity with their surrounding environment (Kark 2013). Keeping in mind biogeochemical cycles as an example. The chemical elements present in mountains and forests' soils, which are used as nutrients by primary producers, may be bleached away by rain and irrigation water, running into the rivers and, ultimately, the sea. However there can also be a nutrient inflow from the sea towards land. For instance, salmon fish are born in rivers and migrate towards the sea, where they will live most of their lives, storing nutrients. At the end, they will return back to the rivers, where they will spawn and die. Their decaying body mass will recycle chemical elements back to the mountains and forests' soils, due to the action of carnivores, scavengers, detritivores and decomposers (Bishop et al. 2017). However, if artificial structures (such as dams) are installed, salmon cannot ascend the rivers, disrupting the biogeochemical cycles. A similar situation occurs also in sandy beaches. These ecosystems are known to foster primary production *in situ* (McLachlan and Brown 2006), being carried out mostly by diatoms (Colombini and Chelazzi 2003). Cross-boundary transportation of nutrients, towards both sea and land, is done by physical and biotic vectors, respectively as wind or water and mobile consumers (Colombini and Chelazzi 2003). Organic beach-cast material (such as seaweeds, seagrasses, mangroves or salt-marsh plants, other stranded macrophytes, fruits and seeds, driftwood, etc.— usually referred to as 'wrack') that washes up on the shore, plus drifting carrion (animal carcasses in decomposition or remaining portions of it), supply important nutrients for oligotrophic beach ecosystems that cannot retain them in the sand. Wrack and carrion are brought to the beach by waves, currents and tides, being an important structural and functional element in this type of ecosystem, as they provide an alternative habitat, food resource and nutrient reservoir (Dugan et al. 2011, MacMillan and Quijon 2012, Schlacher et al. 2013). The drifting organic material deposited on the sand is crucial as a food input for consumers in the beach biological community (Colombini and Chelazzi 2003, Schlacher et al. 2013). It can form extensive accumulation strandlines, sometime as far as into the dunes (Barreiro et al. 2011), which may represent some aesthetical issues. In this context, beach grooming and the installation of artificial coastal structures, such as seawalls, may prevent nutrient input and interrupt the cycle of organic matter in beach ecosystems, causing its degradation and malfunctioning. In the long run, it will result in the loss of beach biodiversity, services and goods (Afghan et al. 2020).

Rainwater infiltrates and recharges the groundwater under dunes. This freshwater layer beneath the dunes is also known as high quality drinking water. Japanese sake breweries, located along the coastlines, used this groundwater to brew a dry and full-bodied sake (Kaneko and Matsushima 2017). In the Netherlands, dunes function not only as a natural embankment for land protection, but also filter river water for public consumption (Arens et al. 2005).

In addition, coastal sand dune systems provide resources (edible herbs, fungi, timber, etc.), recreational opportunities (bathing, swimming, fishing, footpath, etc.) and habitat for diverse living organisms. These dune functions are known as ecosystem services. However, coastal sand dune systems still providing full functionality of these ecosystem services were being lost in Japan, during the last decades (Yura 2014).

JAPANESE TYPES OF COASTLINES

The definitions used to establish the range of a coast varies from country to country. In the case of sandy shores, the Japanese concept of the coast consists on the intertidal zone and the dry sand beach (also called back beach). In other words, it is comprised between the mean low tide sea level and the upper normal storm wave limit line (Fig. 4.4a). Later it follows the hinterlands (also referred as coastal lands), which are about 100 m wide landwards from the inland limit of the coast (Fig. 4.4a). Thus, the coastal area of Japan consists on the coast and the respective hinterlands.

Regarding the modifications operated in the coastlines by humans, the coastal areas are divided into three categories.

1. Natural coast: where there are no artificial structures. The natural coasts are further divided into tidal flats, sandy beaches, gravel beaches, rocky beaches and sea cliffs. Only south islands have coral reefs in Japan.

2. Semi-natural coast: where there are artificial structures present in the coastlines, but these still do look the same as the natural ones. In this case, the intertidal zone is kept under natural conditions, but artificial structures are built on the back beach (e.g., coastal protection infrastructures, such as seawalls and concrete blocks).

3. Artificial coast: is strongly altered by artificial structures, such as coastal protection constructions or developed by reclamations or landfills for industrial, transportation and urban land-use. In this case, both the back beach and the intertidal zone are modified by this kind of artificial structures.

Half of the coastlines in Japan remain natural (Fig. 4.4b), but they correspond mostly to cliffs (26.3%). In contrast, the natural coast lines in the Hokkaido region (in the northern part of Japan) are mainly composed by sandy beaches (26.6%) (Matsushima et al. 2001). The difference is due not only to topographical factors, but also to the development history between regions. Sandy beaches are easily accessed by humans and provide several goods and services that make them very appealing for human populations. Therefore, these ecosystems have been highly explored since time immemorial, whereas coastal cliffs present challenging hurdles. Accordingly, it is comprehensible that natural coasts, especially sandy beaches, are only left undisturbed in areas where human population density is low for some reason. In fact, due to severe weather conditions during winter, the Hokkaido region still has a relatively low population density and a short history, whose full-scale development began only at the end of the 19th century (Walker 2004). For these reasons, Hokkaido still has wide stretches of sandy natural beaches. Nevertheless, the fact that coastal cliffs are classified as strictly protected areas (mainly because of their difficult access), and that sandy beaches are not even properly regulated in terms of their usage, is one of the main reasons for the degradation of the coastal areas in Japan, in particular sand dunes and related sediment transport systems (Matsushima et al. 2001).

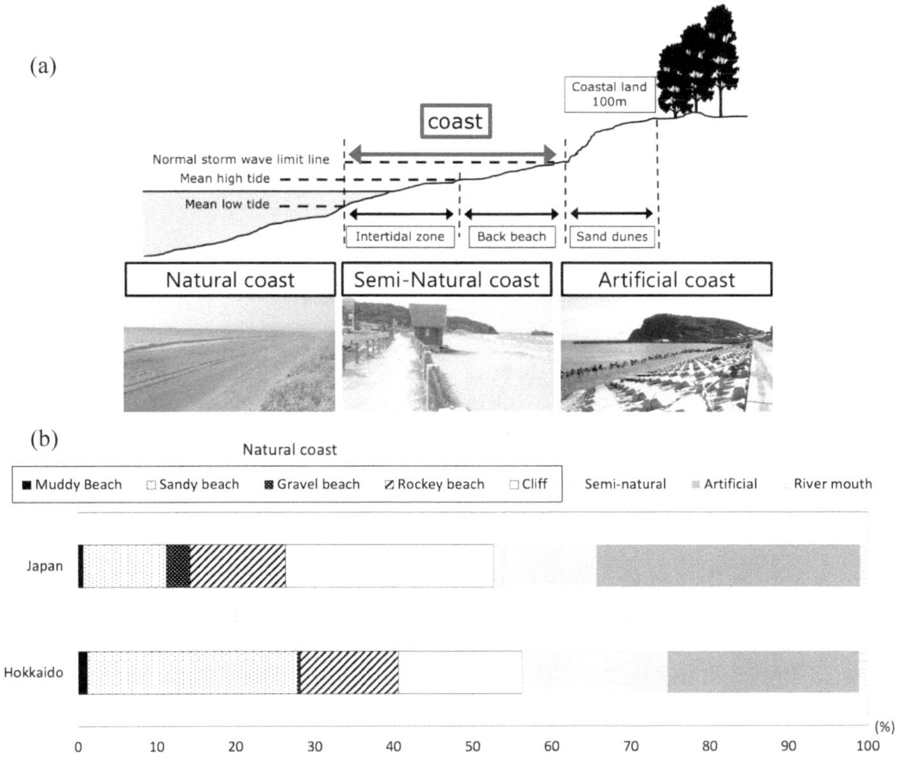

Figure 4.4 Definition of coastal areas and types of coastlines in Japan. (a) The coastal area is composed by the coast and part of the hinterlands (also referred as coastal lands). The coast consists on the intertidal zone and dry sand beach (also known as back beach), which is comprised between the mean low tide sea level and the upper normal storm wave limit line. The hinterlands begin on the upper limit of the coast and extends for 100 m landwards. (b) Comparison between the types of coastlines in Japan and those just found in the Hokkaido prefecture, in the northern part of the country, which still presents well preserved natural coast stretches of sandy beaches (adapted from Matsushima et al. 2001).

The definition of the coast range has another problem. The natural coast still is considered only as the beach itself (intertidal zone and back beach; Fig. 4.4a), without the presence or artificial structures (Matsushima et al. 2014). The adjacent hinterlands are not included. This means that even if a resort hotel or other infrastructures are developed on these strips of land, 100 m behind the back beach, it will still be considered as a natural coast, just as long as the sandy beach keeps its natural functions and conditions. As a result, most dune systems have been destroyed, even on what are considered as 'natural coasts'.

Coastal Development in Japan

Concerning coastal development in Japan, it can be classified into three categories: landfills and urbanization; afforestation; plus coastal protection structures and disconnection.

Landfills and Urbanization

Along with the population increase and rapid economic growth after World War II, the Japanese coastal areas have been extensively converted into urban and industrial zones. Especially in the metropolitan inner bay areas, such as Tokyo Bay (Fig. 4.5a) and Osaka Bay, where natural coasts mainly composed of tidal flats have disappeared due to development. This land reclamation resulted in the loss of ecotones for the beach ecosystems and the deterioration of the inner bay environment, along with the loss of the water purification function of the tidal flats.

Such dense development along coastal areas leads to the loss of beaches and tidal flats as buffer zones between sea and land. It has resulted in a higher disaster risk, due to increased exposure of coastal communities to hazards. While considering that storm surge disasters, caused by extreme weather events, are increasing year by year.

Afforestation

The history of intensive afforestation along the Japanese coast dates back to the beginning of the 17th century, after the *Sengoku Jidai* ('Japanese Warring States' period). These were troubled times in Japan, which started with the Onin civil War in 1467 and finished in 1615, with the *Tokugawa shogunate*, a feudal military government (1603–1867). As the domestic situation stabilized and the Japanese population increased, farmland development was actively promoted, for the purpose of increasing food production. However, as such developments reached close to the coast, they began to suffer losses from damages caused by blowing sand and salt spray. As a result, black pine trees (*Pinus thunbergii* Parl.) have been planted on coastal dunes and beaches, as a bio-shield (both as a windbreak and a sandbreak forest; Fig. 4.5b; Kudoh 1985). Since black pine is resistant to drought and is a fast-growing tree, even in nutrient-poor soils, the afforestation of this species has spread and became popular nationwide, as a Japanese traditional coastal landscape (Suwa 2013, Iwaizumi et al. 2018). Such afforestation was particularly intensive during the reconstruction period of Japan post–World War II (1945–1970). Until the middle of the 20th century, these coastal forests played one other important role for local residents, as a firewood resource, for which they also became popular. However, as the use of firewood has gradually decreased, along with the introduction and increasing distribution of electricity, forest maintenance has been neglected. As a result, pine forests have been left in dense population conditions and the growth of trees has been adversely affected by it (e.g., the leaves from the under layers cannot receive enough light to photosynthesize and the trunks of individual trees cannot grow large). But the more serious problem with the dense monoculture of these trees is pine wilt disease, caused by pine wood nematodes *Bursaphelenchus xylophilus* (Steiner and Buhrer, 1934) Nickle, 1970, which are originally from North America. Most black pine trees could not survive this disease and the forests lost their function as a bio-shield (Mukasyaf et al. 2021).

In addition, planting trees on the dunes has become a factor for losing the resilience inherent to these ecosystems (Uda 2010). The pine forest contributes to stabilize the dunes, as their understory vegetation (natural plants established under the tree canopy) prevents sand particles from moving away by the action of wind and water. Although these ecosystems are highly adapted to disturbance, the shift from an unstable to a stable state inducing habitat changes. In a long run, the dunes will be unable to cope with disturbance, which will increase the vulnerability of these ecosystems to disaster and consequent loss of their biodiversity, functioning, goods and services.

Coastal Protection Structures and Disconnection

Due to advances in civil engineering technology, coastal protections (such as seawalls and concrete blocks) have been installed on the coast, to prevent coastal erosion and storm surges, since the middle of the 20th century. Similar to land fill, such coastal structures provided a stable environment for the hinterlands, which contributes to the expansion of land-use changes and the enhancement of economic activity. On the other hand, it results in a significant loss of the dunes and of their direct and indirect functions.

Most impacts are the loss of coastal dunes itself. In recent years, the structures used for coastal protection have been changing from a wall-like upright breakwater to a gentle slope stepped seawall. This change has improved access to beaches and coastlines, but it has ecological negative repercussions. It results in the loss of beaches and dunes, as the enlarged bottom area of a stepped seawall causes sediment deficit for these ecosystems. The coastal protections also allowed the development of hinterlands and dunes, which have been transformed into residential areas, farmlands and coastal forests (Fig. 4.5c). These land-use changes have increased natural hazard exposure and consequent disaster risk. In fact, the damages caused by extreme weather events have been increasing in number during the recent years. So, the importance of sand dunes and of their role as buffer zones for natural hazards (e.g., salt spray, wind-blown sand, storm surges and wave overtopping) has been recognized and praised, as far as to develop programs to manage and/or restore these ecosystems, as well as to mitigate the impact of anthropogenic pressures.

The indirect impacts consist mostly in ecosystem's changes due the disruption of ecotones. Dune plant communities are subjected to a panoply of environmental characteristics that affect their dynamics, namely some intraspecific parameters, ranging from seed germination to adult growth and reproduction rates (Hesp and Martínez 2007). It includes environmental gradients (namely, salinity, sand

(a)

(b)

(c)

Figure 4.5 Coastal development in Japan: (a) Land-use changes in Chiba city, located along the Tokyo Bay area, visible by comparing aerial photographs from 1945 and 2015 [Source: Konjaku-Map on the web (http://ktgis.net/kjmapw/)]; (b) Afforested black pines in a dune system, known as *Hakusya-seisho* (the beauty of the contrast between flat white sand beaches and vertical green pine trees), one of the most popular coastal scenes in Japan (Kyoto region); (c) Typical development of coastal areas characterized by mobile substrata, (1) in a natural situation (2) and after being intervened in terms of human settlement.

mobility, radiation and nutrient availability in the substrate) and disturbance events that vary in time and space (such as: wave scarping, water intrusion, substrate erosion and sand burial) (Hesp and Martínez 2007). Bearing this in mind, it is impressive that coastal protections provide a stable environment for the hinterlands, but they also hinder the supply of sand, salt spray and nutrients from the sea. Therefore, dune plant communities have been degrading and the nutrient cycle has been disrupted, resulting in deterioration of the entire coastal ecotones.

Thus, coastal development in Japan has greatly contributed to economic growth and population well-being, but it has also resulted in the loss of coastal ecosystems and of their services. This duality raises the question on: how should the coastal areas be developed in the future?

THREATENED LANDSCAPE AND GREEN INFRASTRUCTURE

A catastrophic earthquake with a Mw 9.0 occurred on 11 March 2011, at a depth of approximately 25 km and 130 km offshore of the northeastern part of Japan, called the Great East Japan Earthquake (Fraser et al. 2013). This event had the largest magnitude ever recorded in Japan and it is still the 4th largest one in the world, since 1900. The earthquake also brought on a tsunami disaster, where the inundation area reached about 500 km^2, registering 18,428 dead and missing people, plus 404,893 destroyed buildings (NPA 2020). Yuriage town, in Miyagi prefecture, was one of the most severely damaged coastal village in Japan. Most of all the buildings and other facilities were washed away by the tsunami, with the exception of their basements. The beaches and dunes were also affected, but these last ones were able to regenerate only three months after the tsunami (Fig. 4.6).

After the tsunami disaster in the coastal areas of east Japan—especially on the Pacific side of the Tohoku region, the Japanese government decided to take more active measures to protect the country's inland areas from such natural hazards. So, seawalls were built in an extension of about 400 km along the coastlines, replacing the dunes in a process of self-sustaining recovery (Fig. 4.7). The total cost of constructing these seawalls were over 10 billion US dollars. Additionally, continuous maintenance costs will also be required every year, to assure its integrity and functionality. Although these seawalls ensure a certain level of safety in the coastal areas, it has been pointed out that they will prevent the connectivity of ecotones and degrade the resilience of dune ecosystems (Nishihiro et al. 2014).

To make matters even worse, black pine trees were replanted in the hinterlands of the seawalls, with embankment of mountain soil. Coastal forests have been planted in this area for a long time, but most of them were uprooted and/or washed away by the tsunami (Sakamoto 2012, Tanaka et al. 2013). The damages incurred in the trees resulted from high groundwater levels. These prevent the growth of the plants' root system, restraining the taproot (vertical root) from burrowing deep and developing plate roots (horizontal roots) in thin soil layers (Oda 2001, Hirano et al. 2018). Moreover, the plate root system shows lower resistance to uprooting (Dupuy et al. 2005), for which the black pine trees could not endure the tsunami forces. On account of this, 2 m high embankments with mountain soils were placed over the sand dunes, before re-planting new trees, to ensure enough soil layers as forest beds. This strategy aimed to foment the growth of deeper and vaster root systems, for holding up a tree and soil particles together.

Figure 4.6 Japanese coastal areas damaged by the tsunami disaster of 11 March 2011: (a) a former residential area, in which only the houses' basements remained; (b) an impaired artificial dune; (c) a black pine forest with uprooted and snapped trees, plus destroyed facilities; (d) plants already recovering and rehabilitating the dunes. All photos were taken at the Yuriage village, on 17 June 2011 (3 months after the tsunami).

Figure 4.7 A seawall constructed in place of a dune, in Miyagi, Japan. The photo was taken on 4 November 2014.

Adding mountain soil in the dune systems causes drastic environmental changes. For example, hardening and densification of mountain soil embankments reduces water permeability and promotes the formation of puddles on the surface, which ironically results in a significant inhibition of black pine growth (Ono et al. 2016). Moreover, the seed bank contained in the mountain soil increases the number of inland exotic plants on the coastal areas. As a result, natural habitats on coastal ecosystems have been lost or deteriorated, and their biological communities have suffered changes in species composition.

In order to reduce disaster risk and exposure to natural hazards, it is necessary to set back land-use from the coast and keep buffer zones. Of course, it might be difficult to keep or gain enough space to restore dunes in highly urbanized coastal areas. For instance, residence relocation is not easy, but vacant houses are increasing in urban areas due to population decline. Moreover, fallow lands are also increasing year by year. Therefore, coastal protection and management solutions must be thoroughly pondered.

The Ministry of Land, Infrastructure, Transport and Tourism in Japan reported that Japanese social capital stock management (destined to build new infrastructures and/or maintain the existing ones) would not be sustainable after 2037. This alert was made based on a predictive budget shortage, resultant from increasing maintenance costs of infrastructures, plus population aging and decline. Especially for shrinking cities (a designation for those urban areas becoming uninhabited due to population decline, a concerning phenomenon occurring in Japan), there is no choice rather than using green infrastructure concepts for adapting to recent extreme weather events, caused by climate change.

Although concrete structures—called 'gray infrastructures' (e.g., seawalls) - are useful for a single specific purpose, natural ecosystems called 'green infrastructures' (e.g., coastal sand dunes) are useful for multi purposes, embracing not only an ecological worth, but also scenic, cultural and economic values.

Green infrastructure is one of the key concepts of Nature-based Solutions (NbS). NbS is an umbrella concept that covers a whole range of ecosystem-related approaches [e.g., ecological restoration, Ecosystem-based Disaster Risk Reduction (Eco-DRR), green infrastructure, etc.] (Cohen-Shacham et al. 2019). NbS are defined by the International Union for Conservation of Nature (IUCN) as "actions to protect, sustainably manage, and restore natural or modified ecosystems, that address societal challenges effectively and adaptively, simultaneously providing human well-being and biodiversity benefits" (Cohen-Shacham et al. 2016). In this context, green infrastructure is defined as "an inter-connected network of waterways, wetlands, woodlands, wildlife habitats and other natural areas that support native species, maintain natural ecological processes, sustain air and water resources and contribute to the health and quality of life for communities and people" (Benedict and McMahon 2002). Green infrastructure helps protect and restore naturally functioning ecosystems and provides economic benefits (e.g., increasing property values, decreasing the costs of public infrastructure and services). Investing in green infrastructure is often more cost effective than developing conventional public works projects, such as flood control, water treatment systems and storm water management (Benedict and McMahon 2002). However, the high degree of uncertainty in green infrastructure functions has been regarded as a problem when compared to those known for gray infrastructures (Onuma and Tsuge 2018). Since such descriptions in the past were often unproductive and polarized between gray and green approaches, it is necessary to consider a hybrid infrastructure that complementarily integrates both concepts (Nakamura et al. 2019).

Conclusion

Traditionally, sand dune systems are used to protect hinterlands in Japan, as in other countries. In some cases, sand dunes have been artificially formed and used as natural dikes (Perk et al. 2019). On the same line of reasoning, there is also a hybrid plan that fuses the strength of a seawall with the multiple functions of a sand dune (Matsushima et al. 2019, Almarshed et al. 2020). The aim of this plan is to cover the seawall with sand and turn it into a dune. By doing this, it is expected to ensure the connectivity of ecotones and generate new habitats, in addition to the disaster protection function. The seawall will also be protected from time deterioration effects, namely those caused by wave attack, sand erosion, ultraviolet radiation, corrosion due to sea breeze, sun heating, etc., Consequently, the life span of this coastal protection artificial structure will be extended as well.

The global trend of green infrastructure installation will be inevitable. Conserving healthy ecosystems is the basis of the Sustainable Development Goals (SDGs), adopted by all United Nations Member States in 2015 (Stockholm Resilience Centre 2016). Coastal ecosystems were (and still are) highly adapted to the environmental driving forces, such as tsunamis and storm surges. So, the 2011 tsunami was a great catastrophe for human societies, but it was just another disturbing event for the natural coastal biological communities. Nevertheless, climate changes are already undeniable and they carry along environmental consequences. In the case of Japan, an increase in the number of bright sunshiny days per year, as well as in the annual average temperature and precipitation has been observed; storm surges are becoming more frequent and some with astonishing intensity; besides the fact that the sea-level has risen at a rate of 3.2 $mm.year^{-1}$ from 1993–2010 (Japan Meteorological Agency 2018, Ministry of the Environment 2018, Mukasyaf et al. 2021). As the sea level rise is believed to be an irreversible threat, potential adaptability of coastal ecosystems was pointed out (Kench et al. 2018).

Now that social and natural conditions are deteriorating worldwide, new efforts are needed in the future development of coastal areas. Gray infrastructures present a single purpose, with proven results. Although they are effective in protecting human populations, properties and income activities, gray infrastructures are known to cause changes in habitat and biological communities, loss of ecosystems' resilience, plus the goods and services that they provide, as well as the ecosystems all together. On the other hand, green infrastructures have the premise of multipurpose objectives, which includes long-lasting ecological, economic and social benefits. Although promising, the green infrastructures have not yet proven their worth, discouraging those in doubt to risk it. In other words, to overcome these obstacles, it might be required to combine both infrastructures in a complementary approach, rather than choosing between gray and green infrastructure, depending on the situation. In this context, hybrid protection structures may be a compromising solution to safeguard, mitigate or recover coastal ecosystems, namely coastal sand dunes.

REFERENCES

Acosta, A., S. Ercole, A. Stanisci, V.D.P. Pillar and C. Blasi. 2007. Coastal vegetation zonation and dune morphology in some Mediterranean ecosystems. J. Coastal Res. 23(6): 1518–1524.

Afghan, A., C. Cerrano, G. Luzi, B. Calcinai, S. Puce, T.P. Mantas, et al. 2020. Main anthropogenic impacts on benthic macrofauna of sandy beaches: a review. J. Mar. Sci. Eng. 8(6): 405.

Almarshed, B., J. Figlus, J. Miller and H.J. Verhagen. 2020. Innovative coastal risk reduction through hybrid design: combining sand cover and structural defenses. J. Coastal Res. 36(1): 174–188.

Arens, B., L. Geelen, R. Slings and H. Wondergem. 2005. Restoration of dune mobility in the Netherlands. pp. 129–138. *In*: J.-L. Herrier, J. Mees, A. Salman, J Seys, H.V. Nieuwenhuyse and I. Dobbelaere [eds.]. Proceedings Dunes and Estuaries 2005: International conference on nature restoration practices in European coastal habitats. Flanders Marine Institute (VLIZ) Special Publication 19, Koksijde, Belgium.

Barreiro, F., M. Gómez, M. Lastra, J. López and R. De la Huz. 2011. Annual cycle of wrack supply to sandy beaches: effect of the physical environment. Mar. Ecol. Prog. Ser. 433: 65–74.

Benedict, M.A. and E.T. McMahon. 2002. Green infrastructure: smart conservation for the 21st century. Renew. Resour. J. 20(3): 12–17.

Bishop, M.J., M. Mayer-Pinto, L. Airoldi, L.B. Firth, RL. Morris, LH.L. Loke, et al. 2017. Effects of ocean sprawl on ecological connectivity: impacts and solutions. J. Exp. Mar. Biol. Ecol. 492: 7–30.

Calafat, A., S. Vírseda, R. Lovera, J.R. Lucena, C. Bladé, L. Rivero, et al. 2021. Assessment of the restoration of the remolar dune system (Viladecans, Barcelona): The resilience of a coastal dune system. J. Mar. Sci. Eng. 9(2): 113.

CIA (Central Intelligence Agency). 2020. The world fact book. Washington, D.C., USA.

CIESIN—Center for International Earth Science Information Network, Columbia University. 2012. National aggregates of geospatial data collection: population, landscape, and climate estimates, Version 3 (PLACE III). Palisades, New York, USA: NASA Socioeconomic Data and Applications Center (SEDAC). https://doi.org/10.7927/H4F769GP. Accessed 01 February 2021.

Cohen-Shacham, E., G. Walters, C. Janzen and S. Maginnis [eds]. 2016. Nature-Based Solutions to Address Societal Challenges. International Union for Conservation of Nature and Natural Resources, Gland, Switzerland.

Cohen-Shachama, E., A. Andrade, J. Dalton, N. Dudley, M. Jones, C. Kumar, et al. 2019. Core principles for successfully implementing and upscaling Nature-based Solutions. Env. Sci. Policy 98: 20–29.

Colombini, I. and L. Chelazzi. 2003. Influence of marine allochthonous input on sandy beach communities. Oceanogr. Mar. Biol. Annu. Rev. 41: 115–159.

Dugan, J.E., D.M. Hubbard, H.M. Page and J.P. Schimel. 2011. Marine macrophyte wrack inputs and dissolved nutrients in beach sands. Estuaries Coast. 1–12.

Dupuy, L., T. Fourcaud and A. Stokes. 2005. A numerical investigation into the influence of soil type and root architecture on tree anchorage. Plant Soil 278: 119–134.

Fraser, S., A. Raby, A. Pomonis, K. Goda, S.C. Chian, J. Macabuag, et al. 2013. Tsunami damage to coastal defences and buildings in the March 11th 2011 Mw 9.0 Great East Japan earthquake and tsunami. Bull. Earthq. Eng. 11: 205–239.

Hesp, P. 2000. Coastal sand dunes: form and function. Coastal Dune Vegetation Network Technical Bulletin No. 4. New Zealand Forest Research Institute, Rotorua, New Zealand.

Hesp, P. and M. Martínez. 2007. Disturbance processes and dynamics in coastal dunes. pp. 215–247. *In*: E.A. Johnson and K. Miyanishi [eds]. Plant Disturbance Ecology. Academic Press, Amsterdam, The Netherlands.

Hirano, Y., C. Todo, K. Yamase, T. Tanikawa, M. Dannoura, M. Ohashi, et al. 2018. Quantification of the contrasting root systems of *Pinus thunbergii* in soils with different groundwater levels in a coastal forest in Japan. Plant Soil 426: 327–337.

Iki, S., H. Hirosawa, S. Akiba, M. Isoda, M. Murata, Y. Matsunaga, et al. 2017. Factors and changes of shoreline and landcover in Japan's coast. J. Jpn. Soc. Civil Eng. B3, 73(2):I 492–497.

IPCC (Intergovernmental Panel on Climate Change). 2014. Climate Change 2014: Impacts, adaptation, and vulnerability. Part A: global and sectoral aspects. Contribution of working group II to the fifth assessment report of the intergovernmental panel on climate change. *In:* C.B. Field, V.R. Barros, D.J. Dokken, K.J. Mach, M.D. Mastrandrea, T.E. Bilir, et al. [eds]. Cambridge University Press, Cambridge, UK and New York, USA.

Iwaizumi, M.G., S. Miyata, T. Hirao, M. Tamura and A. Watanabe. 2018. Historical seed use and transfer affect geographic specificity in genetic diversity and structure of old planted Pinus thunbergii population. For. Ecol. Manag. 408: 211–219.

Japan Meteorological Agency. 2018. Climate change monitoring report 2017. Japan Meteorological Agency, Tokyo, Japan.

Kaneko, K. and H. Matsushima. 2017. Coastal sand dune ecosystem services in metropolitan suburbs: effects on the sake brewery environment induced by changing social conditions. Prog. Earth Planet. Sci. 4(28).

Kark, S. 2013. Effects of ecotones on biodiversity. pp. 142–148. *In*: S.A. Levin [ed.]. Encyclopedia of biodiversity. 2nd Ed., Vol. 3. Academic Press, Waltham, Massachusetts, USA.

Kench, P.S., M.R. Ford and S.D. Owen. 2018. Patterns of island change and persistence offer alternate adaptation pathways for atoll nations. Nat. Commun. 9: 605.

Kudoh, T. 1985. Coastal disaster prevention forest in Japan. JARQ 19(1): 55–58.

Kummu, M., H. Moel, G. Salvucci, D. Viviroli, P.J. Ward and O. Varis. 2016. Over the hills and further away from coast: global geospatial patterns of human and environment over the 20th–21st centuries. Environ. Res. Lett. 11: 034010.

MacMillan, M.R. and P.A. Quijon. 2012. Wrack patches and their influence on upper-shore macrofaunal abundance in an Atlantic Canada sandy beach system. J. Sea Res. 72: 28–37.

Marcenò, C., R. Guarino, J. Loidi, M. Herrera, M. Isermann, I. Knollová, et al. 2018. Classification of European and Mediterranean coastal dune vegetation. Appl. Veg. Sci. 21: 533–559.

Matsushima, H., S. Asakawa and T. Aikoh. 2001. On the conservation of coastal area scenic resources in Hokkaido, Japan. J. Jpn. Inst. Land. Archi. 65(1): 633–636. (in Japanese with English summary).

Matsushima, H., H. Arita, H. Naito and S. Sugawara. 2014. Diversity of coastal environment and the efforts toward their conservation in Ishikari coast, Hokkaido. Landsc. Ecol. Manag. 19(1): 41–49. (in Japanese with English summary).

Matsushima, H., A. Suzuki, K. Kimura, X. Zhong and Y. Hirabuki. 2019. Greening on the seawall: challenge to conversion from gray to green infrastructure. Proc. JpGU meeting 2019: H-DS10-03.

McGranahan, G., D. Balk and B. Anderson. 2007: The rising tide: assessing the risks of climate change and human settlements in low elevation coastal zones. Environ. Urban. 19: 17–37.

McLachlan, A. 1991. Ecology of coastal dune fauna. J. Arid Environ. 21: 229–243.

McLachlan, A. and A.C. Brown. 2006. Ecology of Sandy Shores. Academic Press, Amsterdam, The Netherlands.

Merkens, J.-L., L. Reimann, J. Hinkel and A.T. Vafeidis. 2016. Gridded population projections for the coastal zone under the Shared Socioeconomic Pathways. Glob. Planet. Change 145: 57–66.

Ministry of the Environment. 2018. Synthesis report on observations, projections, and impact assessments of climate change, 2018: climate change in Japan and it's impacts. Pacific Consultants, Co., Ltd., Tokyo, Japan.

Mukasyaf, A.A., K. Matsunaga, M. Tamura, T. Iki, A. Watanabe and M.G. Iwaizumi. 2021. reforestation or genetic disturbance: a case study of *Pinus thunbergii* in the Iki-no-Mastubara coastal forest (Japan). Forests 12: 72.

Nakamura F., N. Ishiyama, S. Yamanaka, M. Higa, T. Akasaka, Y. Kobayashi, et al. 2019. Adaptation to climate change and conservation of biodiversity using green infrastructure. River Res. Applic. 36: 921–933.

Nishihiro, J., K. Hara and Y. Hirabuki. 2014. Biodiversity conservation and infrastructure reconstruction after a large-scale disaster: lessons from the coastal regions of southern Sendai Bay. Jpn. J. Conserv. Ecol. 19(2): 221–226. (in Japanese with English summary).

NPA (National Police Agency). 2020. Police measures and damage situation of the Tohoku-Pacific Ocean Earthquake in 2011. Press release on June 10th, 2020.

NSW Department of Land and Water Conservation. 2001. Coastal dune management: a manual of coastal dune management and rehabilitation techniques. Coastal Unit, DLWC, Newcastle, UK.

Oda, T. 2001. Study on the reaction of planted tree root systems to water-logging and its application to developing forests in damp lowlands of coastal sand dunes. Spec. Bull. Chiba Pref. For. Res. Center 3: 1–78. (in Japanese with English summary).

Ono, K., A. Imaya, K. Takahashi and T. Sakamoto. 2016. Evaluation of the berms built on the Restoration of the Mega-Tsunami-Damaged Coastal Forests—Comparison with the effects of soil-scratching as a soil physical correction method among the various types of machinery. Bull. Forestry Forest Prod. Res. Inst. 15: 65–78. (in Japanese with English summary).

Onuma, A. and T. Tsuge. 2018. Comparing green infrastructure as ecosystem-based disaster risk reduction with gray infrastructure in terms of costs and benefits under uncertainty: A theoretical approach. Int. J. Disaster Risk Reduct. 32: 22–28.

Perk, L., L. van Rijn, K. Koudstaal and J. Fordeyn. 2019. A rational method for the design of sand dike/dune systems at sheltered sites; Wadden Sea coast of Texel, The Netherlands. J. Mar. Sci. Eng. 7: 324.

Saito, U. 2004. Hakusya-Seisyo. p. 212. *In:* Japanese Association for Coastal Zone Studies [eds]. Encyclopedia of Coastal Zone Environment. Kyoritsu Shuppan, Tokyo, Japan. (in Japanese).

Sakamoto, T. 2012. Regeneration of coastal forests affected by tsunami. Forestry and Forest Products Research Institute, Tsukuba, Japan.

Schlacher, T.A., S. Strydom and R.M. Connolly. 2013. Multiple scavengers respond rapidly to pulsed carrion resources at the land ocean interface. Acta Oecol. 48: 7–12.

Seino, S. 2010. Problems of coastal environment and expectations of revegetation technology in terms of Satoumi, the nature and human harmonized coastal zone. J. Jpn. Soc. Reveget. Tech. 35(4): 498–502. (in Japanese).

Small, C. and R.J. Nicholls. 2003. A global analysis of human settlement in coastal zones. J. Coastal Res. 19(3): 584–599.

Stockholm Resilience Centre. 2016. Annual Report 2016. Stockholm University, Stockholm, Sweden.

Suwa, R. 2013. Evaluation of the wave attenuation function of a coastal black pine *Pinus thunbergii* forest using the individual-based dynamic vegetation model SEIB-DGVM. J. For. Res. 18: 238–245.

Tanaka, N, J. Yagisawa and S. Yasuda. 2013. Breaking pattern and critical breaking conditions of Japanese pine trees on coastal sand dunes in huge tsunami caused by great East Japan earthquake. Nat. Hazards 65: 423–442.

Uda, T. 2010. Japan's beach erosion: reality and future measures. World Scientific, Singapore.

Vafeidis, A., B. Neumann, J. Zimmermann and R.J. Nicholls. 2011. MR9: analysis of land area and population in the low-elevation coastal zone (LECZ). UK Government's foresight project, migration and global environmental change, Government Office for Science, London, UK.

Valiela, I. 2006. Global coastal change. Blackwell, Oxford, UK.

Walker, B.L. 2004. Meiji modernization, scientific agriculture, and the destruction of Japan's Hokkaido Wolf. Env. History 9(2): 248–274.

Walker, L.R., J. Walker and R. del Moral. 2007: Forging a new alliance between succession and restoration. pp. 1–18. *In:* L.R. Walker, J. Walker, J. and R.J. Hobbs [eds]. Linking restoration and ecological succession. Springer, New York, USA.

Wootton, L., J. Miller, M.M.S. Christopher, M. Peek,A. Williams and P. Rowe. 2016. Dune manual. New Jersey, Sea Grant Consortium, New Jersey, USA.

Yura, H. 2014. Crisis and its factors of coastal dune vegetation. Lands. Ecol. Manag. 19(1): 5–14. (in Japanese with English summary).

Chapter 5

Strategies to Mitigate Coastal Erosion

Carlos Coelho*, Ana Margarida Ferreira and Rita Pombo

RISCO and Civil Engineering Department, University of Aveiro, Campus Universitário de Santiago, 3810-193 Aveiro, Portugal (ccoelho@ua.pt, margarida.ferreira@ua.pt, ritanovo@ua.pt).

STRATEGIES TO MITIGATE COASTAL EROSION

Many shorelines have been retreating at sandy coasts worldwide (Luijendijk et al. 2018), in a redistribution process of sediments resulting from the relationship between sediment availability in the coastal system and their transport capacity by natural hydrodynamic forces (waves, tides and currents). This process is considered a problem, as it endangers activities, infrastructures and human development along the coast. Problems with coastal erosion are complex and their worsening tendency generates frequent criticism among the populations, for whom immediate solutions are necessary. However, coastal erosion mitigation demands effective strategies, planned in long-term perspectives, supported by the identification of risk areas and by cost-effective and cost-benefit analyses. The definition of risk areas allows establishing priorities, in which human occupation can be considered (or should be) forbidden. Cost-effective and cost-benefit analyses may also help decision-makers, showing what is best in a longer perspective, by comparing the costs of the strategies with their physical and economical behavior. To better delineate the strategies to mitigate coastal erosion, an adequate understanding of the causes and consequences of this phenomenon is required, in addition to all the physical,

---

*Corresponding author: ccoelho@ua.pt

environmental, social and economic processes, plus impacts resulting from each intervention or strategic scenario that can be adopted.

Causes of Coastal Erosion

Understanding the causes of coastal erosion is needed to better manage and plan the littoral environment. However, it is difficult to make an important hierarchy of these causes, as they change their relevance from one location to another, being specific at each site. On sandy beaches, coastal erosion is mainly due to the generalized sea-level rise, misruled littoral use and occupation, sometimes along with destruction of its natural defenses, plus external works in harbors that result in perturbations in the littoral system and mainly, the reduction of sediment supply from natural sources (Silva et al. 2007, Coelho et al. 2009).

Globally, sediment supply to coastal areas has decreased drastically in the last decades. In rivers and respective basins, several different human actions can be pointed out as causes of the coastal systems' sediment deficit. Works in rivers have impacted on the sediment volumes transported (Fig. 5.1) and interventions in the river basins affect the production of sediments by soil erosion.

Dams are built for hydroelectric purposes, water supply, irrigation and leisure activities. However, dams' construction creates important water reservoirs updrift, where the flow presents low velocity, without the capacity to transport sediments. It is also where the peak flows are controlled during important rainfalls, decreasing the amount of sediments to be transported during floods (Fig. 5.1a). Water supply for populations, by transferring parts of the flow from natural movable bed channels to piped systems, also decreases the number of sediments to be transported.

Sand extractions (mainly for construction) have consequences in terms of current changes and sediment deficit. Dredging navigation channels also leads to a sediment deficit (Fig. 5.1b). If the removed sediments are not reintroduced somewhere else downdrift, within the hydrographic basin/river course, the deficit will increase in the coastal areas.

Moreover, artificial canals and river bank protections also decrease the amount of sediments available for fluvial transport (Fig. 5.1c). They are used to stabilize margins, or improve water flow conditions, reducing bank degradation and erosion by rivers and streams. Changes in land use, agricultural techniques and groundcover are other anthropogenic actions in the river basins, with impact in the fluvial sediment supply to the coastal systems (Coelho et al. 2009).

Actions in the littoral also represent important factors to increase the coastal erosion problems (Fig. 5.2). Along the coastline, port and harbor infrastructures (breakwaters and jetties) limit the natural sediment dynamics and littoral drift, causing the interruption of longshore sediment transport, which leads to problems downdrift (Fig. 5.2a). At harbor entrances, dredging activities may also contribute to increased erosion in the coastal systems, if no reposition of the sediments is guaranteed.

The coastal protection structures (as groins, longitudinal revetments and detached breakwaters) are built with a specific purpose of intervention, but they do not add sediments to the coastal system. Secondary negative effects of these structures may anticipate sediment deficit downdrift or aggravate wave reflection effects, affecting sediments stability in the region of the intervention (Fig. 5.2b).

(a) (b) (c)

Figure 5.1 Anthropogenic actions affecting sediment transport in rivers: (a) dams; (b) dredging/sand extraction; (c) artificial channels.

Constructions over the dune systems decrease the coasts' natural protections and increase the risk levels of exposure. Anthropogenic actions, such as those related with recreation and leisure activities (Fig. 5.2c), plus urbanization, may cause damages in dune systems, leading to their degradation or destruction. These actions will also induce sediment deficit, diminishing the natural defense capacity of the coastal zones during storm events, resulting in more probable wave overtopping, flooding, damage and destruction of buildings or marginal areas.

(a)

(b)

(c)

Figure 5.2 Anthropogenic actions constraining the natural sediment dynamics at the littoral: (a) breakwaters and jetties (Google Earth image), (b) coastal structures and (c) dune degradation.

Additionally, climate change effects and sea level rise may increase sediment deficit and coastal erosion problems in the future. Climate change effects include, not only the forecasted sea level rise, but also the changes in storms' intensity and frequency, as well as in rainfall regime, which may induce changes in sediment dynamic patterns. The average sea level rise potentiates storm effects over land, causing larger and more frequent wave overtopping and flooding, with full or partial disappearance of sandy beaches or other coastal environments. Depicting the trends for the future and projecting consistent scenarios are important challenges for coastal engineers and planners.

Consequences of Coastal Erosion

The energetic sea actions (waves, tides and currents) have the capacity to transport sediments. Sediment deficit results in general shoreline retreat and loss of territory, corresponding to beach width reduction, which also causes more frequent wave overtopping and flooding events. Thus, the negative consequences of coastal erosion are the degradation or complete destruction of the natural systems (beaches and dunes), plus the increasing number of new coastal zones exposed to wave actions. Damages in coastal defense structures and littoral infrastructures are reported more often. These events lead to conflicts with human and economic activities. Some environmental and/or historic values are also jeopardized. Therefore, the consequences of coastal erosion represent potential harm, determined by the exposure and vulnerability of the sites in which they occur. The vulnerability is dependent on the soil predisposition to erode, the exposure hinges on the quantification of potential land loss and the consequences stand on the value of the territory, considering all the economic, social, patrimonial and/or environmental factors (Samuels and Gouldby 2009, Coelho et al. 2020).

The consequences of coastal erosion are mainly evaluated from an economic point of view, as it is unusual to report casualties among the persons affected by it. However, addressing an economic value to all the potential negative consequences of coastal erosion is not a consensual task, as various stakeholders evaluate the coastal zones in different perspectives. Amongst the groups of people with potential conflict of interests are local inhabitants, tourists, fishermen, aquatic sports practitioners, bar and restaurant owners, etc. When discussing how to preserve the littoral against erosion, it is necessary to include these stakeholders, as they are the first and most affected by its consequences.

Goals and Structure

Considering that the main cause of erosion problems is the lack of sediment in coastal systems, or that the consequences of the shoreline retreat is the loss of territory and damages inflicted to infrastructure and property, two strands of analysis are possible to mitigate them: (1) acting at the cause, by reducing the human activities that decrease the sediment volumes reaching the coastal systems or by bringing sediments into the coasts, by performing artificial nourishments of the dunes and/or beaches; (2) acting at the consequence, by reducing the exposure level (adapting or retreating the infrastructures at urban water fronts), or by protecting the shore by building coastal structures. It should be kept in mind that these protection structures do not add sediments to the coastal system, but allow transferring the erosion problem to a downdrift location, where the negative consequences may present lower impacts.

STRATEGIES TO ACT ON THE CAUSES

Worldwide, coastal zones reveal serious erosion problems related with significant negative sediment budgets. Thus, one pertinent approach to mitigate erosion describes strategies to decrease the deficit of sand in the coastal areas. These mitigation strategies are usually implemented by diminishing the anthropogenic processes that stop sediments from reaching the coastal systems, and/or through artificial nourishment of the erosion affected sites, providing sand to them from other sources. Both levels of action present different challenges, advantages and disadvantages.

Mitigate Human Actions

Dams in rivers, harbors' interference in the longshore sediment dynamics and dunes' degradation are considered the most significant actions that reduce or constrain the quantity of sediments in the littoral.

Dams

Rivers are currently heavily artificialized by the construction of dams. As previously mentioned, these structures are built with different purposes related to hydroelectricity generation, flood control, irrigation, water supply, etc. They change the flow regime of rivers, with consequences on the amount of sediments reaching the littoral. Dams act as a barrier that decrease the river flow velocity, weaken the rivers' sediment transport capacity and consequently trap sediments in updrift areas of the reservoirs (Chen 2005, Coelho 2005, Coelho et al. 2009). However, the negative impacts of dams in the river's sediment transport rate can be mitigated through measures to reduce and control silting in reservoirs, which includes techniques to reduce affluent sediments and their deposition, plus techniques to remove those already deposited.

Techniques to Reduce Deposition at Reservoirs

The techniques to reduce the affluent sediments aim to diminish the amount of these sediments that reaches the dam's reservoir. This solution can be performed through watershed management, upstream sediment retention and sediment derivation (Palmieri et al. 2003). The protection of soils and the land slope control along the river basin are possible strategies often used in watershed management. Sediment retention is carried out through structures built upstream of the reservoirs.

These structures promote sediment deposition in the upstream water line. Sediment derivation consists in the construction of a gallery that transports and diverts the sediments to another location downstream of the dam and the respective reservoir. However, this solution present high costs and operational problems, with accumulation of sediments in the gallery. All these referred techniques improve the performance of the dam's reservoirs, but they do not mitigate coastal erosion, because they do not necessarily increase the volume of sediments crossing the dams.

The techniques to reduce deposition at reservoirs also aim to facilitate the passage of sediments through the dam, before they precipitate and consolidate at the bottom. These strategies allow preserving the reservoir's storage capacity, being carried out in times of floods, when the intensity of the affluent flows is higher. It is estimated that 80% of the annual solid load is present in these flows and about 80% of that solid load may be discharged downstream. So these techniques are implemented through turbidity currents and sluicing.

Turbidity current is a dense and stratified current, with fine particles suspended in the water column. The high concentration of those particles makes the water of the turbidity current denser than the one in the reservoir, which implies that these currents move near the bottom. During the passage through the reservoir, the turbidity currents can deposit or resuspend granular particles (Morris and Fan 1998, Palmieri et al. 2003). Some inherent conditions are necessary for the occurrence of these currents: high slopes of the riverbed and affluent flows with large quantity of sediments, a gorge type reservoir or an embedded valley and adequate hydraulics (great depths and reduced flow velocities). If the earlier conditions are observed, the current may pass the dam by bottom discharges, when the opening of the floodgates is coordinated with the arrival of the current at the dam. If the floodgates are opened before the current arrives, clean water will be released and wasted. If the floodgates are opened too late, the particles present in the current will deposit at the bottom.

Sluicing can be performed in different circumstances, depending on the water level in the reservoir. It can be lowered to the level of flood control and maintained throughout the flood season, or it can be slightly lowered only during the occurrence of floods. The lowering of the water level allows increasing the flow speed, resulting in an enhancement of the capacity to transport sediments. Affluent sediments during the flood season, and sometimes even the ones deposited in the dry season, can be discharged to downstream of the dam. To perform this intervention, flood forecasting is necessary to open the floodgates in the propitious moments. This operation implies lowering the water level even before the entrance of the respective flood flow at the upstream limits of the reservoir. This technique is less efficient than using the turbidity currents, since the flow rate is lower. On the other hand, as the reservoir is kept full, no water is wasted, making it the most economical strategy.

Techniques to Remove Sediments Deposited at the Bottom of the Reservoirs
These are performed through water discharges at the bottom of the reservoir, by lowering the water level, emptying the reservoir, siphoning or dredging. Techniques that include bottom discharge must be carried out essentially in the periods of flood, avoiding the occurrence of silting problems downstream of the dam.

Bottom discharge by lowering the water level implies lowering the amount of water of the reservoir to a minimum level allowed by ecological conditions, keeping it constant throughout the operation. The floodgates are opened quickly, releasing and resuspending the sediments, which will be evacuated by the flow. An erosion cone is formed in the proximity of the discharges and the upstream sediments are eroded, moving towards the dam. However, only the sediments located in the cone are discharged (Morris and Fan 1998). The effectiveness of this solution depends on the topography of the reservoir, the sediments' characteristics, the capacity, elevation, duration and the execution of the discharge itself (Palmieri et al. 2003). This technique is generally applied when the dam's storage capacity is compromised due to sediment retention.

Bottom discharge by emptying the reservoir is usually applied in small systems. It comprises three phases: (1) opening the floodgates, (2) lowering the water level and (3) when it drops below the level of the discharge, the rivers' free flow starts. The water flow causes erosion upstream, next to the discharge, with consequent progressive lowering of the riverbed (Morris and Fan 1998). This technique increases the risk of water supply failure during critical hydrological periods and implies the loss of large amounts of water.

Siphoning is a strategy in which the deposited sediments are hydraulically sucked and removed by pressurized flows, created by suction dredgers. This technique uses the water level difference between the upstream and downstream sections of the dam and works automatically (Morris and Fan 1998). Siphoning presents several advantages: it expends less water to work than other methods; it can be used in varied operating conditions; it has low maintenance and operation costs, as it has great spatial mobility. However, this strategy presents disadvantages related to the clogging of pipes with large sediments, being inefficient in large reservoirs.

Dredging is one of the most popular solutions to remove sediments, being mostly adopted in small reservoirs. Different types of dredging systems (as mechanical dredges) can be used to remove the sediments out of the reservoir. Jets may also be employed to agitate sediments to be later transported downstream by turbidity currents (Morris and Fan 1998).

Breakwaters and Jetties

Protection structures perpendicular to the coastline interrupt the longshore sediment transport, promoting the retention of sediments from the littoral drift. The deposition of sediments updrift of these structures frequently causes problems of erosion in downdrift territory. The purpose of sediment transfer processes is to transpose the sediments from the accumulation zone to areas with deficit, in an attempt to replicate the sediment transport that would occur if these breakwaters and jetties did not exist.

Sediment transfer can be performed by two different approaches: by-pass systems and dredging (Rodrigues 2010, Buisson et al. 2012). By-pass involves the sediment pumping through pipes, from the updrift zone of the structure to downdrift, using a continuous or intermittent operating system. The infrastructure consists of: a motor pump; pump protection structure; mechanical arm, crane or winch that supports the suction system; suction craters; discharge pipes and reinforcement pumps along the pipeline (NRC 1995, Bodge and Rosati 2003).

Likewise, dredging consists of excavating sediments from the bottoms updrifting the coastal structures, by using dredgers, and restoring them to the littoral drift. The selection of the dredging equipment is carried out based on a technical and economic analysis of several factors that condition its performance, namely: distance between dredging and deposition sites, site conditions, site location, depth of water, local traffic, etc. (Bray 1979).

There are several types of dredgers, with different types of mechanism and operation systems, which are classified as mechanical, hydraulic or mixed (mechanical/hydraulic) dredgers. Mechanical dredgers are used to remove gravel, sand and very cohesive sediments, such as clay and highly consolidated silt. Usually, sediments excavated using mechanical dredgers are transported to the deposition site by boats (barges). Hydraulic dredgers use centrifugal pumps that provide the necessary strength to remove and suck sediments together with water, being more suitable for removing poorly consolidated sand and silt. The material dredged by hydraulic dredgers can be transported by boats, pipes or directly unloaded at the deposition site.

Present sediment management practices require that the material dredged in ports and navigation channels must be deposited on beaches located at downdrift zones of their location, but only if they are not contaminated by any type of pollutants. It is a current practice to minimize the impact of breakwaters and jetties. These sediments cannot be used for other purposes, namely those with commercial intents.

Natural Systems

Dunes are natural coastal protection systems that act like a barrier to the propagation of energetic maritime events, preventing populations and goods from being affected during storms. Coastal management entities should promote rules to help the stabilization of these systems and prevent their destruction. Recreational activities that may damage the dunes should be avoided or prohibited. Planting vegetation and building wood palisades along the dunes are techniques that allow promoting sand retention and stabilization of these ecosystems (Fig. 5.3a). Walkways should be implemented in order to facilitate beach access, preventing people from walking on the dunes and trampling its vegetation (Fig. 5.3b). These types of measures are already being performed along several coastal stretches worldwide (Buisson et al. 2012).

Information to the Population

Coastal areas are parts of territory used for different types of activities. The general perception and opinion about coastal management is that many times it is conditioned by people's interests and expectations (Housley 1996). Information and efficient communication about coastal processes are tools to raise awareness levels and thus, mobilize civil society to adopt actions that contribute to the preservation of coastal systems. For this purpose, information about coastal processes should be clear and easily accessible, integrating the available knowledge. Communication of relevant information is crucial for implementing measures and actions, besides fostering the interaction with communities and citizens (Santos et al. 2014).

(a)

(b)

Figure 5.3 Strategies to promote sand retention and stabilization of dunes: (a) wood palisades and (b) walkways.

Artificial Nourishments

Artificial nourishment consists of human-induced addition of large volumes of sand to a coastal area presenting a sediment deficit. The concept of this coastal intervention strategy is based on an attempt to replicate the natural processes of beach and dune formation: sand is deposited in the coastal system (beaches, dunes or littoral drift) and nature takes care of its distribution, by the action of waves, currents and winds (Tondello et al. 1998, Kamphuis 2000, Taal et al. 2016). This solution presents an element different from other coastal defense strategies, which is the fact that it acts directly on the cause of erosion, trying to contour the deficit of sediments in coastal areas.

The first artificial beach nourishment was performed in 1922, on Coney Island (United States of America) (Hamm et al. 2002). In Europe, the first intervention was carried out in 1950, at Tamariz beach (Portugal), with a volume of 15 thousand cubic meters of sand (Martins and Veloso-Gomes 2011). Since then, this type of intervention has gained notoriety, become more common and has been adopted by several countries (Buisson et al. 2012, Pinto et al. 2018). In Europe, an average of 27.5 beach nourishment projects are performed per year, while in the United States it is around 30 annual projects (Hamm et al. 2002). The Sand Engine was the most recent nourishment project of great magnitude, executed in the Netherlands. This project consisted of a concentrated beach nourishment in space and time of 21.5 million cubic meters of sand (Schipper et al. 2016).

By performing artificial nourishments, beach profiles are reinforced and consequently, beaches will be able to maintain their protective functions in scenarios of energetic maritime events, as well as their recreational and tourism functions. This supports the opinion that a beach nourishment project can be a good coastal erosion mitigation strategy (Tondello et al. 1998, Marinho et al. 2018a). However, artificial nourishments are not a permanent solution. The removal of sediments, which occurs naturally from the deposition site, requires renourishment over time to maintain its designed function, often leading to the misinterpretation that the project and the intervention were a failure (Clark 1995). Periodic renourishment intervals range from 2 to 10 years and depend of several factors related to initial design, namely: wave climate, sand used, frequency and type of storms (NRC 1995, ASBPA 2007, Schipper et al. 2016).

When beach nourishment is performed, it should be clear that the sediments will be removed from the deposition site. The project success should not be assessed by the time taken for filling material to run out from the deposition site, but by its ability to reduce erosion rate and to prevent ecosystems, located behind the dunes, of being affected by storm events (ASBPA 2007). Furthermore, the sediments will benefit the downdrift coastal stretches over time, after being mobilized from the deposition site.

In order to increase the longevity of beach nourishment projects, an integrated system can be achieved through a combination of beach nourishment interventions and hard structures, such as groins or detached breakwaters (NRC 1995, Clark 1995, Saponieri et al. 2018). The potential sediment retention of the groins allows reducing the sediment transport capacity and increasing the longevity of the project (Fig. 5.4a). The breakwaters allow dissipating the waves energy, promoting the formation of a salient or tombolo in the sheltered area (Fig. 5.4b), keeping the sediments at the nourished site (Kamphuis 2000).

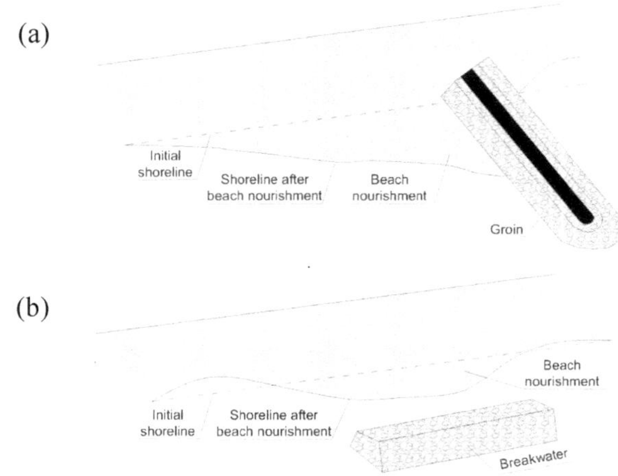

Figure 5.4 Integrated system that combines beach nourishment with coastal structures (based on Kamphuis 2000): (a) groin and (b) detached breakwater.

Sediment Sources

As stated earlier, beach nourishment has become a common and important strategy to mitigate coastal erosion, being adopted worldwide. However, this strategy requires the existence of large volumes of sediments, since beach nourishment projects involve values ranging from thousands to millions of cubic meters of sand. The availability of sediments is a serious issue in a beach nourishment project, since its source should be near to its destination. Otherwise, the costs to transport sand can make the project into an impracticable solution. The selection of the excavation source is dependent on factors related to the availability of sediments, compatibility between native and filling material, plus costs. The costs are related to the equipment that is used to collect and to transport the sediments into the deposition site (dredges, trucks, pipelines, boats, etc.; Fig. 5.5).

The compatibility between native sediments and filling material is based on the grain size, sediment color, composition and texture of the materials (ASBPA 2007). The dimension of the filling material should match the local sediments, since those with larger dimensions are more stable and will promote a different dynamic at the deposition site, inducing erosion downdrift, while sediments with smaller dimensions are less stable and easily eroded (Hearon et al. 2002, Haney et al. 2007, Gravens et al. 2008).

The sources of nourishment sediments may include extraction from quarries, reprocessing of quarry/mining waste, fossil beach deposits, port and harbor dredging activities, lagoonal or barrier island coasts, offshore deposits or derived from downdrift sediments and fluvial supplies (French 2001).

According to Dean (2002), the main advantage of using offshore sources is the fact that it is easy to find large stretches of adequate sand, usually 1 to 20 km away from the deposition site. However, offshore dredging must be carried out outside the active beach profile. Otherwise, the sediments will be just removed from one place to the other, in the same dynamic coastal system.

The material dredged from ports and navigation channels represents an economically interesting sediment source, since dredging is imperative in these places, in order to guarantee navigation conditions. In this case, these sediments can be reused downdrift for beach nourishment, which will be advantageous for relatively short distances for transportation and the lower costs associated with it (Haney et al. 2007, Marinho et al. 2018b).

(a)

(b)

Figure 5.5 Dredged materials from water channels to be used in beach nourishments: (a) a dredger collecting sediments and (b) sand deposition on a beach.

Deposition Sites

The nourishment filling can be deposited in different positions on the beach profile, namely: in the dune, in the dry beach, in the entire beach profile or in a nearshore bar (Hearon et al. 2002, Campbell and Benedet 2006). The selection of the deposition site depends on the level of protection required and the deposition method (ASBPA 2007). For example, if offshore sand banks are used as borrowing sources, the placement of the sand in a submerged part of the beach can be more cost-effective, since its placement in the dry zone involves the use of additional equipment to transport the sand, such as trucks, pipelines or boats (Fig. 5.6).

The choice for the deposition site may not be consensual among all the entities involved in the coastal management zone, since the perception of a beach nourishment success is usually strongly conditioned by the time that the sand remains at the intervened location, as mentioned earlier.

Figure 5.6 Deposition of filling material in a submerged zone of the beach profile.

Dune nourishment allows increasing its height and/or reinforcing its cross-shore profile (Fig. 5.7a). Usually, this type of intervention is used after storm events that compromised the dune's natural function as a protection barrier. The time in which the sand remains in the dune can be considerable, since it is only affected by wave action during storm periods (Marinho, 2018). However, this solution does not increase the width of the dry beach and thus, it is not adequate when the objective of the project also includes improving beach recreational opportunities (Hearon et al. 2002, ASBPA 2007).

Nourishment of the dry beach immediately allows increasing its width, representing new areas available for protection and recreation (Fig. 5.7b). However, the natural hydrodynamics (waves and currents) promote sediment redistribution across the entire beach profile, resulting in a progressive decrease of the dry beach width, causing the sensation of loss and failure of the intervention (Hearon et al. 2002, ASBPA 2007).

The profile nourishment consists in placing the filling material across the entire active profile of the beach (dry beach, intertidal and submerged zones). This solution tries to rebuild the beach in a stable configuration, which leads to small redistribution of sediments and reduced changes in the width of the dry beach (Fig. 5.7c). The main disadvantages of this solution are related to the fact that its execution is more complex and it does not offer great protection in the case of a storm. This lower protection results from the non existent of extra sand reserves on the beach, contrary to the situation in which sand is deposited in the dunes or in the dry zone (Hearon et al. 2002, ASBPA 2007).

In the nearshore bar nourishment scheme, sand is placed in a submerged zone of the active beach profile (Fig. 5.7d). The sand is distributed by wave and current actions, increasing the beach width through time. But the required time for sediments to reach the dry beach varies, depending on maritime agitation (Hearon et al. 2002, ASBPA 2007).

All the options described for cross-shore profile deposition sites will also represent downdrift benefits along time, as the added sediments will be in the active beach profile and will reinforce the littoral drift, during a period of time that will depend on the longshore sediment transport rate.

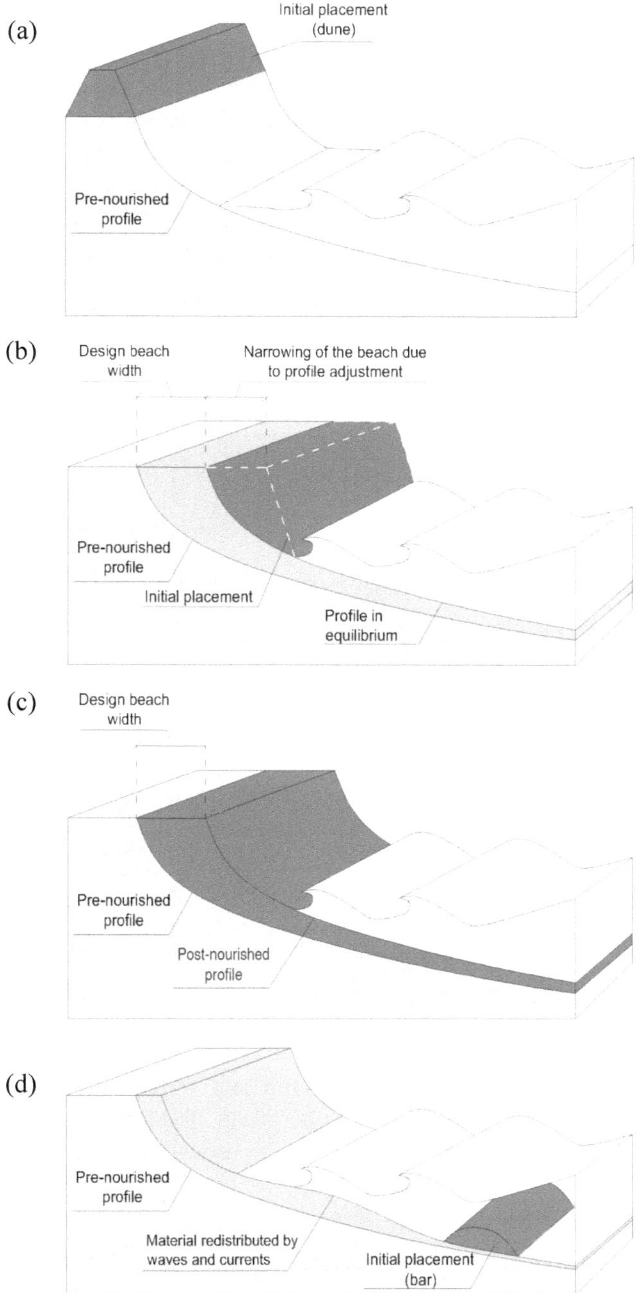

Figure 5.7 Design of artificial sand nourishment schemes (based on Hearon et al. 2002): (a) dune, (b) berm, (c) profile and (d) nearshore bar (offshore zone of the active profile).

STRATEGIES TO ACT ON THE CONSEQUENCES

At historic eroding coasts, it must be expected that erosion will continue diminishing the beaches width, due to a weak sediment supply. This fact has been increasing the costs of coastal protection. Nowadays, sediment deficit has become a widespread phenomenon and it will continue worsening the problem of coastal erosion. When mitigation measures acting solely on the causes of coastal erosion are no longer enough, the main question is: defend or retreat?

Mitigating the consequences of coastal erosion will not add sediments on the shores. Coastal protection with defense structures will shift the problem elsewhere downdrift, benefiting the area to be protected and jeopardizing another one, less valuable or vulnerable. In comparison with beach nourishment, that is considered a soft alternative, coastal defense structures are hard engineering solutions of shore protection. Soft and hard engineering alternatives are often combined to reduce downdrift impacts. For instance, when having the purpose of slowing down the loss of sand placed by beach nourishment projects (Williams et al. 2017). Accordingly, a continuous monitoring process should always be considered, while a possible relocation of infrastructures and goods is discussed, for reducing permanent exposure to direct wave climate.

Coastal Fronts Exposure Reduction to Erosion Effects

To allow natural retreating of the shoreline position is to assume land loss, which has direct negative impacts on entailing economic activities, environmental values, cultural and historical legacy, not to mention that it is dependent on the acceptability of many stakeholders. This is a wide-ranging discussion, since it involves the social and economic components of relocating people (including subsequent compensations), when firmly assuming territory loss. Elevated structures, flood proofing measures, zoning restrictions, storm warning and evacuation planning are some coastal adjustment methods that might be considered when facing it (Basco 2008).

Comprehensive coastal planning and management include policies concerning public life, economic and social development. Coastal management tools should be adequate to help legally competent entities classifying areas of higher erosion risk, so as to constrain land use. These tools may present a strategic and operational framework to be applied on coastal areas, helping to prevent and reduce risks, including vulnerability to climate change, by establishing the ways on how to operate retreat measures. They also may limit and forbid new constructions in critical areas exposed to erosion, overtopping and flooding events, anticipating equilibrium conditions between natural shoreline retreat and sediment deficit.

Infrastructure Adaptation

Coastal urbanizations were developed in areas of high natural sediment dynamics, leading to the deterioration and weakening of those systems, thus themselves becoming more exposed to damage. The accommodation strategy advocates the flexible adaptation of infrastructures and activities in littoral areas, in order to deal with the risks of erosion and consequent impacts. The adoption of resilient urban solutions includes: restraining the use of land, promoting seasonal occupation, filling the gaps in the urban water fronts (e.g., with parking areas), creating solutions for wave energy dissipation, building mechanisms that may conduct overtopped volumes of water elsewhere, plus building and rehabilitating structures to endure wave action.

Adapting buildings to the local reality is then a priority for preventing hazardous occurrences. Typical coastal constructions were developed throughout the years, to overcome coastal challenges near the beach or even on the sandy dunes. Examples of these constructions include houses being usually made of wood, supported by foundation piles, which allow the natural sand drifting below the building, as well as water draining/flowing in case of wave overtopping (Fig. 5.8). The access to the main entrance is made by stairs, which could be expanded at any time, with wooden boards, if the sand drifts away. There used to be vegetation planted around the house to help anchor the dune, such as pines trees, which could later be used for wood construction, firewood, as well as for pulp and paper industry.

Figure 5.8 Typical beach house on the dune system.

Managed Retreat

Retreat is a permanent evacuation or abandonment of a coastal infrastructure subjected to high erosion rates and flooding damages (Fig. 5.9). In a managed retreat situation, coastal communities may be relocated and settled elsewhere inland, as an accommodation measure to manage natural hazard risks (Hino et al. 2017). There is a wide range of expenses and constraints related to this alternative, starting in the broad array of challenges that the victims are subjected to, ending with the costs of designing and building the relocation site. A managed retreat project should always be fully assessed, namely in what concerns the ways to qualify services and infrastructures, in a context of social, touristic and landscaping valorization. Moreover, it should contribute to support local natural, fishing and bathing characteristics (if they exist). So far, the benefits of a managed retreat on exposed sandy shores can only be presented in conceptual terms, until real demonstration projects provide tangible answers and the values of the new features can be appreciated (Nordstrom et al. 2015, Hino et al. 2017).

(a)

(b)

Figure 5.9 Examples of managed retreat: (a) infrastructures damaged by extreme events and (b) house construction for fishermen relocation.

In general, the practice of managed retreat is still limited, as there is an important social resistance to it. Additional difficulties are related to the price to pay to compensate people for the loss of their houses, goods or activities. The decision is also dependent on the dominant stakeholders, usually with vested interests that conflict with the native goals of coastal management, and on the inconsistency of policies, resultant from changes in the dominant parties of a legislature (Neal et al. 2017).

Protection with Coastal Structures

The decision of safeguarding coastal heritage and patrimony, with defense structures, implies giving up on aesthetical and leisure values, assuming the consequent impacts on the intervention vicinity. The coastal structures are intended to constrain the sediment dynamics or to reduce the wave energy, and set the shoreline position in a specific location. Such options will not add sediments into the coastal system. The coastal structures will instead protect/promote accretion of sediments in the most valuable areas and transfer the erosion problems to less vulnerable and valuable ones downdrift. Longitudinal revetments and detached breakwaters' main goal is to dissipate wave energy, while groins mainly interfere with the littoral drift.

Several structures can be combined to mitigate the negative effect of downdrift erosion. It is common to have longitudinal revetments at downdrift of groins. Groins and detached breakwaters can also be considered to retain sand from artificial nourishments.

The coastal structures have an initial cost, related to its planning and construction, plus the cost of their maintenance. The high values involving coastal structures is justified when flooding and wave action in low-lying areas substantially threaten human investment. Thus, coastal structures are more common in urbanized waterfronts. Nontraditional technologies (e.g., beach drains and geotextile bags) can also be implemented to try to reduce the structures' costs. However, as they are more recent solutions, their performance is not yet proved, mainly in more dynamic environments.

Groins and Jetties

Groins and jetties are barrier structures placed perpendicular to the coastline and extended into the surf zone. Groins are short barriers usually built in straight portions of the shoreline (Bush et al. 2001). Their main purpose is to retain sand and promote updrift accretion, giving rise to a new beach, or to expand the existing one (Fig. 5.10a). However, a downdrift sediment deficit may be anticipated when these structures are built (Fig. 5.10b). They also control longshore currents. A groin can work as a standalone solution or can be designed to work together as a groin field (creating a so-called sawtoothshaped shoreline), stabilizing a wider coastal stretch. Its efficiency is, however, dependent on the longshore sediment transport and it is especially vulnerable to wave direction turnover. If there is no longshore sediment transport, the groins do not retain sediments. Moreover, if erosion is anticipated downdrift, wave actions from the most vulnerable side can have larger impacts. Groins also need maintenance in the long run. They are most commonly constructed of rubble mound (quarry stone). As other rubble mound protective structures, groins can be colonized by algae and invertebrates, provide shelter also for fish, acting as an alternative habitat for several life forms (Yozzo et al. 2003).

The main purpose of jetties is not to prevent coastal erosion in a specific location, but usually end up having the same type of impact (Bush et al. 2001; Smith, 2016). They are often very long structures, usually built at the banks of tidal inlets and river mouths. Their intent is to stabilize one or both banks from a shifting position, ensuring water flow and preventing channel siltation (Bush et al. 2001). Therefore, jetties are a long-term solution to maintain channel navigability, by reducing costs with dredging operations, and therefore benefiting shipping, commerce, industry and trade (Bush et al. 2001). Sometimes, it may be required to bypass sediments for the downdrift side of a jetty, as to ameliorate erosion problems along the downdrift shorelines, and prevent future sediment inflow into the inland body of water (CTNC 2017).

(a)

(b)

Figure 5.10 Groin impacts on longshore sediment transport: (a) updrift sediment accretion and deposition, plus (b) downdrift erosion.

Longitudinal Revetments

.The longitudinal revetments are deployed parallel to the coastline. They are designed to set a fixed position of the shoreline, as well as to reflect and dissipate wave energy, absorbing the direct impact of waves and reducing the wave run-up (Fig. 5.11a). From an aesthetical point of view, they create a rocky wall appearance along the seacoast. They should only be used in situations where reflected wave energy can be tolerated. In the long-term, the reflective effect of this type of structure, combined with the pre-existent sediment deficit, may deepen the bottom and gradually diminish the forefront beach width, hence enabling higher waves to reach the structure. In consequence, it might create instability of the foundations, fostering potential damages, with the increasing frequency of overtopping and floods events. To mitigate these aspects, maintenance actions are sometimes required, like deepening the structure foundation (Fig. 5.11b) or increasing the structure crest level (which will reduce the view from the landside). Increasing the structures' dimension (as building a berm) will be more effective in reducing the wave run-up and overtopping, but will also decrease the beach width. The longitudinal revetments' weakest points are their boundary limits, which may need more frequent repairing works. This type of solution is sometimes performed with geosynthetic materials (geotubes, bags or containers), which protect the cliff from erosion in a more economic and environmental-friendly way. Notwithstanding, this option will require more maintenance interventions.

(a)

(b)

Figure 5.11 Rocky longitudinal revetments fixing the shoreline position: (a) profile section of a longitudinal revetment and (b) rehabilitation works (specifically, a foundation deepening).

Detached Breakwaters

The detached breakwaters are a long-lasting solution for the seashore stabilization, since they mitigate partially and permanently the energetic wave actions that would reach the coast, namely by reducing the transmitted wave height. They also promote waves breaking in a further distance from the coastline, reducing wave energy and potential longshore sediment transport capacities. Therefore, a sheltered area behind the structure is created, where a salient (Fig. 5.12a) or tombolo (Fig. 5.12b) is shaped by the local wave regime. The accumulation of sediments in this sheltered area boosts the sediment deficit downdrift. If a tombolo is formed, the connection between the shoreline and the detached breakwater will make this structure act like a groin, constraining the littoral drift. It will promote more accumulation of sediments at the updrift side and will anticipate erosion at the downdrift one.

(a)

(b)

Figure 5.12 Detached breakwaters and sediment deposition in sheltered areas: (a) salient and (b) tombolo.

An offshore structure has higher costs of construction and maintenance than an onshore implanted one, as in the case of longitudinal revetments and groins. The increased expenses are related with the operational costs to transport the material offshore and also with the potential low stability of the foundations in sandy coasts.

Like longitudinal revetments and groins, detached breakwaters can work together as a field of these structures, protecting a wider coastal stretch. There are two major types of detached breakwaters: submerged or emerged. Low freeboards are preferred for aesthetic reasons, smaller dimensions and consequent lower construction costs. If they are built offshore, they are usually made of a rock and rubble mound, comprising various layers of stone.

Detached breakwaters can even be considered artificial reefs, if they are intended to improve ecological stability and protect ecosystems. Nature-based measures also offer other environmental and ecosystems services beyond coastal protection, which would further increase these coastal structures' benefit-to-cost ratio (Reguero et al. 2018). In this situation, a broad-crested submerged structure is built, but instead it can simply be used a sunk ship or a geotube. Geotubes act as filters, separating the sand accumulated within from the water passing through it. It appears to be a low-cost solution, but it demands high maintenance. The detached breakwaters may secondly promote water sport activities (like surfing, snorkeling and scuba diving), if the right wave conditions are created.

FINAL REMARKS

Several strategies to mitigate coastal erosion were briefly described, referring to the main goals of each approach, plus their positive and negative impacts. In fact, there are no perfect solutions to mitigate coastal erosion and thus, for each strategy adopted, it is necessary to balance and weigh all aspects and impacts. These aspects have different preponderance in each location and thus, coastal management alternatives should describe distinct levels of functional effectiveness, structural robustness, sedimentary and morphological consequences on the environment and landscape, construction difficulties, maintenance needs and cost (Coelho et al. 2020).

Mitigating the causes of sediment deficit in coastal systems should always be kept in mind by decision makers, as well as trying to reduce the influence of river interventions and breakwaters, including preserving the natural systems. After implementing all these possible measures, the discussion and decision on artificial nourishments (to mitigate the causes of coastal erosion), relocation of populations or protection of the coastline (both to decrease the level of exposure) should be supported by cost and benefit assessments, with medium to long-term perspectives. The cost assessment of a coastal erosion mitigation strategy involves several aspects, not only related to the type of intervention, but also with the location, function, economic conjecture, material availability, intervention dimension, etc. The dimension and depth at which the works will be deployed can lead to more difficulties and thus, increasing costs. These also depend on the materials used, the maintenance frequency and the degree of exposure to which the works are subjected. They must also consider economic, social, cultural, environmental, etc., aspects at the same time (Coelho et al. 2020).

Cost and benefit analyses are fundamental to support the decisions of medium to long-term management and planning strategies, often being demanded (Costa et al. 2009). Cost–benefit studies provide insight on the measures/strategies that might provide more net benefits. Coastal zone managers should rely on cost–benefit analyses when defining protection, adaptation and/or retreat strategies (Nicholls and Tol 2006). Finally, strategies to mitigate coastal erosion need to be thoroughly evaluated, as they represent an interference with the coastal environment and lead to multiple, divergent and local specific impacts, which imply large investment, as well as maintenance costs (Coelho et al. 2020).

Acknowledgments

This work was financially supported by the project 'Integrated Coastal Climate Change Adaptation for Resilient Communities', INCCA-POCI-01-0145-FEDER-030842, funded by FEDER, through 'Competividade e Internacionalização' in its FEDER/FNR component and by national funds (OE), through FCT/MCTES.

REFERENCES

ASBPA (Shore and Beach Preservation Association). 2007. Shore protection assessment: how beach nourishment projects work. Shore Protection Assessment, US Army Corps of Engineers, Vicksburg, Mississippi, USA.

Basco, D. 2008. Coastal, project planning and design. Shore protection projects (Part V, Chapter 3). Coastal Engineering Manual 1110-2-1100. US Army Corps of Engineers, Washington, DC, USA.

Bodge, K. and J.D. Rosati. 2003. Coastal, project planning and design. Sediment management at inlets (Part V, Chapter 6). Coastal Engineering Manual 1110-2-1100. US Army Corps of Engineers, Washington, D.C., USA.

Bray, R. 1979. Dredging: A Handbook for Engineers. Edward Arnold, London, UK.

Buisson, P., A. Rousset, ANCORIM and J. Massey. 2012. ANCORIM—Overview of Soft Coastal Protection Solutions. BRGM/ONF, Bordeaux, France.

Bush, D.M., O.H. Pilkey and W.J. Neal. 2001. Human impact on coastal topography. pp. 581–590. *In* J.H. Steele [ed.]. Encyclopedia of Ocean Sciences, 2nd Ed. Academic Press, Cambridge, Massachusetts, USA.

Campbell, T. and L. Benedet. 2006. Beach nourishment magnitudes and trends. J. Coast. Res. SI39: 57–64.

Chen, J. 2005. Dams, effect on coasts. pp. 357–359. *In*: M.L. Schwartz [ed.]. Encyclopedia of Coastal Science. Encyclopedia of Earth Science Series. Springer, Dordrecht, The Netherlands.

Clark, J.R. 1995. Coastal Zone Management Handbook. Lewis Publishers, Boca Raton, USA.

Coelho, C. 2005. Riscos de exposição de frentes urbanas para diferentes intervenções de defesa costeira. Ph.D. Thesis. University of Aveiro, Aveiro, Portugal.

Coelho, C., T. Conceição and B. Ribeiro. 2009. Coastal erosion due to anthropogenic impacts on sediment transport in Douro River—Portugal. p. 15. *In*: M. Mizuguchi and S. Sato [eds]. Proceedings of Coastal Dynamics 2009: Impacts of Human Activities on Dynamic Coastal Processes, Paper 72. World Scientific Publishing, Singapore, Republic of Singapore.

Coelho, C., P. Narra, B. Marinho and M. Lima. 2020. Coastal management software to support the decision-makers to mitigate coastal erosion. J. Mar. Sci. Eng. 8: 37.

Costa, L., V. Tekken and J. Kropp. 2009. Threat of sea level rise: costs and benefits of adaptation in European union coastal countries. J. Coast. Res. 56: 223–227.

CTNC (Climate Technology Centre and Network). 2017. Jetties. Technology Compendium. CTNC, Copenhagen, Denmark.

Dean, R.G. 2002. Beach nourishment theory and practice. Advanced Series on Ocean Engineering 18. World Scientific, Singapore, Republic of Singapore.

French, P. 2001. Coastal Defences: Processes, Problems and Solutions. Routledge, London, UK.

Gravens, M., B. Elbersole, T. Walton and R. Wise. 2008. Coastal project, planning and design. Beach fill design (Part V, Chapter 4). Coastal Engineering Manual. Coastal Engineering Manual 1110-2-1100. US Army Corps of Engineers, Washington, DC, USA.

Hamm, L., M. Capobianco, H.H. Dette, A. Lechuga, R. Spanhoff and M. Stive. 2002. A summary of European experience with shore nourishment. Coast. Eng. 47: 237–264.

Haney, R., L. Kouloheras, V. Malkoski, J. Mahala, J. and Y. Unger. 2007. Beach nourishment: MassDEP's guide to best management practices for projects in Massachusetts. Massachusetts Department of Environmental Protection and Massachusetts Office of Coastal Zone Management, Boston, Massachusetts, USA.

Hearon, G., C. Ledersdorf and P. Gadd. 2002. California beach restoration study, Part II: beach nourishment. Department of Boating and Waterways and State Coastal Conservancy, Sacramento, California, USA.

Hino, M., C.B. Field and K.J. Mach. 2017. Managed retreat as a response to natural hazard risk. Nat. Clim. Change 7: 364–370.

Housley, J.G. 1996. Justification for beach nourishment. 25th International Conference on Coastal Engineering, Orlando, Florida, USA.

Kamphuis, J. 2000. Introduction to coastal engineering and management. Advanced Series on Ocean Engineering 16, World Scientific, Singapore, Republic of Singapore.

Luijendijk, A., G. Hagenaars, R. Ranasinghe, F. Baart, G. Donchyts and Stefan Aarninkhof. 2018. The state of the world's beaches. Sci. Rep. 8(6641): 1–11.

Marinho, B. 2018. Artificial nourishments as a coastal defense solution: monitoring and modelling approaches. Ph.D. Thesis, University of Aveiro, Aveiro, Portugal.

Marinho, B., C. Coelho, M. Larson and H. Hanson. 2018a. Short- and long-term responses of nourishments: Barra-Vagueira Coastal Stretch, Portugal, J. Coast. Conserv. 22: 475–489.

Marinho, B., C. Coelho, M. Larson and H. Hanson. 2018b. Monitoring the evolution of nourished beaches along Barra-Vagueira coastal stretch, Portugal. Ocean. Coast. Manag. 157: 23–39.

Martins, H. and F. Veloso-Gomes. 2011. Alimentação artificial de praias em ambientes energéticos intermédios. 6.as Jornadas de Hidráulica, Recursos Hídricos e Ambiente, Faculty of Engineering, University of Porto, Portugal: 29–42.

Morris, G.L. and J. Fan. 1998. Reservoir Sedimentation Handbook. McGraw-Hill Book Co., New York, USA.

Neal, W., O. Pilkey, J. Cooper and N. Longo. 2017. Why coastal regulations fail. Ocean Coast. Manag. 156: 21–34.

Nicholls, R.J. and R.S.J. Tol. 2006. Impacts and responses to sea-level rise: a global analysis of the SRES scenarios over the twenty-first century. Philos. Trans. A. Math. Phys. Eng. Sci. 364(1841): 1073–1095.

Nordstrom, K., C. Armaroli, N. Jackson and P. Ciavola. 2015. Opportunities and constraints for managed retreat on exposed sandy shores: examples from Emilia-Romagna, Italy, Ocean Coast. Manag. 104: 11–21.

NRC (National Research Council). 1995. Beach nourishment and protection. The National Academies Press Washington, DC, USA.

Palmieri, A., F. Shah, G.W. Annandale and A. Dinar. 2003. Reservoir conservation volume I: the RESCON approach—economic and engineering evaluation of alternative strategies for managing sedimentation in storage reservoirs. A contribution to promote conservation of water storage assets worldwide. The World Bank, Washington, DC, USA.

Pinto, C., T. Silveira and S. Teixeira. 2018. Alimentação artificial de praias na faixa costeira de Portugal continental: enquadramento e retrospetivas das intervenções realizadas (1950–2017). Relatório Técnico. Agência Portuguesa do Ambiente, Amadora, Portugal.

Reguero, B., M. Beck, D. Bresch, J. Calil and I. Meliane. 2018. Comparing the cost effectiveness of nature-based and coastal adaptation: a case study from the Gulf Coast of the United States. PLoS One 13(4) e0192132: 1–24.

Rodrigues, L. 2010. Gestão de sedimentos na zona costeira—alimentações artificiais. Master Thesis, University of Aveiro, Portugal.

Santos, F., A. Lopes, G. Moniz, L. Ramos and R. Taborda. 2014. Gestão da zona costeira—o desafio da mudança. Relatório do Grupo de Trabalho do Litoral, Portugal.

Samuels, P. and B. Gouldby. 2009. Language of risk: project definitions, 2nd Ed. FloodSite Project, Document Number: T32: 04–01.

Saponieri, A., N. Valentini, M. Di Risio, D. Pasquali and L. Damiani. 2018. Laboratory investigation on the evolution of a sandy beach nourishment protected by a mixed soft-hard system. Water 10: 1171.

Schipper, M., S. Vries, G. Ruessink, R. Zeeuw, J. Rutten, G. Gelder-Maas, et al. 2016. Initial spreading of a mega feeder nourishment: observations of the sand engine pilot project. J. Coast. Eng. 111: 23–28.

Silva, R., C. Coelho, F. Taveira-Pinto and F. Veloso-Gomes. 2007. Dynamical numerical simulation of medium-term coastal evolution of the West Coast of Portugal. J. Coast. Res. SI50: 263–267.

Smith, P.E. 2016. Types of marine concrete structures. pp. 17–64. *In*: M.G. Alexander [ed.]. Marine Concrete Structures. Woodhead Publishing, Sawston, UK.

Taal, M., M. Loffler, C. Vertegaal, J. Wijsman, L. Van der Valk and P. Tonnon. 2016. Development of the sand motor: concise report describing the first four years of the monitoring and evaluation programme (MEP). Deltares—Enabling Delta Life. Retrieved January 27, 2021, from https://www.dezandmotor.nl/uploads/2016/09/usability-report-sand-motor-final1juli.pdf

Tondello, M., P. Ruol, M. Sclavo and M. Capobianco. 1998. Model tests for evaluating beach nourishment performance. pp. 3096–3109. *In*: B.L. Edge [ed.]. Coastal Engineering 1998. ASCE—American Society of Civil Engineers, Reston, Virginia, USA.

Yozzo, D., J.E. Davis and P.T. Cagney. 2003. Coastal, project planning and design. Coastal Engineering for Environmental Enhancement (Part V, Chapter 7). Coastal Engineering Manual 1110-2-1100. US Army Corps of Engineers, Washington, DC, USA.

Williams, A., N. Rangel-Buitrago, E. Pranzini and G. Anfuso. 2017. The management of coastal erosion. Ocean Coast. Manag. 156: 4–20.

Chapter 6

Exploring the Sun and the Sea—Vulnerabilities and Mitigation Tools Regarding Touristic and Recreational Activities on Coastal and Beach Systems

João Paulo Jorge[1,2*], Paulo F.C. Lourenço[1,2], Verónica N. Oliveira[1,2] and Ilaha Guliyeva[1]

[1]School of Tourism and Maritime Technology, Polytechnic of Leiria, Campus 4, Rua do Conhecimento, 2520–614 Peniche, Portugal, (emails: jpjorge@ipleiria.pt, paulo.lourenco@ipleiria.pt, veronica.oliveira@ipleiria.pt, 4150472@my.ipleiria.pt).

[2]Centre for Tourism Research, Development and Innovation (CITUR), Portugal.

INTRODUCTION

Tourism pressure, climate change and the resulting increase in sea levels heighten beaches' vulnerabilities, already depleted by erosion. From a socioeconomic point of view beach tourism transforms a natural resource into a social and economic value, since it is founded on the consumption and use of biophysical elements. According to Fraguell et al. (2016) the investment in beaches is estimated to produce a direct rate of return of 700% in relation to tourist spending. On the

*Corresponding author: jpjorge@ipleiria.pt

other hand, tourism activities' demands, coupled with the need to satisfy tourists' satisfaction with experiences, are dependent on multiple factors like: tourist profile (age bracket, income level, education levels, among others), their level of involvement in tourism's own recreational activities, tourism supply available in the surrounding territory and its legal and managing regulation.

Managing a beach in tourism terms is inherently complex, and it all begins with its formal definition, since the beach has no judicial definition. From the geomorphological point of view, definitions conform to a transition from a water environment to a terrestrial environment. The Blue Flag's beach definition points to the beachfront and associated water plain.

Beaches are formed by wave-deposited accumulations of sediment located at the shoreline. According to Short and Woodroffe (2009), "the beach extends from wave base where waves begin to feel bottom and shoal, across the nearshore zone, though the surf zone to the upper limit of wave swash". The water plain has an extension equal to the beachfront, and extend about 100 yards into the ocean, including a swimming area and the appropriate channels for sport and leisure activities. The ISO 13009 norm defines the beach as a natural or artificial area formed by sand, gravel, pebbles, boulders or other materials. It also characterizes beaches as leisure activities' areas, served by beach operators who provide services to visitors. Pedestrian paths, strands, parking lots and similar amenities are not part of the beach but surround it.

Over the past 50 years, coastal and beach systems have been subjected to two global phenomena: on the one hand growing human occupation accompanied by intense utilization as tourism activity generators; on the other hand, a biophysical system loss as a consequence of increasing pressure derived from human activities, deterioration of water quality, generalized erosion and climate change processes, which lead to a natural surrounding degradation. These phenomena synergistically put pressure on the beaches' natural systems, conditioning their environmental functions and utilization capacity (Jorge 2019).

The combined effects of population increase, associated activities, infrastructure development and generalized erosion and degradation processes, cause a situation of never-ending stress which demands a fast and coordinated response.

TOURISM AND CLIMATE CHANGE IMPACTS ON COASTAL AND BEACH SYSTEMS

Tourism and Climate Change Connection

Tourism is one of the largest global economic sectors growing enormously since the 1970s and contributing to economies both at local and national level. According to WTTC (2020), tourism contributed 10.3% of the global Gross Domestic Product (GDP), and supported 330 million jobs in 2019. The United Nations Environment Programme (2009) emphasize that "A phenomenon of such magnitude could not remain without consequences for the climate on account of the greenhouse gas emissions generated by trips and stays". However, the relationship between tourism and climate change is complex, as tourism is a victim of climate change, and at the same time, a significant contributor to this change (McKercher et al. 2010). For instance, climate is an important resource for tourism, which can influence destination choice and timing of travel, as well as, travel experience, while tourism can contribute to climate change through greenhouse gases (GHG) emissions and by using natural and environmental resources.

Climate is expected to significantly influence the future of coastal tourism sector, as it can dramatically affect the competitiveness and sustainability of coastal destinations. Furthermore, climate change can reduce the economic value of this sector, through reducing tourism potential and attractiveness of the area, which makes tourism climate-sensitive in a larger extent. Among all types of tourism, coastal/beach tourism is expected to be more affected by climate change impacts, as this type of tourism significantly depends on climate and weather conditions. Table 6.1 provides a qualitative summary of climate-related changes on the various socioeconomic sectors of the coastal zone.

Table 6.1 Summary of climate-related impacts on coastal socio-economic sectors

Coastal socio-economic sector	**Temperature rise (air and seawater)**	**Extreme events (storms, waves)**	**Floods (sea level, runoff)**	**Rising water tables (sea level)**	**Erosion (sea level, storms, waves)**	**Salt water intrusion (sea level, runoff)**	**Biological effects (all climate drivers)**
Freshwater resources	•	•	•	•	–	•	@
Agriculture and forestry	•	•	•	@	–	•	@
Fisheries and aquaculture	•	•	@	–	@	•	•
Recreation and tourism	•	•	@	•	•	–	•
Biodiversity	•	•	•	•	•	•	•
Settlements/infrastructure	•	•	•	•	•	•	–

(Note: • strong; @ weak; – negligible or not established)
Source: Adapted from Nicholls et al., *Coastal systems and low-lying areas* (Cambridge: Cambridge University Press, 2007, 315–356).

It was found that 1°C increase in temperature, is expected to shift tourist destinations gradually to the north and up to the mountains, which would affect the preferences of 3S (sun, sea and sand) tourists from western and northern Europe (Giannakopoulos et al. 2011). Besides, research on the Mediterranean, one of the most popular beach tourism destinations worldwide, show that increase in temperature due to climate change is expected to gradually decrease tourism attractiveness during peak season (summer), however, a gradual increase is expected during shoulder seasons (spring and autumn) as a result of a more favorable climatic suitability of the area (Amelung et al. 2007, Nicholls and Amelung 2008, Hein et al. 2009, Rutty and Scott 2010, Amengual et al. 2014).

In addition, it is very important for tourists to feel safe during their holidays, therefore, extreme weather events such as heat waves or tropical storms (Becken et al. 2011) can negatively influence tourists' safety. The potential changes to this attraction due to the global warming can negatively influence tourists' destination choice (Scott and Lemieux 2010). Accordingly, climate change is expected to influence destination choice, as changes in climate patterns of the destinations will push tourists to make changes in their choices and travel to the alternative destinations with more favorable climate conditions. Besides, Hamilton and Tol (2007) found that due to climate change, an attractiveness of domestic climate will increase, which will result in people from these countries choosing domestic holidays instead of taking holidays abroad. Furthermore, strengthening extreme weather events due to climate change will negatively influence the safety of tourists, which can lead to negative consequences, such as cancellation of trips, or never returning to the same destination (Nyaupane and Chhetri 2009). Taking into consideration all the above-mentioned potential impacts of climate change that can influence the tourism sector, as well as, economic and social value of tourism, and its important role in sustainable development, WTO and UNEP (2008) stress that tourism is one of the most emerging sectors that needs to be adapted to the potential impacts of climate change and global warming.

Moreno (2010) argues that climate can influence the time of the year when tourists travel, as well as, can define the environmental context that certain tourism interests stem from. In her turn, Gomez-Martin (2005) claims that climatic conditions can be promoted as the main attraction of the destination, since other parameters (besides number of sun hours and high temperatures), such as wind, traditionally rejected, can be appraised as an important resource. She found out that very strong winds during the whole year can focus the tourism sector on gliding and wind surfing in Tarifa, Spain. Furthermore, climate change can dramatically influence the competitiveness and sustainability of tourist destinations, especially, climate-sensitive destinations. Scott and Lemieux (2010) argue that every destination is climate-sensitive to some extent. Thus, the impacts of climate change will differ depending on the location or the adapting capacity of destinations and so on.

Researchers point out that climate change will have critical impacts on beach destinations (Buzinde et al. 2010, Klint et al. 2012). This is because weather and climate are determinant factors in outdoor tourism activities and making these destinations highly vulnerable to the climate change (Zaninovic and Matzarakis 2009). For example, Sovacool (2012) points out that the Maldives are exceptionally vulnerable to climate change due to the rise of sea level which threats not only tourism in the islands, but also the existence of the islands.

According to UNEP (2009), tourism is estimated to create about 5% of total carbon emissions, primarily due to transport (mainly by air transport). Unfortunately, according to Intergovernmental Panel on Climate Change (IPCC 2013, IPCC 2019), even if one stops GHG emissions today, climate will continue to change for many decades due to the past emissions. Lenzen et al. (2018) underline that global carbon emissions related to tourism are currently not well quantified. In their work they quantified tourism-related global carbon flows regarding 160 countries, and their carbon footprints under origin and destination accounting perspectives.

"between 2009 and 2013, tourism's global carbon footprint has increased from 3.9 to 4.5 $GtCO_2$, four times more than previously estimated, accounting for about 8% of global greenhouse gas emissions. Transport, shopping and food are significant contributors. The majority of this footprint is exerted by and in high-income countries. The rapid increase in tourism demand is effectively outstripping the decarbonization of tourism-related technology. We project that, due to its high carbon intensity and continuing growth, tourism will constitute a growing part of the world's greenhouse gas emissions." (Lenzen et al. 2018).

However, researchers address other contributors such as beach-front hotels which can increase beach erosion and as a result can lead to rising sea levels. Thus, on-site tourism activities are also significant contributors to climate change as they can be an accelerator of this change. Williams and Ponsford (2009) point out that tourists are consuming a large amount of water, energy and other resources at the tourist destinations more than at home. Besides, to provide tourists with comfort and luxury services, tourism accommodation establishments are overusing environmental resources, especially, water resources and this leads to the harmful impacts on the natural environment.

Overall, to cope with climate change impacts, it is important for all societies and economic sectors to get involved and work together. Several sectors, including tourism have already begun to take measures to adapt to climate change. Mukogo (2014) argues that due to its rapid growth and being a core driver of global economy, the tourism sector has the ability to tackle negative impacts of climate change and can lead the way by adopting green practices and emphasizing sustainable development.

Climate Change Impacts on Coastal and Beach Systems

Throughout history coastal areas have always been centers of human activity due to their rich variety of ecosystems and habitats, which provide a range of goods and services critical to human sustenance and well-being, especially food production, raw materials and transportation options (World Bank 2017). Being the heart of major socio-economic activities, coastal zones and especially coastal cities will be affected by a range of climate change impacts (Nordhaus 2010). For instance, climate change will have direct impacts on coastal zones such as sea level rise, and indirect impacts including coastal erosion, land loss, obstructed drainage and flooding (Nicholls et al. 2007, Hunt and Watkiss 2011). According to World Bank (2017), "Coastal areas are particularly vulnerable because exposure to hazards comes both from the sea and from the land, and because of their high socioeconomic and naturalistic value".

According to IPCC (2013), the global mean sea levels will rise by between 9 and 88 cm by 2100, implying a rate of increase between two and four times greater than during the 20th century. Regarding oceans, IPCC projects that sea level will rise more than about 95% of the ocean area. Besides, it is predicted that about 70% of the coastal areas globally will experience sea level change.

Nevertheless, Hunt and Watkiss (2011) argue that coastal areas and cities will be affected not only by sea level rise, but also by its secondary effects. For instance, the research done by Hunt and Watkiss (2011) in major cities found out that Mumbai in India is expected to experience structural instability due to coastal shifting. Besides, Sherbinin et al. (2007) evaluated climate change in Rio de Janeiro in Brazil noting that the city would experience sea level rise, along with coastal erosion and storm surges.

Besides, extreme climate events can lead to a huge amount of land loss (McGranahan et al. 2007, IPCC 2019), especially, in beaches and port cities where the frequency of flooding is expected to increase due to the climate change. According to Nguyen et al. (2016) "the importance of the coastal zone will further intensify in future, due to the ever-increasing number of people who live there". Adger et al. (2005) indicate that 1.2 billion people, 23% of the world's population, now live within 100 km of the coast, and about 50% of the world's population are likely to do so by 2030".

All the above-mentioned reasons emerge adaptation strategies in coastal and beach systems to cope with climate change impacts. As stated by World Bank (2017), "adaptation to climate change in the context of coastal areas is defined as a policy process entailing decisions on policy and technological interventions that aim at reducing the vulnerability of the system to climatic changes". In order to identify the coastal areas that are under the threat of climate-change impacts, it is needed to perform vulnerability assessments by using reliable scientific tools which will help to set up proper development and land-use strategies in those areas. It is necessary to examine the sensitivity of coastal areas to the changes, their adaptive capacity and other factors that may influence these components, and after this, certain adaptation options can be suggested which can reduce sensitivity to climate change and can promote the development of adaptive capacity of coastal areas.

For instance, according to USAID (2009), Laukkonen et al. (2009) and Michailidou et al. (2016), the adaptation measures that can help to minimize climate-change impacts on coastal zones are coastal wetland protection and restoration (acting as buffer against extreme weather events, storm surge, erosion, and floods; limits salt water intrusion), payment for environmental services (provides incentives to protect critical habitats that defend against damages from flooding and storm surges as well as coastal erosion), beach and dune nourishment (protects shores and restores beaches; serves as a 'soft' buffer against flooding, erosion, scour and water damage), coastal-watershed management (preserves estuaries, which act as storm buffers and protect against coastal groundwater salinization), integrated coastal management (provides a comprehensive process that defines goals, priorities, and actions to address coastal issues, including the effects of climate change). However, even while not considering climate change, coastal areas face a wide range of problems, such as water pollution, population growth, habitat change, degradation. It is expected that climate change will accelerate those problems, which means that there is a need of urgent actions for implementing coastal-adaption strategies to reduce the impacts of climate change, which is considered one of the most important challenges of the 21st century.

BEACH TOURISM: PLANNING, MANAGEMENT AND TOOLS FOR SUSTAINABILITY

Beach Tourism

Regarding the importance of coastal zones for tourism, it is clear that coastal zones have vital importance for tourism in the form of sandy beaches or spectacular cliffs. According to UNEP (2009), beach tourism is one the 'most common' tourism types based on the unique combination of land and sea resources such as beaches, infrastructure and marine biodiversity.

Hall (2001) defines coastal tourism as "the full range of tourism, leisure, and recreationally oriented activities, that take place in the coastal zone and the offshore waters. These include coastal tourism development (accommodation, restaurants, food industry and second homes) and the infrastructure supporting coastal development (retail businesses, marinas, and activity suppliers)". This definition is very important as it recognizes various elements involved in the tourism sector, from demand to supply.

Mieczokowski (1990) divided the coastal zone into four areas that are relevant to tourism—the marine zone, the beach, the shore land and the hinterland. The research considering beach tourism the most important of all four, because of the main tourist activities taking place on the beach (Mieczokowski 1990). According to Moreno and Amelung (2009), although the typical activities include 3S (sun, sand and sea), the development of coastal tourism has led to the diversity of activities which can be divided in shore-based leisure activities and off-shore recreation. Shore-based leisure activities include sunbathing, land-based wildlife watching, nature appreciation, etc. Diving, surfing, underwater photography, cruising, recreational fishing, yacht trips belong to offshore recreation activities.

Although all tourism activities are to a certain extent sensitive to weather and climate conditions, outdoor activities in coastal environments are among the most weather sensitive ones. As the enjoyment and safety of tourist activities in coastal environments are highly dependent on weather and climate conditions, these environments have been identified as highly vulnerable to a changing climate (IPCC 2007). Climate change has various direct and indirect impacts on coastal tourism such as sea level rise, flooding, coastal erosion, changes in precipitation, strong winds, rise in temperatures (air and water), extreme weather events, storms, physical damage to infrastructure, environmental changes etc. (UNEP, 2008). As reported by IPCC (2019), about 24% of the world's sandy beaches are currently eroding by rates faster than 0.5 m per year.

Sea level rise might affect coastal tourism in different ways, for example, sea level rise can lead to the reduced size of beaches, which are of vital importance for recreational activities related to beach tourism in the form of sunbathing and swimming, as well as can cause damages to the tourist infrastructure on the coast. For instance, 1 m sea level rise would cause damage to 49–60% of tourist properties, such as loss or damage of 21 airports, inundation of land in nearly 35 ports in Caribbean resorts, as nearly a third of these resorts are less than 1 m above the high-water mark (IPCC, 2013). Moore (2010) argued that coastal erosion which is the result of sea level rise, might cause damage to infrastructures such as hotels, restaurants, roads and others. In many coastal destinations, these tourism infrastructures and facilities have been built very near to the coastline in order to fulfil visitors' preferences concerning the proximity to the sea.

Being very vulnerable to climate change, it is vitally important for coastal zones to implement adaptation strategies. WTO and UNEP (2008) suggested that for adapting coastal zones to climate-change impacts, the strong seasonality of beach tourism has to be taken into consideration, as it can be exacerbated by climate change. In consonance with the previously mentioned IPCC (2019) underlines that the responses from most countries are often driven by the recreational value of beaches and the high economic benefits associated with beach tourism.

Adaptation strategies related to sea level rise and its secondary effects for coastal tourism can be (1) minimizing built structures close to the beaches; (2) managing adaptive access points to the beaches; (3) building soft (sand bags, beach nourishment) and hard (seawalls, dikes, floodgates) structures for beach protection; (4) developing risk management plans in tourist accommodation establishments to prevent flooding and extreme climate events.

Overall, due to the high sensitivity of coastal tourism and recreation to weather conditions, the high level of exposure of coastal zones to climate change and the limited adaptation capacity of coastal destinations, climate change is expected to jeopardize the sustainability of many coastal tourist destinations across the globe (Moreno, 2010).

According to Freitas (2001) there are three aspects of climate relevant for tourism. First, the physical aspect which facilitates or complicates tourist activities through rain, wind or snow. For instance, sunbathing will be impossible if there is rain or wind. Second, the aesthetic aspect of climate such as the quality of light that can affect the appearance of tourist surroundings. Finally the thermal aspect, which describes how comfortable the tourist feels. Climate and weather are important factors for tourism industry as they influence tourists' destination choice (Becken 2010), as well as, their activity participation, satisfaction and safety (Becken and Wilson 2013) in the chosen destination. Becken and Wilson (2013) noted that "climate and weather are key ingredients of a destination's geography as they influence tourist flows and have significant on-site impacts on the tourism resource base". Furthermore, climate and weather parameters determine when, where and why tourists travel (Scott et al. 2009), and the quality of the tourism/ recreation experience (Gössling et al. 2012, Moreno 2010).

Despite several studies regarding the strong relationship between tourism and weather, the importance of weather and climate for destination choice decision-making has not been highly researched and there is a lack of empirical evidence. As stated by Dubois et al. (2016), Falk and Hagsten (2018) and Matthews et al. (2019), there are some knowledge gaps that need to be filled in: (1) literature mainly focused on climate-change impacts on tourism demand, visitation level and international arrivals, and there are likely limited studies which have explored climate-change impacts on destination choice; (2) studies on climate change impacts on coastal destination choice have limitedly been explored; (3) studies on climate change impacts on destination choice are not enough.

Beach Management and Planning for Sustainable Tourism

The challenges involved in planning and implementing management processes surrounding beaches meant for tourism utilization are many and complex. Overall, these processes can be seen as a coastal resource management continuum, based on an integrated and holistic approach, with the intent of meeting current challenges and problems, through the setting up of some ground principles, a spatial strategy and specific support methods. The main obstacle is the satisfaction of the generalized interests, achieving a shared vision of the future, making these spaces operational, as well as generating the necessary availability on the part of stakeholders, both individuals and structures, who have to be able to bring a long-term managing plan to life.

In fact, one of the greatest concerns when pushing for tourism development is the need to control its impact over the natural and socio-cultural environments, reaffirming the necessary synergy between good practices and preserving and managing natural and cultural heritage.

Achieving sustainability goals through tourism planning and destination development requires the institutional capacity to integrate diverse areas, taking into account a variety of natural, human and cultural assets for a long period of time, as well as implementing appropriate instruments. A beach destination cannot successfully implement a sustainable tourism strategy without a set of assessment instruments (statistical indicators, carrying capacity, among others), economical instruments (eco-fees and financial incentives), command and control instruments (laws, regulations, plans, activity licensing) and voluntary instruments (voluntary certification, codes of conduct) providing a framework, as well as processes that encourage voluntary adherence from the interested parties to sustainable practices and approaches.

Ariza et al. (2010), within the framework of their Beach Quality Index (BQI), which allows an evaluation of beach quality and supports decision-making, propose three interrelated components accounting for main functions supported by beaches (Table 6.2): Recreational Function Partial Index (RFI), the Natural Function Partial Index (NFI) and the Protective Function Partial Index (PFI). All of them are highly complex, grouping 13 sub-indices comprehending: the socioeconomical dimension, which sees the beach as a unit of production, supporting leisure and recreational activities, therefore satisfying both tourists and local residents; the biophysical dimension, meaning all ecological functionalities, assessed by the quality of the natural systems present in the beach, including the monitoring of pollution and the effect of human changes on the physical properties of beaches; and the spatial planning dimension, that assumes the beach as the most important element of protection against extreme climatic events and as an example of resilience, implying different aspects of management efforts. Al three dimensions have to be strengthened coherently so that all functionalities can be fulfilled without conflict or overlap.

Table 6.2 Structure synthesis of the Beach Quality Index (BQI)

Components	Partial indices	Description
RFI Monitors processes related to the recreational experience of users	Microbiological Water Quality (α)	Provides criteria for evaluating *Coliforms* and *Streptococcus*
	Beach Crowding (IC)	Measure of quality of use considering optimum and crowdedness thresholds
	Environmental Quality (IEQ)	Integrated measure of the aesthetic and hygienic environmental quality
	Services and Facilities (ISerF)	Evaluation of 11 components. Differences for urban and urbanized beaches
	Activities (IAct)	Evaluates annoying and other types of undesirable behaviour
	Access and parking (IAcPar)	Measure of accessibility to surrounding areas, signposting, access to the beach and parking and transportation
	Comfort Quality (IComf)	Evaluation of aspects of the beach structure and climatic conditions that affect users' experience: 8 factors
	Surrounding Area Quality (IS)	Evaluates landscape and aesthetic quality
	Beach Safety (IBS)	Integrated measure of the safety and rescue services
NFI Monitors processes related to beach biophysical condition	Natural Conditions (IN)	Assess quality of the natural systems in the wind-controlled upper part of the beach (vegetation representation, surface coefficient and development of the habitat)
	Water–Sand Pollution (IWSP)	Monitors effects of pollution events on different natural communities
	Physical Quality (IPQ)	Represents the effect of human changes on the physical properties of beaches (grain size, surface and wave regime)
PFI Monitors gains and losses of sediments in relation to protection of coastal facilities	Protection (IP)	Represents the importance of beaches in protecting coastal features in the study area

Source: Adapted from Ariza et al., "Proposal for a Beach Integral Quality Index for urban and urbanized beaches". *Environmental Management* 45 (2010): 998–1013.

It becomes evident that coastal tourism not only represents the largest and constantly growing segment of the tourism industry, but also among the most important (and fastest growing) economic activities taking place at the sea. At the same time—depending on the type of activity—beach and marine tourism can have a varying nature, ranging from alternative and 'eco' activities to activities with a rather mass character. Consequently, in some sun and sea destinations, tourism has grown at a huge rate, becoming a mass phenomenon. Regarding spatial patterns, coastal tourism often causes increased environmental impacts and pressure to the coastal and marine ecosystems, characterized by a very high biodiversity (European Commission 2014).

Tourism depends on a healthy environment and the sustainable use of natural capital, but activities are often concentrated in already densely populated areas, leading to vast increases in water demand, more waste and emissions from air, road and sea transport at peak periods, more risks of soil sealing (the covering of the ground by an impermeable material) and biodiversity degradation (from infrastructure developments), eutrophication and other pressures.

Environmental impacts generated by tourism can adversely affect competitiveness of tourism destinations, not only through the reduction in the quality of their tourism inputs, but also through a potential fall in demand as a consequence of the emergence of 'environmentally sensitive' tourists. Despite the reputation of tourism as being unsustainable, the sector has been a pioneer in the search for impact-reducing alternatives by introducing environmental accreditations (Fraguell et al. 2016). In fact, the first tourism resources to be targeted were beaches, via the Blue Flag campaign, launched in France in 1987 (Lucrezi et al 2015). Blue Flag is considered the predecessor of all quality and environmental accreditations for sustainable tourism management in coastal areas. Its promotional campaigns over 30 years have made it one of the well known environmental awards among tourists and businesspeople alike. Its prestige inspires confidence and generates wide media coverage (Nelson et al. 2000, Kozak and Nield 2004, Aliraja and Rughooputh 2005, McKenna et al. 2011, Lucrezi et al. 2015).

BLUE FLAG Eco-label: Sustainable Beach Management Tool

According to the International Standards Organization (ISO 2018), the aim of an environmental label is to encourage the demand and the offer of products that cause less pressure on the environment throughout their life cycle, through the communication of verifiable, reliable and not misleading information on the environmental aspects of the products and services.

Over the last two decades the evolution of environmental labeling has been remarkable. The European Network for Sustainable Tourism Development (ECOTRANS) contained in its data base, in 2001, more than 50 ecolabels and awards and over 300 examples of good practices by tourism companies (Hamele 2002). According to ECOTRANS (2016), for tourism services (covering a wide variety of elements such as hospitality, culture, recreation, sport, hygiene and others) there was more than 150 quality labels worldwide. Weston et al. (2018)

estimated that in Europe there are up to 100 labels, such as Blue Flag, Green Key, European Ecolabel, and Green Globe 21 among others. According to these authors, "there is considerable fragmentation and diversity in the criteria applied, principles, management and governance of the labels. Just as with other types of labels, quality labels are susceptible to market saturation and as such, consumers may find it difficult to distinguish more reputable labels from others."

Certification, when based on transparent, objective and measurable procedures, provides a reliable basis for identifying tourism businesses and destinations and their products and services that meet sustainability criteria. It provides guidance for travelers and other purchasers to encourage sustainable consumption patterns and green purchasing. It gives a checklist and targets for businesses to work towards when creating sustainable products and services and supply chains and is used to improve innovation and market access. Therefore, the marketing and branding of certified tourism has become an important feature of sustainable tourism development. Environmental certifications, when implemented along with spontaneous initiatives geared towards environmental management, may represent effective ways of improving the environmental quality, the competitiveness of tourism destinations, and of enhancing the territory's touristic brand.

One popular tool used for beach management throughout the world is the use of beach awards and eco-labels, such as the Blue Flag. Beach awards have been said to bridge the gap between recreation and conservation and have become a beach management tool. The Blue Flag is one of the world's most recognized voluntary eco-labels awarded to beaches, marinas and sustainable boating tourism operators. It is awarded annually by the Foundation for Environmental Education (FEE) to coastal and inland beaches and marinas, which comply with a set of requirements relating to the following four categories:

(1) Environmental education and information,
(2) Water quality,
(3) Environmental management,
(4) Safety and services.

Despite criteria having evolved from year to year, becoming ever stricter, the number of beaches and marinas distinguished has been increasing in a significant manner. In 2020, 4614 beaches, marinas and sustainable boating tourism operators were distinguished in 47 countries from Europe, Africa, America and Oceania. Figure 6.1 charts the total number of Blue Flag awards for beaches, marinas and boats over the past 15 years.

The Blue Flag Program was launched on a European scale in 1987, as part of the European Year for the Environment. The environmental education program, authored by the Foundation for Environmental Education (FEE), resulting in this eco-label, had as its goals creating awareness among regular citizens, and also government officials, for the need to protect marine and coastal environments, and encouraging actions that lead towards solving existing issues.

In the mid-1980s, a number of factors converged to arouse coastal destinations' interest in obtaining Blue Flag accreditation for tourist beaches (Nelson et al. 2000). These included the incipient obsolescence of the traditional sun-and-sand tourism model; the need to compete with new tourist areas with more attractive landscapes; new demands for environmental quality; the deterioration of the coastal landscape; and the depletion of basic natural resources. Blue Flag accreditation enabled destinations to recover lost prestige and to project a better image of beaches complying with quality standards (Nelson et al. 2000, Aliraja and Rughooputh 2005, McKenna et al. 2011, Zielinski and Botero 2019).

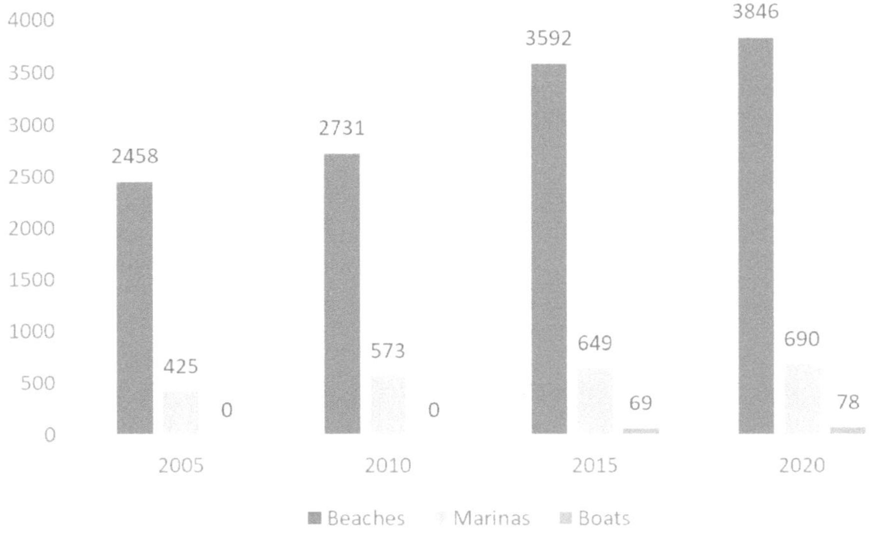

Figure 6.1 Number of Blue Flag awards from 2005 to 2020. Source: Blue Flag (n.d.), retrieved from https://www.blueflag.global/

The Blue Flag Campaign promotes the sustainable development of coastal areas, encourages cooperation between the tourism and environmental sectors (especially locally) and good environmental practices among users of beaches, marinas and boats, provides codes of conduct which are respectful of the environment.

Furthermore, one could say that it is an eco-labeling that demands the fulfilment of a set of requirements related to the Quality of Life (QoL): some of them referring directly to the environmental quality, and other to the additional comfort and services that tourists and residents can enjoy. A large part of the requirements relates to environmental indicators that affect QoL, such as: the absence of wastewater discharges, the separate waste collection in the area or the promotion of sustainable transport. Another part of the requirements demanded by the Blue Flag refers to more general indicators of QoL, such as safety and surveillance, cleaning, accessibility to the beach or the availability of drinking water in it. Table 6.3 shows the Blue Flag criteria specifically related to the environmental management category.

Table 6.3 Criteria used through Blue Flag environmental management category

• The local beach operator should establish a beach management committee • The local beach operator must comply with all regulations affecting the location and operation of the beach • Sensitive area management • The beach must be clean • Algae vegetation or natural debris should be left on the beach • Waste disposal bins must be available at the beach in adequate numbers • Facilities for the separation of recyclable waste materials should be available • An adequate number of toilet facilities must be provided	• The toilet or restroom facilities must be kept clean • The toilet or restroom facilities must have controlled sewage disposal • There should be unauthorised camping, driving or dumping of waste on the beach • Access to the beach by dogs and other domestic animals must be strictly controlled • All buildings and beach equipment must be properly maintained • Coral reefs in the vicinity of the beach must be monitored • A sustainable means of transportation should be promoted in the beach area

Source: Foundation for Environmental Education, *Blue Flag Beach Criteria and Explanatory Notes* (Copenhagen: Foundation for Environmental Education, 2017)

However, literature points out some negative and less favorable aspects with regard to this certification, which must be considered and improved. Mir-Gual et al. (2015) stated that although the Blue Flag lead to excellent water quality, the security of the services and facilities and environmental education, this certification does not evaluate all environmental aspects of the beach. For example, Davenport and Davenport (2006) and Fraguell et al. (2016) refer that the beach cleaning mandatory by Blue Flag involves the removal of unsightly natural debris as well as the litter generated by the growing number of tourists, sometimes requiring mechanized operations that remove all seaweed and other natural debris washed onto the beach by the tide. Consequently, these mechanical cleaning measures, ecologically aggressive, decrease the sandy shore biodiversity, affecting the coastal morphology, wildlife protection, natural vegetation and increase the erosion process (Boevers 2008, Defeo et al. 2009, Mir-Gual et al. 2015). In order to avoid these consequences Davenport and Davenport (2006) stated that it "requires control over access (to control demand), effective planning that pays due attention to the values of ecological services, and rigorous environmental legislation, properly enforced."

Even though some criteria regarding environmental management need crucial improvements and control, with consequences for biodiversity recovery (e.g., beach cleaning), in general, according to Oliveira and Soares (2018), the benefits from the Blue Flag Program to coastal areas' environmental quality are evident. This eco-label is a valuable tool to local authorities and other agents for a sustainable beach management, offering visitors and residents quality assurance parameters in order to achieve sustainable tourism goals (Nahman and Rigby 2008, Fraguell et al. 2016). Thus, one can highlight positive impacts for the resident population, who saw investments in:

- Urban solid and liquid waste management, industrial and agricultural waste processing, in order to improve beach and seaside water quality;

- Social, recreational, dune protection, and sensitive areas infrastructure improvement, accessibilities, environmental education facilities;
- Youth educational initiatives, aiming at recycling and reutilization awareness;
- Educational and scientific initiatives focusing on such themes as climatic change, coastal erosion, sustainable urbanism, preservation of biodiversity;
- Local economy initiatives, generating multiplying effects from surge in small businesses related to leisure, sports, food, culture, etc.;
- Improving tourism attractiveness through focus on Quality Beach Areas.

It can be said through the Blue Flag eco-label, several improvements related to quality of life, environmental health, wellness, landscape improvement, citizenship and environmental education, tourist attractiveness increases, innovative initiative promotion, etc., were achieved.

These days environmental policy is closely related to the aim of increasing the QoL. One of the most important political and societal problems is how to improve the quality of life of the population while living in the carrying capacity of the natural environment and without compromising the long-term human, economical and ecological capital of the future. That is, how to balance economic wellbeing with environmental wellbeing.

In the case of tourism, the policies should aim at promoting sustainable tourism practices that minimize the negative impacts of tourism on the environment, while the positive economic impacts in the quality of life are kept for the residents of the tourist destinations (job creation, access to infrastructure, social and cultural services). There is a bidirectional relationship between the activities of tourism and the environment in the sense that the environmental impacts they generate may adversely affect the competitive position of the entire tourism destination and, therefore, the QoL of residents.

The reason is not only the reduction of the quality of tourism inputs, but also the potential decrease in consumption due to the existence of segments of 'environmentally sensitive' tourists, who take into account issues such as environmental quality or sustainability in their choice of destination. In particular, the degradation in quality of the destination devalues the quality of the tourist's experience.

The beach management scenario surrounding many coastal destinations before implementing Blue Flag was mostly unsustainable. In the best cases, only minimal management was in place, covering clean water and sand and the requirements for recreational services and first aid. Gaping inefficiencies were evident in the planning of beach services and uses, controlling attendance, providing information and environmental awareness among beach stakeholders and users, as well as the nonexistence of emergency plans, maintenance actions, and the recovery of ecological systems.

Fostering a more respectful and sustainable use of such a fragile ecosystem as the beach—which from the economic and tourist point of view is a scarce spatial unit, since tourism use is regulated by a set of institutional planning instruments, such as the spatial plans of the coastal zones, land-use management and beach concession regulations—, Blue Flag remedies some of these shortcomings. It reports high degrees of satisfaction among tourists who make use of a beach awarded the Blue Flag, and it promotes their adoption of sustainable tourism practices (Fraguell et al. 2016).

In the interests of sustainability, Blue Flag focuses its interest on environmental management by promoting compliance with current environmental regulations, the development of a coastal management plan, responsible management of waste collection, beach cleanliness and sustainable transport. In water quality management, safety and services, it aims to ensure the comfort and safety of swimmers by ensuring an excellent quality of bathing water, applying control measures to prevent the uncontrolled discharge of pollutants on the beach, and improvements to beach accessibility. In informing and awareness of, bathers and the local population, it develops environmental education activities regarding the need to protect sensitive ecosystems and protected areas located in the coastal area (Fraguell et al. 2016).

Conclusions

There is today an international awareness of the importance of placing limits to the development of activities occurring in coastal areas. These areas are characterized by its high population, service and infrastructural density, resulting in higher consumption of natural resources (water, air, soil), and consequential associated negative impacts: waste disposal issues, air pollution, soil erosion, loss of biodiversity, etc. In addition, the impacts of climate change exacerbate pressures on these areas.

Coastal tourism is based on the use and consumption of biophysical factors, turning a natural resource into a social and economic value. The interface between the sea and the land is the preferred zone for recreational activities, with the beach as the main space and the focal point for a whole range of tourist attractions (Fraguell et al. 2016).

In order for tourism to be part of coastal areas development strategies, its own development will have to adopt integrated planning actions, based on economic, socio-cultural and environmental goals, and incorporating quantifying instruments providing limits to the use of resources, as well as increasing of tourism quality, like certification and eco-labels. One popular tool used for beach management throughout the world is the use of beach awards and eco-labels. Beach awards have been said to bridge the gap between recreation and conservation and have become as a beach management tool.

Blue Flag is a world-renowned ecolabel trusted around the globe. The mission of Blue Flag is to promote sustainability in the tourism sector, through environmental education, environmental protection and other sustainable development practices (FEE 2019). It aims to ensure that a beach can be promoted for its sustainable management, cleanliness and safety. Blue Flag identifies beaches that meet a set of requirements relating to four aspects: (1) quality of bathing water, (2) environmental management of the area, (3) information and environmental education for tourists and residents, and (4) security, services and facilities.

The Blue Flag eco-label is not static, rather it attempts to adapt to the changes and new requirements of the fragile but dynamic ecosystem that is the coastline. The criteria for gaining accreditation are periodically updated at an international level.

The evolution of the Blue Flag program and the continued renewal of the criteria demonstrates a willingness to approach the principles governing the paradigm of sustainable tourism. It aims to gradually adopt an increasingly more integrated perspective on planning and management for the whole tourism system covering the beach as a resource. Therefore, controlling socio-cultural and environmental negative impacts, as well as increasing residents' quality of life, should be a major concern in tourism development of coastal areas.

REFERENCES

Adger, N.W., T.P. Hughes, C. Folke, S.R. Carpenter and J. Rockstrome. 2005. Social-ecological resilience to coastal disasters. Science. 309: 1036–1039.

Aliraja, S. and S.D. Rughooputh. 2005. Towards introducing the Blue Flag eco-label in SIDS: The case of Mauritius. IRFD World Forum on Small Islands Developing States: Challenges, Prospects and International Cooperation for Sustainable Management. Mauritius: 1–15.

Amelung, B., S. Nicholls and D. Viner. 2007. Implications of global climate change for tourism flows and seasonality. J. Travel Res. 45: 285–296.

Amengual, A., V. Homar, R. Romero, C. Ramis and S. Alonso. 2014. Projections for the 21st century of the climate potential for beach-based tourism in the Mediterranean. Int. J. Climatol. 34: 3481–3498.

Ariza, E., J.A. Jiménez, R. Sarda, M. Villares, J. Pintó, R. Fraguell, et al. 2010. Proposal for a Beach Integral Quality Index for urban and urbanized beaches. Environ. Manage. 45: 998–1013.

Becken, S. 2010. The Importance of Climate and Weather for tourism. Land Environment and People. Miscellaneous Publications. Lincoln University. Canterbury, New Zealand.

Becken, S., J. Wilson and K. Hughey. 2011. Planning for Climate, Weather and Other Natural Disasters—Tourism in Northland. Land Environment and People. Research Paper N° 1. Lincoln University. Canterbury, New Zealand.

Becken, S. and J. Wilson. 2013. The impacts of weather on tourist travel. Tour. Geogr. 15: 620–639.

Blue Flag. n.d. https://www.blueflag.global/

Boevers, J. 2008. Assessing the utility of beach ecolabels for use by local management. Coast. Manage. 36: 524–531.

Buzinde, C.N., D. Manuel-Navarrete, E. Yoo and D. Morais. 2010. Tourists' perceptions in a climate of change: eroding destinations. Ann. Tour. Res. 37: 333–354.

Davenport, J. and Julia L. Davenport. 2006. The impact of tourism and personal leisure transport on coastal environments: a review. Estuar. Coast. Shelf Sci. 67: 280–292.

Defeo, O., A. McLachlan, D.S. Schoeman, T. Schlacher, J. Dugan, A. Jones, et al. 2009. Threats to sandy beach ecosystems: a review. Estuar. Coast. Shelf Sci. 81: 1–12.

Dubois, G., J.P. Ceron, S. Gössling and C.M. Hall. 2016. Weather preferences of French tourists: lessons for climate change impact assessment. Clim. Change 136: 339–351.

ECOTRANS. 2016. Sustainability in Tourism A Guide through the Label Jungle. Naturefriends International, Vienna.

European Commission. 2014. A European Strategy for more Growth and Jobs in Coastal and Maritime Tourism. EC—Com 86, Brussels.

Falk, M. and E. Hagsten. 2018. Winter weather anomalies and individual destination choice. Sustainability 10: 2630.

Foundation for Environmental Education. 2017. Blue Flag Beach Criteria and Explanatory Notes. Foundation for Environmental Education (FEE), Copenhagen.

Foundation for Environmental Education. 2019. Blue Flag and Sustainable Development Goals. Foundation for Environmental Education (FEE), Copenhagen.

Fraguell, R., C. Martí, J. Pintó and G. Coenders. 2016. After over 25 years of accrediting beaches, has Blue Flag contributed to sustainable management. J. Sustain. Tour. 24: 882–903.

Freitas, C.R. 2001. Theory, concepts and methods in climate tourism research. International Society of Biometeorology Proceedings of the First International Workshop on Climate, Tourism and Recreation, Neos Marmaras, Greece, 3–20.

Giannakopoulos, C., E. Kostopoulou, V.K. Varotsos, K. Tziotziou and A. Plitharas. 2011. An integrated assessment of climate change impacts for Greece in the near future. Reg. Environ. Change 11: 829–843.

Gomez-Martín, M.B. 2005. Weather climate and tourism a geographical perspective. Ann. Tour. Res. 32: 571–591.

Gössling, S., D. Scott, C.M. Hall, J.P. Ceron and G. Dubois. 2012. Consumer behaviour and demand response of tourists to climate change. Ann. Tour. Res. 39: 36–58.

Hall, C.M. 2001. Trends in ocean and coastal tourism: the end of the last frontier? Ocean Coast. Manag. 44: 601–618.

Hamele, H. 2002. Eco-labels for tourism in Europe: moving the market toward more sustainable practices pp. 187–210. *In:* M. Honey [ed.]. Ecotourism and Certification: Setting Standards in Practice. Island Press, Washington DC, USA.

Hamilton, J.M. and R.S. Tol. 2007. The impact of climate change on tourism in Germany, the UK and Ireland: a simulation study. Reg. Environ. Change 7: 161–172.

Hein, L., M.J. Metzger and A. Moreno. 2009. Potential impacts of climate change on tourism: a case study for Spain. Curr. Opin. Environ. Sustain. 1: 170–178.

Hunt, A. and P. Watkiss. 2011. Climate change impacts and adaptation in cities: a review of the literature. Clim. Change 104: 13–49.

IPCC. 2007. Climate change 2007: The physical science basis. summary for policymakers. working group i contribution to the fourth assessment report of the intergovernmental panel on climate change. pp. 93–129. *In*: S. Solomon, D. Qin, M. Manning, Z. Chen, M. Marquis, K.B. Averyt, et al. [eds]. Cambridge University Press, Cambridge, UK and New York, USA.

IPCC. 2013. Climate change 2013: The physical science basis. working group I contribution to the fifth assessment report of the intergovernmental panel on climate change. pp. 121–158. *In*: T.F. Stocker, D. Qin, G.-K. Plattner, M. Tignor, S.K. Allen, J. Boschung, et al. [eds]. Cambridge University Press, Cambridge, UK and New York, USA.

IPCC. 2019. Special report on the ocean and cryosphere in a changing climate. pp. 73–114. *In*: H.-O. Pörtner, D.C. Roberts, V. Masson-Delmotte, P. Zhai, M. Tignor, E. Poloczanska, et al. [eds]. In press.

ISO. 2018. Contributing to the UN Sustainable Development Goals with ISO standards. International Organization for Standardization, Geneva, Switzerland.

Jorge, J.P. 2019. Gestão de recursos costeiros. pp. 22–27. *In*: P. Almeida [ed.]. Manual de Boas Práticas e Sustentabilidade no Turismo. AIRO, Oeste Portugal: Tourism Startup Program, Caldas da Rainha, Portugal.

Klint, L.M., E. Wong, M. Jiang, T. Delacy, D. Harrison and D. Dominey-Howes. 2012. Climate change adaptation in the Pacific Island tourism sector: analysing the policy environment in Vanuatu. Curr. Issues Tour. 15: 247–274

Kozak, M. and K. Nield. 2004. Role of quality and eco-labelling systems. J. Sustain. Tour. 12: 138–148.

Laukkonen, J., P.K. Blanco, J. Lenhart, M. Keiner, B. Cavric and C. Kinuthia-Njenga. 2009. Combining climate change adaptation and mitigation measures at the local level. Habitat Int. 33: 287–292.

Lenzen, M., Y. Sun, F. Faturay, Y. Ting, A. Geschk and A. Malik. 2018. The carbon footprint of global tourism. Nat. Clim. Chang. 8: 522–528.

Lucrezi, S., M. Saayman and P. Van der Merwe. 2015. Managing beaches and beachgoers: Lessons from and for the Blue Flag award. Tour. Manag. 48: 211–230.

Matthews, L., D. Scott and J. Andrey. 2019. Development of a data-driven weather index for beach parks tourism. Int. J. Biometeorol. doi: 10.1007/s00484-019-01799-7. Epub ahead of print.

McGranahan, G., D. Balk and B. Anderson. 2007. The rising tide: assessing the risks of climate change and human settlements in low elevation coastal zones. Environ. Urban 19: 17–37.

McKenna, J., A.T. Williams and J.A. Cooper. 2011. Blue Flag or red herring: do beach awards encourage the public to visit beaches? Tour. Manag. 32: 576–588.

McKercher, B., B. Prideaux, C. Cheung and R. Law. 2010. Achieving voluntary reductions in the carbon footprint of tourism and climate change. J. Sustain. Tour. 18: 297–317.

Michailidou, A.V., C. Vlachokostas and N. Moussiopoulos. 2016. Interactions between climate change and the tourism sector: multiple-criteria decision analysis to assess mitigation and adaptation options in tourism areas. Tour. Manag. 55: 1–12.

Mieczokowski, Z. 1990. World Trends in Tourism and Recreation. Peter Lang Publishing, New York.

Mir-Gual, M., G.X. Pons, J.A. Martín-Prieto and A. Rodríguez-Perea. 2015. A critical view of the Blue Flag beaches in Spain using environmental variables. Ocean Coast. Manag. 105: 106–115.

Moore, W.R. 2010. The impact of climate change on Caribbean tourism demand. Curr. Issues Tour. 13: 495–505.

Moreno, A. and B. Amelung. 2009. Climate change and coastal and marine tourism: review and analysis. J. Coast. Res. 56: 1140–1144.

Moreno, A. 2010. Climate change and tourism: impacts and vulnerability in coastal Europe. Ph.D. Thesis, Maastricht University, Maastricht, Netherlands.

Mukogo, R. 2014. Greening of the Tourism Sector an Effective Mitigation Measure Against Climate Change. International Institute for Peace through Tourism, Vermont, USA.

Nahman, A. and D. Rigby. 2008. Valuing Blue Flag status and estuarine water quality in Margate. S. Afr. J. Econ. 76: 721–737.

Nelson, C., R. Morgan, A.T. Williams and J. Wood. 2000. Beach awards and management. Ocean Coast. Manag. 43: 87–98.

Nguyen, T., J. Bonetti, K. Rogers and C. Woodroffe. 2016. Indicator-based assessment of climate-change impacts on coasts: a review of concepts, methodological approaches and vulnerability indices. Ocean Coast. Manag. 123: 18–43.

Nicholls, R., P. Wong, V. Burkett, J. Codignotto, J. Hay, R. McLean, et al. 2007. Coastal systems and low-lying areas. pp. 315–356. *In*: M.L. Parry, O. Canziani, J.P. Palutikof, P.J. van der Linden and C. Hanson [eds]. Climate Change 2007: Impacts, Adaptation and Vulnerability. Contribution of Working Group II to the fourth assessment report of the Intergovernmental Panel on Climate Change, Cambridge University Press, Cambridge, UK.

Nicholls, S. and B. Amelung. 2008. Climate change and tourism in northwestern Europe: impacts and adaptation. Tour. Anal. 13: 21–31.

Nordhaus, W.D. (2010). Economic aspects of global warming in a post-Copenhagen environment. Proc. Natl. Acad. Sci. USA 107: 11721–11726.

Nyaupane, G. and N. Chhetri. 2009. Vulnerability to Climate Change of Nature-Based Tourism in the Nepalese Himalayas. Tour. Geogr. 11: 95–119.

Oliveira, N. and J. Soares. 2018. O Lado Verde da Bandeira Azul—Estudo Piloto dos Benefícios Económico-Ambientais em Seis Municípios Portugueses. ABAE-Associação Bandeira Azul da Europa, Lisboa.

Rutty, M. and D. Scott. 2010. Will the Mediterranean become 'too hot' for tourism? A reassessment. Tour. Plan. Dev. 7: 267–281.

Scott, D., S. Gössling and C.R. Freitas. 2009. Preferred climates for tourism: Case studies from Canada, New Zealand and Sweden. Clim. Res. 38: 61–73.

Scott, D. and C. Lemieux. 2010. Weather and climate information for tourism. Procedia. Environ. Sci. 1: 146–183.

Sherbinin, A., A. Schiller and A. Pulsipher. 2007. The vulnerability of global cities to climate hazards. Environ. Urban 19: 39–64.

Short, A.D. and C.D. Woodroffe. 2009. The Coast of Australia. Cambridge University Press, Melbourne, Australia.

Sovacool, B.K. 2012. Perceptions of climate change risks and resilient island planning in the Maldives. Mitig. Adapt. Strateg. Glob Chang. 17: 731–752.

United Nations Environment Programme. 2009. UNEP 2008 Annual Report. Retrieved from: https://wedocs.unep.org/handle/20.500.11822/7742

USAID. 2009. Adapting to coastal climate change: a guidebook for development planners. U.S. Agency for International Development, Washington DC, USA.

Weston, R., H. Hamele, M. Balas, R. Denman, A. Pezzano, G. Sillence, et al. 2018. Research for TRAN Committee—European Tourism Labelling. European Parliament—Policy Department for Structural and Cohesion Policies, Brussels.

Williams, P.W. and I.F. Ponsford. 2009. Confronting tourism's environmental paradox: transitioning for sustainable tourism. Futures 41: 396–404.

World Bank. 2017. Adaptation to climate change in coastal areas of the ECA Region: a contribution to the umbrella report on adaptation to climate change in ECA. World Bank Group: Washington DC, USA.

WTTC. 2020. Travel & Tourism: Recovery Scenarios 2020 & Economic Impact from COVID-19. World Travel and Tourism Council, London.

Zaninović, K. and A. Matzarakis. 2009. The bioclimatological leaflet as a means conveying climatological information to tourists and the tourism industry. Int. J. Biometeorol. 53: 369–374.

Zielinski, S. and C.M. Botero. 2019. Myths, misconceptions and the true value of Blue Flag. Ocean Coast. Manag. 174: 15–24.

Chapter 7

Validation and Use of Biological Metrics for the Diagnosis of Sandy Beach Health

Jenyffer Vierheller Vieira[1,2*] **and Carlos Alberto Borzone**[2]

[1]Santos Basin Beach Monitoring Project (PMP-BS), University of the Region of Joinville (UNIVILLE), Rodovia Duque de Caxias, nº 6365 CEP 89240-000, Iperoba, São Francisco do Sul, SC, Brazil (jenyffervvieira@gmail.com), current affiliation.

[2]Sandy Beach Ecology Laboratory, Center for Marine Studies, Federal University of Parana, Av. Beira Mar, s/n CEP 83255-000, Pontal do Sul, Pontal do Paraná, PR, Brazil (capborza@gmail.com).

INTRODUCTION

In the 19th century, English royalty created the prevailing custom of sea bathing, which culminated with the great coastal occupation in the 20th century (Angulo 2004). Since then, the popularity of beaches used for recreational and tourism purposes is growing at a fast pace, and these environments are increasingly being submitted to a large variety of anthropic disturbances (Brown and McLachlan 2002, Davenport and Davenport 2006, McLachlan and Defeo 2013, Schlacher et al. 2014a).

Among these threats, urbanization of the coastal zone in some countries of Latin America has been considerably swift, following the regional economic boom with consequent 'sun and beach tourism' (González et al. 2014). In addition, use of

*Corresponding author: jenyffervvieira@gmail.com

solid structures to contain coastal erosion, pollution via domestic and/or industrial sewage, manual or mechanical cleaning practices, and recreational activities have endangered the integrity of the beach ecosystem (Defeo et al. 2009, McLachlan and Defeo 2013). Such disruptions vary locally and can act as temporary (pulse) or permanent (press) disturbances and, consequently, cause negative impacts on the natural system (physical and/or biological) at different spatial and temporal scales (Schlacher et al. 2007a, Schlacher et al. 2014a).

Since sandy beaches are affected by both terrestrial and marine environments, they are among the most vulnerable ecosystems (Amaral et al. 2016). Challenges in the management of sandy beaches arise from a duality of purposes: beaches need to function as places of recreation (leisure) and other human finalities and, at the same time, represent unique ecosystems and habitats that require subsidies for conservation in face of such excessive anthropic use (McLachlan et al. 2013). Till date an anthropocentric vision predominates in beach management, in which actions are generally directed toward human well-being. Despite the warnings regarding the need to maintain the ecological functions of beach ecosystems, few initiatives have been adopted for this purpose (Brown and McLachlan 2002, Micalleff and Williams 2002, Schlacher et al. 2008, Harris et al. 2014).

On the other hand, it is impossible to measure all the environmental variables (biotic and physical) and appropriately integrate a large amount of information to ultimately evaluate the health of an ecosystem and collaborate with the decision-making process for its conservation (Barros 2001, Williams and Micallef 2009). The selection of metrics to assess the ecological effects of anthropic activities on sandy beaches can be a useful tool for both the management and conservation of the ecosystem (Schlacher et al. 2014b). In this context, the term 'indicator' (synonymous herein of 'metric') is often employed in the science and management interface. Basically, the environmental quality indicators of beaches, monitored by environmental agencies, are limited to quantifying the presence of fecal coliforms in the waters and sand (WHO 2003, Andraus et al. 2014). Although these metrics are considered good quantitative measures, their results only reflect public health and human utility issues, inferring little about the ecological integrity of the beach environment (James 2000, Schlacher et al. 2014b).

Knowledge on sandy beaches has advanced significantly in the past decades, especially from the perspective of anthropic disturbances (Nel et al. 2014). Numerous scientific publications from around the world have documented the consequences of anthropic impacts in these environments (Brown and McLachlan 2002, Schlacher et al. 2008, Defeo et al. 2009, Gonçalves et al. 2013, González et al. 2014, Schlacher et al. 2014a, Schlacher et al. 2016, Bessa et al. 2017 and many others). Consequently, several biological metrics at the community level (occurrence, species richness, diversity index) and population level (distribution, total abundance, biomass, physiological tolerance, behavioural plasticity) have been used as ecological indicators of beach ecosystem health (reviewed in Schlacher et al. 2014b).

Unfortunately, biological metrics have not yet been incorporated into beach planning and management tools (represented in Brazil by the *Projeto de Gestão*

Integrada da Orla Marítima—Projeto Orla) (Oliveira and Nicolodi 2012). In general, scientific and technical knowledge have little permeability within government agencies responsible for the management of these environments (Scherer et al. 2009). This scenario is associated with a set of considerations, including conflicts of interest; lack of communication between researchers and public administrators (Schlacher et al. 2008, Noriega et al. 2012); little environmental awareness on the part of society and public authority; nominations for public office in environmental organizations often based only on political criteria; and mainly the scarcity of local bio-ecological information on each beach (Amaral et al. 2016). Although science and management are distinct disciplines, they are integrated or should interact, intimately. Therefore, scientific knowledge (regional and local) should be seen as one of the primary foundations for the implementation of management actions, thus avoiding the use of inappropriate practices (Scherer et al. 2009, Souza et al. 2017).

In order to promote the implementation of studies to evaluate the health of sandy beaches (local, regional and global), Brazilian researchers, in association with The Coastal Benthic Habitat Monitoring Network (ReBentos) (www.rebentos.org), have developed standardized, low-cost and easy-to-use collection protocols for continuous and permanent beach monitoring (Amaral et al. 2016). Among the resident beach fauna, the populations of *Ocypode quadrata* (Crustacea: Decapoda) and *Bledius* spp. (Insecta: Coleoptera) have been recognized as some of the key groups for assessing alterations caused by climate change (Borzone et al. 2015, Rosa et al. 2015). Additionally, a considerable amount of literature has proposed the use of these species also as indicators of anthropogenic impact (Vieira et al. 2012, Lucrezi and Schlacher 2014, Vieira 2015, Vieira and Borzone 2017, Souza et al. 2017, Costa and Zalmon 2019).

In Brazil, *O. quadrata* has received special attention as an ecological indicator since it is the only species of the genus in the country, inhabiting beaches (including the entire morphodynamic spectrum) from open to the ocean to low energy estuarine ones (Turra et al. 2005, Rosa and Borzone 2008, Pombo and Turra 2013). In spite of some shortcomings (Pombo and Turra 2013, Silva and Calado 2013), the non-destructive indirect method of counting and measuring the diameter of crab burrows is strongly recommended in health assessments of the beach ecosystem (Souza et al. 2017, Costa and Zalmon 2019). In practice, the large amount of older specimens (adults) on beaches with low numbers of people indicates a better-quality habitat (Vieira 2015, Souza et al. 2017).

Historically, insects are the most neglected taxonomic group in studies concerning sandy beaches due mainly to methodological and taxonomic limitations (Harris et al. 2014). These organisms, particularly coleopterous staphylinids, represent a conspicuous component of upper beach zones around the world (McLachlan and Brown 2006), and their presence is easily detected by surface traces in the sediment from their activity (Herman 1986, Gandara-Martins et al. 2010, Vieira and Borzone 2017).

Although representatives of the genus *Bledius* are considered to be an excellent key group for beach monitoring (Rosa et al. 2015), Vieira and Borzone (2017) highlighted that the impact of recreational use should be evaluated at the species level. In southern Brazil (coast of Paraná), for example, *Bledius hermani* is recognized as the most sensitive species to anthropic disturbances due to its exclusive occurrence in the upper and lower midlittoral regions, which are primarily occupied by tourists (Vieira 2015, Vieira and Borzone 2017).

Considering that high levels of recreational activity and urbanization can also influence beach morphodynamics and, consequently, affect beach fauna in some situations, it would be difficult to differentiate natural changes from anthropogenic ones (McLachlan et al. 2013, Amaral et al. 2016). In this case, it is recommended that ecosystem health assessments also analyze the physical aspects together (particularly wave and tidal regime and sedimentological characteristics), thus allowing reliable and robust interpretations, as well as the identification of the main factor accountable for the responses of the beach fauna.

In view of this scenario, the objective of the present study was to validate the use of *O. quadrata* (adults) and *B. hermani* populations as biological metrics of beach ecosystem health, illustrating their application in the physical and anthropic contexts of the ocean beaches on the coast of Paraná, southern Brazil. To this end, the relationships between the biological metrics, physical parameters and anthropic characteristics (represented by the Recreational Potential Index—RPI) of the beaches were analyzed. The investigation was guided by the following hypothesis: regardless of the physical differences between the beaches of the Paraná coast, it is expected that species variability (biological metrics) be related to the anthropic factor (RPI). From this, we were able to (1) identify the most appropriate type of management (priority for conservation, priority for recreation or multiple uses) for each beach analyzed and (2) propose guidelines for the conservation and sustainable use of the beach environments along the coast of Paraná (southern Brazil).

MATERIALS AND METHODS

Study Area

The coast of the state of Paraná, located in southern Brazil, extends for approximately 126 km and is limited to the north by the Barra do Ararapira (25°18'S) and south by the Barra do Saí (25°58'S) (Angulo et al. 2016). This region presents a humid subtropical climate, a semidiurnal micro-tide regime, and a maximum amplitude of 1.7 m (Lana et al. 2001, Amaral et al. 2016), in addition to facing east, towards the Atlantic Ocean.

Ocean beaches can be found along the entire Paraná coast and are interrupted by some rocky outcrops and the mouths of Paranaguá Bay (north) and Guaratuba Bay (south). Thus, the ocean coast is divided into three distinct stretches: (1) Coastal Plain of Superagui (14 km), north of the Paranaguá Bay; (2) Coastal Plain of Praia de Leste/Matinhos (31.2 km) in the south-north stretch of the Paranaguá and Guaratuba Bays, respectively, and (3) Coastal Plain of Brejatuba and Saí (11.2 km), south of Guaratuba Bay (Angulo et al. 2016 and references therein).

Field expeditions were carried out in July 2013, in which 12 beaches were selected: seven on the Coastal Plain of Praia de Leste (comprising the beaches of Guapê, Shangri-lá, Guarapari, Albatroz, Perequê, Saint Etienne, and Brava) and five on the Coastal Plain of Brejatuba and Saí (including the Grande, Eliane, Vô Fredo, Coroados 1, and Coroados 3 beaches)—(Fig. 7.1). The geographical coordinates of each beach are available in the first column of Table 7.2.

The beaches in Paraná are a frequent destination for second home tourism in search of sun, sand and beaches, especially with people coming from the state capital (Curitiba) and the countryside (Battistuz and Zardo 2014). In Paraná, the use of beaches for recreational purposes began in 1920 decades (Angulo et al. 2016). Despite the significant increase in population growth rates of permanent residents in the last two, heterogeneous urban occupation with a few non-urbanized beaches can still be observed, interspersed with beaches undergoing medium to high urbanization processes (Vieira 2015).

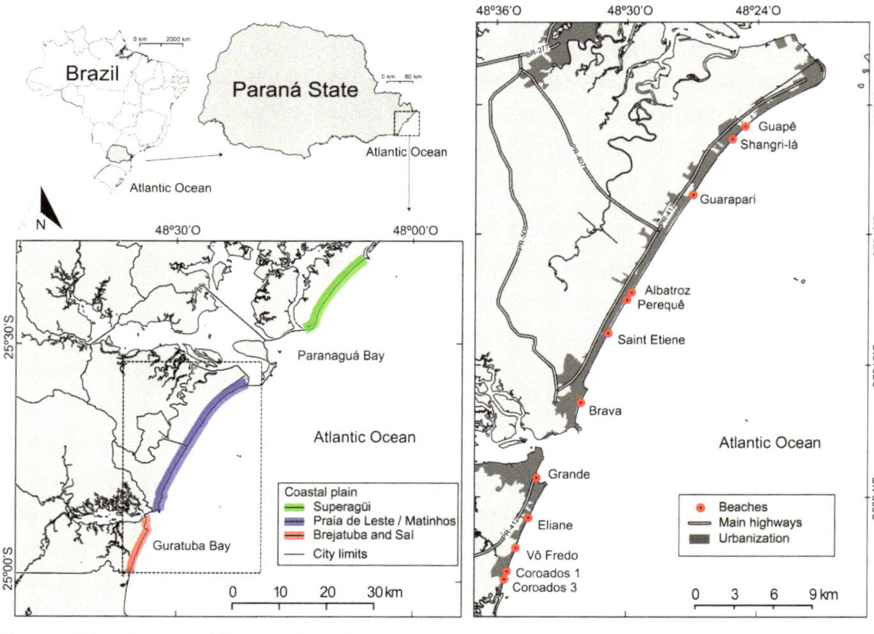

Figure 7.1 Geographic location of the 12 beaches situated along the coastal plain of the state of Paraná (southern Brazil). The geographic pointer indicating north applies to all maps. Satellite image (2003) extracted from Google Earth.

Temporal patterns of recreational use of the Brazilian coast are considerably irregular and closely related to climatic conditions (Veloso et al. 2006, 2008). For instance, beaches of the south coast are exposed to high-intensity human pressure only during a short summer period (January to early February, called Summer—High Season) (Vieira et al. 2012, Vieira and Borzone 2017).

Government initiatives in the State of Paraná, such as the well-known *Operação Verão Paraná*, are carried out during the summer season, the period of the year when the greatest tourist reception takes place, in which coastal municipalities receive social incentives for the development of supervised recreational activities (sports, dance, and gymnastics, among others) (Vieira et al. 2012) and beach cleaning (Krelling et al. 2017).

Beach frequenters usually focus on the beach stretches that provide a combination of facilities (infrastructure, recreational installations, health and safety support) and other tourist services on the waterfront (nightclubs, commercial centers, musical events, etc.) (Breton et al. 1996, De Ruyck et al. 1997, Santana 2003), and, therefore, where recreational impacts are spatially concentrated (Defeo et al. 2009, Schlacher and Thompson 2012). Thus, the choice of the beaches contemplated in the present assessment aimed to reliably reflect such diversity of occupancy scenarios, including beaches with low occupation status (Guapê, Guarapari, Vô Fredo) to strongly densified (particularly Grande and Brava).

Physical Parameters and the Recreational Potential Index (RPI)

During the summer season of 2013/2014, sampling was conducted directed to the physical and anthropic characterization (RPI) of each beach.

The beach slopes were estimated using Emery's profiling technique (Emery, 1961), and their width was measured as the distance between the base of the dune and the lower swash level. In order to determine the morphodynamic state of the beach, the Dean parameter (Ω) (Short and Wright 1983) was employed; three samples of sediment were collected from the upper, middle and lower zones of each beach. A particle size analyzer (Microtrac S3500) was used to estimate the sand fractions, and the sedimentation rate of the particles was evaluated based on tables from Gibbs et al. (1971). The wave period was determined with a stopwatch, and corresponded to 1/10 of the total elapsed time for the passage of 11 consecutive crests at a fixed point in the surf zone. Wave height was visually recorded by the same observer.

The recreational potential index—RPI (modified from McLachlan et al. 2013) was calculated on each beach, based on three criteria: (1) available infrastructure; (2) safety and health, and (3) physical carrying capacity. Each criterion presents a different weight in the index calculation.

The '**available infrastructure**' takes the following parameters into account: (1) access to the beach, (2) recreational facilities, (3) trash bins, (4) lighting, and (5) public toilet; thus, the score for this criterion varies from 0 to 5. The '**safety and health**' criterion is evaluated according to the presence of lifeguards, bathing water quality (available to the public on plates), and the presence of beach cleaning activity (removal of solid waste along with organic material); its score varies from 0 to 3. The third criterion '**physical carrying capacity**' is closely related to the characteristics of the beach profile (width and slope) available to users, and its scores vary from 0 to 2 (Table 7.1).

The RPI value obtained from each beach results from the sum of the individual scores of each criterion and, consequently, can vary from 0 to 10. The score

Table 7.1 Description of the items related to the application of the Recreational Potential Index—RPI (modified from McLachlan et al. 2013)

Criteria and Parameters	Condition and score		
Infrastructure			
Access to the beach	0 – Absence of seaside, only access by the beach	0.5 – Unpaved waterfront, access to the beach by the dune vegetation	1 – Paved waterfront, access to the beach by the dune vegetation
Recreational facilities	0 – Absent	0.5 – Street vendors	1 – 'Operação Verão Paraná', restaurants, parkings, street vendors
Trash bins	0 – Absent	0.5 – Present/few	1 – Present/many
Lighting	0 – Absent	1 – Present	
Public toilet	0 – Absent	1 – Present	
Safety and health			
Lifeguard	0 – Absent	0.5 – Present/few	1 – Present/many
Bathing water quality	0 – Polluted water	1 – Free water pollution	
Beach cleaning	0 – Absent	1 – Present	
Carrying capacity			
Beach profile	0 – Short and steep	1 – Intermediate	2 – Extensive and smooth
Total score	**minimum = 0** **maximum = 10**		

0.5 was adopted for some parameters (accessibility to the beach, presence of recreational facilities, trash bins, lifeguards) where semi-quantitative evaluations were possible. All this information was attained through direct observations in the field. In the present study, RPI values ≤ 4 indicated low recreational potential, $4 < \text{RPI} < 7$ moderate recreational potential, and RPI ≥ 7 intense recreational potential. Details on the scoring of the parameters contemplated in each criterion are shown in Table 7.1.

Biological Metrics

Fauna sampling took place after the summer high season of 2013/2014 (March), with all beaches being sampled only once and during the same period in order to exclude the effects of temporal variability on the abundance of the evaluated species (Souza et al. 2017). *O. quadrata* samplings consisted of counting the openings of active burrows (Barros 2001, Moss and McPhee 2006, Schlacher et al. 2007b, Noriega et al. 2012, Borzone et al. 2015), which was always conducted during low spring tide periods and under good weather conditions. During neap tides, the sea level does not rise enough to eliminate all signs of activity and close the old and inactive burrows of the beach profile (Borzone et al. 2015).

In this evaluation, only active burrows of adult individuals were counted. According to Alberto and Fontoura (1999), burrows above 20.8 mm in diameter are considered adults. The other age structures (recruits and juveniles) were not included in the assessment since the main period of recruitment of this species occurs in summer (Alberto and Fontoura 1999, Negreiros-Fransozo et al. 2002). In addition, *O. quadrata* is able to settle on beaches submitted to different levels of anthropic disturbance (Souza et al. 2017). Thus, the inclusion of these individuals could confuse the interpretation of the real effect of recreational activity. At each beach, 10 perpendicular transects were randomly placed within 50 m along the shore extension. Burrows were quantified inside a square measuring 1.0 m^2 placed on the intertidal beach: from the beginning of the dune vegetation to the last burrow found until the upper swash limit.

The *B. hermani* staphylinid samplings were carried out in parallel with those of *O. quadrata*. Its larvae were not analyzed due to difficulties in identification at the species level (Vieira and Borzone 2017). Six of the 10 transects used in the indirect sampling of *O. quadrata* were randomly selected to sample *B. hermani*. Adverse climatic conditions (heavy rains, undertows or strong winds) were avoided, given they interfere in the level of organism activity (Rosa et al. 2015).

Ten samples were taken from each transect from the border with the embryonic dune (supralittoral region) until the upper midlittoral area, covering the entire distribution range of the staphylinids (Gandara-Martins et al. 2010). A total of six samples were collected at each transect, spaced apart by 2 to 3 m, with the aid of a cylindrical sampler measuring 15 cm in diameter. Prior to the contact between the sampler and the sediment surface, the upper part of the cylinder was sealed with a plastic bag to prevent leakage of the organisms (Ruiz-Delgado et al. 2014, Rosa et al. 2015, Vieira and Borzone 2017).

In the laboratory, the samples were fixed in 10% neutralized formalin solution and washed with the aid of a sieve (mesh opening = 500 μm). The individuals were counted under a stereoscopic microscope and identified using specialized literature (Caron and Ribeiro-Costa 2007).

Data Analysis

Possible differences in the linear abundance of *O. quadrata* (adults) and density of *B. hermani* staphylinids among the beaches (factor) were analyzed by one-way analysis of variance (ANOVA). When significant differences were found, they were discriminated using the Student-Newman-Keuls (SNK) multiple means comparison test. Data normality and the homoscedasticity of the variances were verified by the Shapiro and Cochran tests, respectively (Underwood 1997). When these assumptions were not met, the data underwent transformation.

Correlations between the biological metrics (linear abundance of *O. quadrata* adults and density of *B. hermani* staphylinids) and the physical characteristics (i.e., beach width and slope, mean grain diameter, morphodynamic stage) and the Recreational Potential Index (RPI) of the beaches were assessed by Spearman's correlation test (Quinn and Keough 2002).

All analyses and graphs were generated with R software, version 2.13.0 (R Core Team 2012), using the GAD packages (Sandrini-Neto and Camargo 2012) and Sciplot (Morales 2012).

RESULTS

Beach Environment and Recreational Level

The analyzed beaches presented distinct topographic profiles. Guapê, Shangri-lá, and Guarapari exhibited an extensive profile and a gentle slope, whereas the other beaches were characterized by narrower profiles and moderate slopes (Table 7.2). The average grain size corresponded to a fine sand fraction in all beaches, except for Albatroz and Perequê, where the sediment was composed of fine/medium-sized sand (Table 7.2). The morphodynamic stage of the beaches was classified as intermediate ($1 < \Omega < 6$, Table 7.2). Guapê, Shangri-lá, and Guarapari presented the highest values of omega and were characterized as intermediate-dissipative environments. In contrast, beaches Albatroz, Perequê, and Saint Etienne, which retained values close to the reflective limit, represent intermediate-reflective environments (Table 7.2).

The values obtained from the calculation of the Recreational Potential Index (RPI) ranged from 3.5 to 10 (Table 7.3). The lowest values (RPI $\leq$4.0) were recorded in Guapê, Guarapari, Albatroz, Perequê, Vô Fredo, and Coroados 3, while Shangri-lá, Brava, and Grande were the ones with the highest recreational potential values (RPI $\geq$7).

Table 7.2 Brief physical, sedimentological and morphodynamic characterization of the beaches analyzed along the coast of the state of Paraná (southern Brazil) regarding beach width and slope, mean grain diameter, wave period and height, and the omega index (Dean Ω)

Beach	Localization	Beach width (m)	Slope (°)	Mean sand grain size (mm)	Wave period (s)	Wave height (cm)	Dean's parameter
Guapê	25°37'01.4"S 48°24'34.7"W	80	1.43	0.191	6	70	5.00
Shangri-lá	25°37'36.8"S 48°25'10.4"W	100	1.20	0.191	5	60	5.09
Guarapari	25°40'09.5"S 48°26'57.2"W	90	1.72	0.210	6	80	5.02
Albatroz	25°44'40.0"S 48°29'46.6"W	50	2.87	0.255	12	65	1.59
Perequê	25°45'00.4"S 48°29'59.3"W	55	2.69	0.262	11	55	1.42
Saint Etiene	25°46'33.8"S 48°30'50.2"W	55	2.57	0.247	11	55	1.53
Brava	25°49'42.4"S 48°32'05.2"W	65	2.10	0.216	7	50	2.59
Grande	25°53'08.0"S 48°34'06.0"W	70	2.35	0.210	7	50	2.69
Eliane	25°54'56.9"S 48°34'29.1W	65	2.11	0.241	9	75	2.63
Vô Fredo	25°56'19.1"S 48°35'03.6"W	55	2.63	0.243	10	100	3.12
Coroados 1	25°57'24.0"S 48°35'28.3W	65	2.10	0.215	8	85	3.89
Coroados 3	25°57'45.6"S 48°35'35.1"W	60	2.38	0.221	8	80	3.53

Biological Metrics: *Ocypode quadrata* and *Bledius hermani*

The linear abundance of adult *O. quadrata* varied significantly between beaches ($F_{11,108} = 23.81$, $p < 0.001$). Based on a *posteriori* Student-Newman-Keuls (SNK) test, Guapê and Guarapari presented the highest abundances, while Shangri-lá, Eliane, and Coroados 1 exhibited lower levels of this parameter. No burrows were observed in the Grande beach (Fig. 7.2a; Table 7.4).

The density of *B. hermani* staphylinids also showed significant differences between beaches ($F_{11,60} = 22.55$, $p < 0.001$). Half of the analyzed beaches (Albatroz, Coroados 3, Guarapari, Saint Etienne, Coroados 1, and Vô Fredo) presented the highest densities. At the other extreme, the Shangri-lá and Eliane beaches recorded the lowest values (Fig. 7.2b; Table 7.5). The presence of staphylinids was not detected in the Grande beach (Fig. 7.2b).

Table 7.3 Values obtained concerning the parameters used in the calculation of the Recreational Potential Index—RPI (modified from McLachlan et al. 2013, Table 7.2) and the respective score per beach analyzed along of the State of Paraná (southern Brazil)

	Infrastructure					**Safety and health**			**Carrying capacity**	
Beach	**Access to the beach**	**Recreational facilities**	**Trash bins**	**Lighting**	**Public toilet**	**Lifeguard**	**Bathing water quality**	**Beach cleaning**	**Beach profile**	**RPI**
Guapê	0	0	0.5	0	0	0	–	1	2	3.5
Shangri-lá	1	1	1	1	1	1	1	1	2	10
Guarapari	0.5	0	0.5	0	0	0	–	1	2	3
Albatroz	1	0	0.5	0	0	0	–	1	1	3
Perequê	1	0	0.5	0	0	0	–	1	1	3
Saint Etiene	1	0.5	1	1	0	0.5	–	1	1	5
Brava	1	1	1	1	1	1	1	1	1	9
Grande	1	1	1	1	1	1	1	1	1	9
Eliane	0.5	0.5	1	1	1	0.5	–	1	1	5
Vô Fredo	1	0	0	0	0	0	–	1	1	3
Coroados 1	0.5	0.5	1	1	1	0.5	–	1	1	5
Coroados 3	0.5	0	0.5	0	0	0.5	–	1	1	2

Table 7.4 Summary of the multiple comparison test of the Student-Newman-Keuls (SNK) averages, contrasting the linear abundance of *O. quadrata* adults among the 12 beaches analyzed from the state of Paraná (southern Brazil). $^{ns}p > 0.05$; $*p < 0.05$; $**p < 0.01$; $***p < 0.001$

	Grande	**Shangri-lá**	**Eliane**	**Coroados 1**	**Albatroz**	**Coroados 3**	**Vô Fredo**	**Brava**	**Perequê**	**Saint Etiene**	**Guapê**	**Guarapari**
Grande												
Shangri-lá	ns											
Eliane	ns	ns										
Coroados 1	ns	ns	ns									
Albatroz	*	ns	ns	ns								
Coroados 3	*	ns	ns	ns	ns							
Vô Fredo	***	**	**	**	ns	ns						
Brava	***	**	**	**	ns	ns	ns					
Perequê	***	***	***	***	ns	ns	ns	ns				
Saint Etiene	***	***	***	***	**	**	ns	ns	ns			
Guapê	***	***	***	***	***	***	***	***	***	***		
Guarapari	***	***	***	***	***	***	***	***	***	***	ns	

Table 7.5 Summary of the multiple comparison test of the Student-Newman-Keuls (SNK) averages, contrasting the density of *B. hermani* among the 12 beaches analyzed from the State of Paraná (southern Brazil). $^{ns}p > 0.05$; $*p < 0.05$; $**p < 0.01$; $***p < 0.001$

	Grande	**Shangri-lá**	**Eliane**	**Brava**	**Guapê**	**Perequê**	**Vô Fredo**	**Coroados 1**	**Saint Etiene**	**Guarapari**	**Coroados 3**	**Albatroz**
Grande												
Shangri-lá	Ns											
Eliane	*	ns										
Brava	***	***	*									
Guapê	***	***	*	ns								
Perequê	***	***	*	ns	ns							
Vô Fredo	***	***	***	ns	ns	ns						
Coroados 1	***	***	***	ns	ns	ns	ns					
Saint Etiene	***	***	***	**	*	*	ns	ns				
Guarapari	***	***	***	**	**	*	ns	ns	ns			
Coroados 3	***	***	***	***	***	**	ns	ns	ns	ns		
Albatroz	***	***	***	***	***	***	ns	ns	ns	ns	ns	

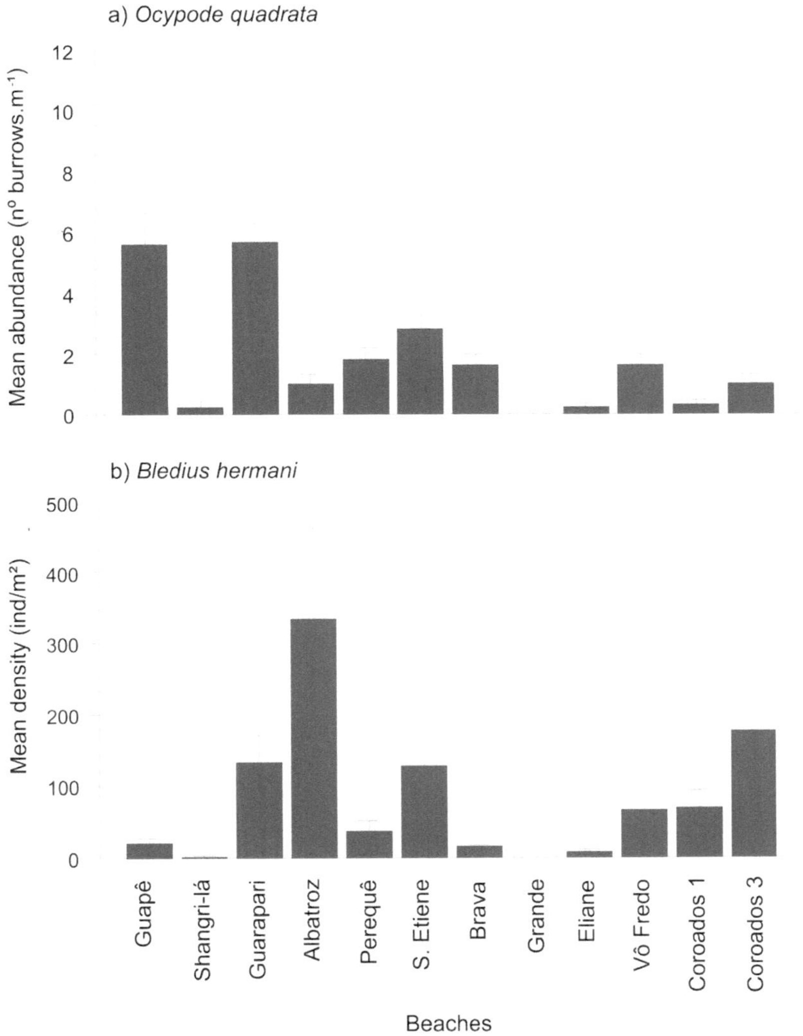

Figure 7.2 Mean values (± standard error) of the linear abundance of adult *O. quadrata* (a) and the density of *B. hermani* staphylinids (b) recorded at the beaches along of the state of Paraná (southern Brazil).

Relationship between the Biological Metrics and the Physical and Anthropic Parameters

The two analyzed biological metrics (linear abundance of *O. quadrata* adults and the density of *B. hermani* staphylinids) showed a significant negative correlation with the recreational potential index (Table 7.6). *O. quadrata* did not present a significant association with the other parameters. With low correlation values, the density of *B. hermani* was negatively related to beach width and the omega morphodynamic index, and positively to the beach slope (Table 7.6).

Table 7.6 Values of the Spearman correlation coefficient (R) between the analyzed biological metrics, linear abundance of adult *O. quadrata* and density of *B. hermani* staphylinids and the physical parameters (beach width and slope, mean grain diameter and the morphodynamic index – Dean) and the Recreational Potential Index (RPI) of the evaluated beaches along of the State of Paraná (southern Brazil)

	Independent variables				
	Beach width	**Slope**	**Mean sand grain size**	**Dean's parameter**	**RPI**
Linear abundance of adult *O. quadrata*	(+) 0.06^{ns}	(–) 0.04^{ns}	(+) 0.05^{ns}	(–) 0.01^{ns}	(–) 0.44^{***}
Density of *B. hermani*	(–) 0.39^{***}	(+) 0.41^{***}	(–) 0.13^{ns}	(–) 0.40^{***}	(–) 0.60^{***}

$^{ns}p > 0.05$; $*p < 0.05$; $**p < 0.01$; $***p < 0.001$

DISCUSSION

Responses of the Biological Metrics to the Physical and Anthropic Factors

Changes in the physical environment of the beaches were evident along the ocean coast of the state of Paraná (southern Brazil), with morphological and sedimentological characteristics consistent with the morphodynamics stage exhibited. Those located on the coastal plain of Praia de Leste presented contrasting morphodynamic conditions between the dissipative stage (Guapê, Shangri-lá, and Guarapari) and the reflective stage (Albatroz, Perequê, and Saint Etienne). The morphodynamic evolution of the beach arch of the Praia de Leste plain occurs as a function of two primary distinct sediment sources, represented by the Paranaguá and Guaratuba Bays (fine and medium sand, respectively) (Borzone et al. 1996). Such granulometric variations reflect a gradual shift between the beach morphodynamic stages in the north-south direction (Angulo et al. 2016 and references therein), also attested in the present study. The Brava beach (located in the municipality of Matinhos) and the beaches on the Brejatuba and Saí coastal plains (Grande, Eliane, Vô Fredo, Coroados 1, and Coroados 3) were classified as intermediate, corroborating previous morphodynamic evaluations (Borzone et al. 1998, Barros et al. 2001, Gandara-Martins et al. 2010).

Despite the physical variability (morphological and morphodynamic) detected among the beaches evaluated in this study, the abundance of adult *O. quadrata* showed no significant correlation with any of the analyzed physical factors, corroborating the tested hypothesis. On the other hand, a significant negative correlation with the Recreational Potential Index (RPI) shows the sensitivity of these organisms to anthropic pressure. In subtropical beaches, anthropic pressure due to recreational use is recognized as an acute disturbance (Defeo et al. 2009). Human trampling, the primary form of disturbance in the recreational beaches of southern Brazil, occurs mainly during the summer, and, therefore, its effects are detected for weeks to months later. In theory, recovery is possible after pulse disturbances, but, in practice, these types of disturbances are often repeated, causing recovery to be incomplete (Schlacher et al 2014a). Considering that the temporal extension of these effects depends on the intensity of the disturbance, Vieira (2015) stated that the pressure resulting from recreation on Parana´s beaches assumed a recurring character, since such disruption is repeated summer after summer.

Currently, there is abundant evidence that the species *O. quadrata* and its congeners respond negatively to some anthropic pressures, such as those directly related to recreational activities (human trampling and traffic of motor vehicles), or indirectly, including mechanical beach cleaning and urbanization, among others (Magalhães et al. 2009, Lucrezi and Schlacher 2014, Schlacher et al. 2016 and references therein). All these stressors probably occur synergistically to impair ghost crab populations, but it is only possible to confirm the intensity of each impact through the adoption of manipulative experimental designs (Glasby and Underwood 1996). On the Brazilian coast, the use of *O. quadrata* as a proxy for beach health has been frequently discussed (Neves and Bemvenuti 2006, Souza et al. 2008, Silva and Calado 2011, Silva and Calado 2013, Pombo and Turra 2013, Borzone et al. 2015, Souza et al. 2017, Costa and Zalmon 2019).

Differences in the density of *B. hermani* staphylinids were correlated to both the physical and anthropic variables, although more significantly to the latter. The higher densities of these individuals were associated with beaches with narrow profiles, marked slopes and lower morphodynamic index values, typical characteristics of intermediate to reflective environments. In contrast, Gandara-Martins et al. (2010) reported elevated densities of *B. hermani* in dissipative beaches of Paraná (extensive and smooth profile). In the present study, the dissipative conditions were only observed in the Guapê, Shangri-lá, and Guarapari beaches, the latter of which presented similar densities to the environments described as intermediate (Vô Fredo, Corados 3) and reflective (Albatroz, Saint Etienne). Such observations indicate that this species shows high plasticity to morphodynamic conditions, a premise supported by the results obtained by Gandara-Martins et al. (2010).

The low density of *B. hermani* in the Shangri-lá beach confirms the negative influence of recreational pressure on the spatial distribution of these organisms, as recently reported by Vieira and Borzone (2017). The complete absence of *Bledius* spp. was evidenced by Barros et al. (2001) on beaches with distinct morphodynamics located on the southern coast of Paraná, including the Grande beach. The authors attributed this fact to some anthropic disturbances, such as the destruction of dunes for road construction and human trampling. In view of the aforementioned issues, the physical characteristics probably do not explain the total absence of staphylinids (Grande and Eliane, respectively), nor the low densities recorded in Shangri-lá, Eliane, Saint Etienne, Brava, and Grande. Therefore, the influence of the anthropic factor appears to be dominant.

Among the species of *Bledius* described on the coast of Paraná (Caron and Ribeiro-Costa 2007, Gandara-Martins et al. 2010, Vieira and Borzone 2017, Castro et al. 2016, Bortoluzzi and Caron 2019), *B. hermani* seems to be the most exposed to recreational pressure (via human trampling), considering that the spatial distribution of users overlaps the region occupied by the species, that is, the upper midlittoral area of the beach (Vieira and Borzone 2017). According to Vieira and Borzone (2017), recreational pressure affects the populations of *B. hermani* in two ways: (1) directly, causing mortality by crushing individuals (larvae and adults); (2) indirectly, interfering in functional activities like superficial foraging, due to the collapse of their burrows. Human trampling causes lethal damage to the larval stages, as the individuals are still unfit for movement (flight) and have delicate body structures (Herman 1986), and the adults, who are buried in the more superficial layers of the sediment (Veloso et al. 2008). In addition, destruction of the burrows exposes these staphylinids to natural adversities, including variations of temperature and humidity (desiccation) and the action of winds and the tide (flooding) (Vieira and Borzone 2017).

It is worth noting that, unlike the other species described on the Brazilian coast, *B. hermani* occurs in practically the entire coastline (south, southeast and northeast) (Castro et al. 2016 and references therein). This distribution enables comparisons between beach environments located in different latitudes and that are exposed to different scenarios of anthropic impact, extending the possibilities of using this species as an ecological indicator of beach health.

Sandy beaches are considered the leading tourist resource on the coast of many countries, environments which are responsible for the flow of hundreds of thousands of people who, every year, stimulate the local economy during the holiday season. Despite the undeniable importance for the touristic and economic development of coastal municipalities, beaches are subject to a substantial variety of threats that endanger the integrity of the ecosystem (Brown and McLachlan 2002, Defeo et al. 2009). One of the major challenges of managing coastal ecosystems, particularly sandy beaches, is to reconcile the development of sustainable use and conservation (McLachlan et al. 2013). In this sense, this chapter suggests the incorporation of biological metrics to current diagnoses, aimed for more complete evaluations and, indeed, more efficient management strategies. From this perspective, one can also recommend the use of the concepts proposed by McLachlan et al. (2013): priority conservation, priority recreation and multiple uses.

Implications for Beach Management: Case Report on the Coast of Paraná (southern Brazil)

The adverse effects of anthropic disturbances (namely recreational activities) on beach fauna can only be inhibited or minimized by planning and adopting management approaches (McLachlan et al. 2013). The various arrangements of human society (residents, tourists, scientists, managers) may present different values and distinct expectations of how a beach should be managed (Jedrzejczak 2004). Efforts to conserve biodiversity often clash with the needs of human use.

The diagnosis of several beaches of the Paraná coast, conducted by the technical staff of the Projeto Orla in the municipalities of Pontal do Paraná, Matinhos, and Guaratuba, is based on physical determinants (types of coastal exposure) and socioeconomic aspects (types of use and occupation). Damage to beach fauna is utterly neglected.

From the ecological point of view, some beaches of low recreational potential (Guapê, Guarapari and Vô Fredo) have endured high abundances of both of the studied species (*O. quadrata* and *B. hermani*) and, therefore, were considered high-priority conservation sites. Coincidentally, these beaches are located in areas of low occupancy and lack of infrastructure (boardwalks, seaside paving, waste bins, showers/toilets, lighting, recreational facilities, etc.), and their frequenters are adept to taking walks and recreational fishing.

At first hand, it is recommended that such beaches be maintained as semi-natural environments, and preferably supported by formal preservation initiatives (Ex. conservation units) (Harris et al. 2015). On the coast of Paraná, the ocean beach located on the Superagui coastal plain (Praia Deserta) is the only stretch of beach from this study included within a federal integral protection unit (Superagüi National Park) (Angulo et al. 2016). When compared to the tourist ocean beaches of the Paraná coast, the Praia Deserta retains elevated abundances of *B. hermani* staphylinids (Gandara-Martins et al. 2010) and *O. quadrata* crabs (pers. observation). Alternatively, the adoption of spatial zoning is also suggested, that is, low recreation sites interspersed with areas of intense recreation along the coast (McLachlan and Defeo 2013). In the early 90s, the coast of Paraná presented a heterogeneous landscape related to the evolution of shoreline occupation (Angulo et al. 2016). Fortunately, after more than two decades, this panorama still prevails as showed by the results of the Recreational Potential Index (RPI) obtained in the present evaluation regarding the analyzed beaches.

Considering the negative effects of human trampling on the populations of *O. quadrata* and *B. hermani* already described in the literature (Vieira 2015, Vieira and Borzone 2017), it is recommended that those beaches of priority conservation undergo an urgent adoption of control measures regarding the number of beach users. Some sites already have features that naturally limit the number of visitors and, therefore, should be maintained. For example, in Guapê, access to the beach is carried out primarily by walking due to the absence of a paved access road to the seashore. Reducing beach accessibility can be a powerful control tool for visitation.

Beach cleaning (manual and mechanical) should be entirely prohibited in these beaches due to the immeasurable negative ecological (physical and biological) damage inherent to debris removal (reviewed in Zielinski et al. 2019). Such cleaning removes not only the solid residue (waste), but also the sand, organic matter, and all associated fauna (Colombini et al. 2011, Defeo et al. 2009, Amaral et al. 2016). The organic detritus is of variable origin and may consist of marine macrophytes, macroalgae, mangrove propagules and dead animals (Ruiz-Delgado et al. 2014). From the beach user's point of view, the debris is considered dirty and without aesthetic value, a perception that is generally related to the strong odor and presence of flies attracted by decomposition (Mclachlan and Brown 2006), mainly in beaches that receive a high contribution of debris. The actual 'waste' has no function, although the natural detritus represents an essential element for the maintenance of beach biodiversity and functioning (Harris et al. 2014). When an entire beach is cleaned, a common practice in many coastal regions of Brazil (and other countries), the effects of wrack removal are more severe (Vieira et al. 2016). The adoption of spatial zoning results in the maintenance of uncleaned zones, which may act as species reservoirs (Colombini et al. 2011).

In addition, other simple measures could be applied, such as the environmental awareness of beach users and the prohibition of civil engineering, as well as the traffic of vehicles on beaches and dunes. The adoption of user signalling can aid in the delimitation of accesses and, at the same time, act as an instrument of environmental awareness. Undoubtedly, civil construction and vehicular traffic can only be restrained by the active supervision of competent organizations throughout the year. McLachlan et al. (2013) warned that the efficiency of preservation strategies is conditional to the maintenance of a minimum stretch of beach. In the present study, the beaches considered as a priority for conservation (Guapê, Vô Fredo, and Guarapari) have similar beach extensions (1– 2 km), although the actual effectiveness of these locations as preservation areas can only be evaluated after applying the aforementioned measures and the long-term monitoring of biological metrics.

In contrast, in the category of priority recreation management, the Shangri-lá, Grande, and Eliane beaches were submitted to a high level of environmental stress, signalled by the absence or low values of biological metrics (*O. quadrata* and *B. hermani*). These beaches are located near urban centers and are characterized by the presence of street vendors, showers, waste bins, lifeguard observation posts, and paving alongside the waterfront access, as well as social mobility items (public transportation, waterfront access roads: 'boardwalk') and several gastronomic options (bars and restaurants) along the waterfront.

In some countries, stimulating the use of beaches with high recreational potential is one of the suggested management strategies (Breton et al. 1996, Harris et al. 2015). In this sense, future actions carried out in the Shangri-lá, Grande, and Eliane beaches should be directed exclusively to the users' well-being, given that the ecological role of these sites has already been compromised. Several aspects are known to influence the return of visitors to a particular beach: beach length, shoreline characteristics, scenery, landscape, water quality, agglomerations and amenities (Krelling et al. 2017). Therefore, installations and activities that ensure the health and safety of beach frequenters, such as the presence of lifeguards, beach cleaning and bathing water quality, should be strongly encouraged on these beaches. Considering the selection of beaches for recreation, in particular, marine litter is a critical aspect taken into account by visitors (Santos et al. 2005). Krelling et al. (2017) reported that in the municipality of Pontal do Paraná, the presence of garbage can potentially (15 to 39.1%) reduce local tourism income, representing losses of US$3.2 to 8.5 million per year.

Studies regarding the risks and safety of beaches on the coast of Paraná are practically non-existent. The lack of familiarity with the beach environment and/or false sense of security renders visitors unaware or causes them to underestimate the natural dangers on the beaches, as well as how to avoid them, enhancing the probability of accidents related to sea bathing (Angulo et al. 2016 and references therein). According to McLachlan et al. (2013), the condition of bathing water quality should be readily available to its users, as should the temporary closure of the beach be indicated when under inappropriate circumstances. In general, coastal municipalities lack an efficient sewage treatment system (Scherer et al. 2009). Thus, a large part of the household waste is still carried to the sea via washout (small bodies of water that drain the water resulting from the pluviometric accumulation of the post-dune region—Gandara-Martins et al. 2015 and references therein), causing the contamination of the beach environment. In practice, fecal bacterial populations can persist in both seawater and beach sand (Halliday and Gast 2011, Andraus et al. 2014). The primary microbiological risk to human health found in sand results from contact with excreta from domestic animals, particularly from dogs (WHO 2003).

Other alternative forms of management of these priority recreation areas are available, including beach certification programs (reviewed in Zielinski et al. 2019). In Brazil, the Programa Bandeira Azul (internationally known as 'Blue Flag') has been implemented since 2006 and is monitored by the Instituto Ambientes em Rede (a non-governmental organization based in Florianópolis, SC, Brazil). The program promotes improvements in beach management, which are certified by compliance with 34 criteria; currently, only 14 beaches present the certification seal (www.bandeiraazul.org.br). As mentioned by Scherer et al. 2009, the low adherence of Brazilian beaches to the program is related to several reasons, such as: lack of effective control of the uses and activities carried out in the sand strip; inefficient basic sanitation; existence of conflict between competences in beach management, and lack of support and financial resources.

Ultimately, the multiple-use management approach was identified in the other analyzed beaches (Albatroz, Perequê, Saint Etienne, Brava, Coroados 1, and Coroados 3), where the biological metrics showed intermediate values. Although urban occupation is already consolidated in these areas, a limited number of facilities are available to the residents/tourists. Here managers should strive to ensure that human use does not compromise the ecosystem or related ecological processes (Harris et al. 2015). Mechanical cleaning, as motor vehicle traffic, should be prohibited for two reasons: (1) such activities, which are accountable for the crushing of organisms (especially *O. quadrata*; pers. observation), are not consistent with beach conservation, and at the same time (2) may compromise the recreational experience of those non-motorized users. The allowed cleaning activity consists of the manual disposal of solid waste. In fact, the analyzed beaches receive little input of organic detritus, and the permanence of such material on the beach should not cause an annoyance to users. If possible, the dune system should be recovered (Noriega et al. 2012) and, therefore, access to the beach should occur exclusively via suspended walkways, in order to prevent direct trampling on the dunes. Similar to the priority recreation beaches, aspects such as safety and health of the frequenters cannot be neglected. On the other hand, improvements in infrastructure (kiosks, parking, lighting) and the development of scheduled public recreational activities are not recommended, given that such actions attract large concentrations of people and, consequently, intensify human trampling. If such recommendations are not implemented, there is no prospect for the maintenance of crab and staphylinid populations in the near future and, thus, the environmental damage already detected will extend all over the Parana´s beaches.

FINAL CONSIDERATIONS

In summary, our results demonstrated that the level of recreational use associated with the sandy ocean beaches of the coast of Paraná (southern Brazil) causes a decrease in the abundance of *Ocypode quadrata* crabs and *Bledius hermani* coleopterous staphylinids as the recreational potential index increases, to the point of disappearance in beaches with a high degree of human intervention. Therefore, the adoption of ecological indicators of environmental stress is strongly recommended given that such measures reflect the health of the beach ecosystem.

The integrated analysis of the biological metrics and Recreational Potential Index of the beaches (exemplified by the case report of the beaches on the coast of Paraná, southern Brazil) enables managers to identify the most appropriate type of management approach for each stretch of the beach, thus ensuring the sustainable use of the ecosystem as a whole. Closer interactions between scientists and decision makers will also be essential if efficient management and conservation strategies are to be formulated and applied to reduce anthropogenic impacts on sandy beaches.

Acknowledgements

Special thanks to all participants in the fieldwork for their valuable efforts. We especially wish to acknowledge Leonardo Sandrini-Neto for statistical support, Fabiano Grecco de Carvalho for graphical edition, Pablo Damian Borges Guilherme for map elaboration and Eurico de Paula Arruda for English revision. J.V. Vieira was supported by a postgraduate grant from the 'Coordenação de Aperfeiçoamento de Pessoal de Ensino Superior', CAPES, of Brazil (CAPES/DGU n° 206/09) during the execution of the present study.

REFERENCES

Alberto, R.M.F. and N.F. Fontoura. 1999. Distribuição e estrutura etária de *Ocypode quadrata* (Fabricius, 1787) (Crustacea, Decapoda, Ocypodidae) em praia arenosa do litoral sul do Brasil. Rev. Bras. Biol. 59: 95–108.

Amaral, A.C.Z., G.N. Corte, J.S.R. Filho, M.R. Denadai, L.A. Colling, C.A. Borzone, et al. 2016. Brazilian Sandy beaches: characteristics, ecosystem services, impacts, knowledge and priorities. Braz. J. Oceanogr. 64: 5–16.

Andraus, S., I.C. Pimentel and J.A. Dionísio. 2014. Microbiological monitoring of seawater and sand of beaches Matinhos, Caiobá e Guaratuba-PR, Brazil. Rev. Estud. Biol. Amb. Divers. 36: 43–55.

Angulo, R.J. 2004. Aspectos físicos das dinâmicas de ambientes costeiros, seus usos e conflitos. Rev. Des. Meio Ambiente. 10: 175–185.

Angulo, R.J., C.A. Borzone, M.A. Noernberg, C.J.L. Quadros, M.C. Souza and L.C. Rosa. 2016. The state of Paraná beaches. pp. 419–464. *In*: A.D. Short and A.H.F. Klein [eds]. Brazilian Beach Systems. Springer, Florida, USA.

Barros, F. 2001. Ghost crabs as a tool for rapid assessment of human impacts on exposed sandy beaches. Biol. Conserv. 97: 399–404.

Barros, F., C.A. Borzone and S. Rosso. 2001. Macroinfauna of six beaches near Guaratuba bay, Southern Brazil. Braz. Arch. Biol. Technol. 44: 351–364.

Battistuz, G.Z. and E.D. Zardo. 2014. Paraná—Estudo Estatístico 20 anos de Turismo. Paraná, Brazil.

Bessa, B., S. Felicita, T.M.B. Cabrini and R.S. Cardoso. 2017. Behavioural responses of talitrid amphipods to recreational pressures on oceanic tropical beaches with contrasting extension. J. Exp. Mar. Biol. Ecol. 486: 170–177.

Bortoluzzi, S. and E. Caron. 2019. *Bledius hyalinus* sp. nov. (Coleoptera: Staphylinidae: Oxytelinae), of the forcipatus group, first recorded from coastal Brazil. Zootaxa. 2: 391–395.

Borzone, C.A., L.C. Rosa, P.D.B. Guilherme and J.V. Vieira. 2015. Monitoramento de populações de *Ocypode quadrata* (Crustacea: Decapoda). pp. 244–249. *In*: A. Turra and M.R. Denadai [eds]. Protocolos para o monitoramento de habitats bentônicos costeiros—Rede de Monitoramento de Hábitats Bentônicos Costeiros—ReBentos [on line]. Instituto Oceanográfico da Universidade de São Paulo, São Paulo, Brazil.

Borzone, C.A., J.R.B. Souza and A.G. Soares. 1996. Morphodynamic influence on the structure of inter and subtidal macrofaunal assemblages of subtropical sandy beaches. Rev. Chil. Hist. Nat. 69: 565–577.

Borzone, C.A., Y.A.G. Tavares and F.C.R. Barros. 1998. Beach morphodynamics and distribution of *Mellitaquin quiquiesperforata* (Leske, 1778) on sandy beaches of southern Brazil. Proc. 9th Int. Echinoderm Conf. San Francisco, California, USA. 581–586.

Breton, F., J. Clapés, A. Marquès and G.K. Priestley. 1996. The recreational use of beaches and consequences for the development of new trends in management: the case of the beaches of the Metropolitan Region of Barcelona (Catalonia, Spain). Ocean Coastal Manage. 32: 153–180.

Brown, A.C. and A. McLachlan. 2002. Sandy shore ecosystems and the threats facing them: some predictions for the year 2025. Environ. Conserv. 29: 62–77.

Caron, E. and C.S. Ribeiro-Costa. 2007. *Bledius* Leach from southern Brazil (Coleoptera, Staphylinidae, Oxytelinae). Rev. Bras. Entomol. 51: 452–457.

Colombini, I., M. Fallaci and L. Chelazzi. 2011. Terrestrial macroinvertebartes as key elements for sustainable beach management. J. Coast. Res. 61: 24–35.

Castro, J.C., E. Caron and L.C. Rosa. 2016. Update on the Brazilian coastal species of *Bledius* Leach (Coleoptera: Staphylinidae: Oxytelinae) with the description of two new species. Zootaxa. 2: 145–157.

Costa, L.L. and I.R. Zalmon. 2019. Multiple metrics of the ghost crab *Ocypode quadrata* (Fabricius, 1787) for impact assessments on sandy beaches. Estuarine Coastal Shelf Sci. 218: 237–245.

Davenport, J. and J.L. Davenport. 2006. The impact of tourism and personal leisure transport on coastal environments. Estuarine Coastal Shelf Sci. 67: 280–292.

De Ruyck, A.M.C., A.G. Soares and A. McLachlan. 1997. Human recreational patterns on beaches with different levels of development. Trans. R. Soc. Afric. 52: 257–276.

Defeo, O., A. McLachlan, D.S. Shoeman, A. Schlacher, J. Dugan, A. Jones, et al. 2009. Threats to sandy beach ecosystems: a review. Estuarine Coastal Shelf Sci. 81: 1–12.

Emery, K.O. 1961. A simple method of measuring beach profiles. Limnol. Oceanogr. 6: 90–93.

Gandara-Martins, A.L., C.A. Borzone, L.C. Rosa and E. Caron. 2010. Ocorrência de três espécies do gênero *Bledius* Leach, 1819 (Coleoptera, Staphylinidae, Oxytelinae) nas praias arenosas expostas do Paraná, Brasil. Braz. J. Aquat. Sci. Technol. 14: 23–30.

Gandara-Martins, A.L., C.A. Borzone, P.D.B. Guilherme and J.V. Vieira. 2015. Spatial effects of a washout on sandy beach macrofauna zonation and abundance. J. Coastal Res. 316: 1459–1468.

Gibbs, R.J., M. D. Matthews and D. A. Link. 1971. The relationship between sphere size and settling velocity. J. Sediment. Petrol. 41: 7–18.

Glasby, T.M. and A.J. Underwood. 1996. Sampling to differentiate between pulse and press perturbations. Environ. Monit. Assess. 42: 241–252.

Gonçalves, S.C., P.M. Anastácio and J.C. Marques. 2013. Talitrid and Tylid crustaceans bioecology as a tool to monitor and assess sandy beaches' ecological quality condition. Ecol. Indic. 29: 549–557.

González, S.A., K. Yànez-Navea and M. Muñoz. 2014. Effect of coastal urbanization on sandy beach coleoptera *Phaleria maculate* (Kulzer, 1959) in northern Chile. Mar. Pollut. Bull. 83: 265–274.

Halliday, E. and R.J. Gast. 2011. Bacteria in beach sands: an emerging challenge in protecting coastal water quality and bather health. Environ. Sci. Technol. 45: 370–379.

Harris, L., E.E. Campbell, R. Nel and D. Schoeman. 2014. Rich diversity, strong endemism, but poor protection: addressing the neglect of sandy beach ecosystems in coastal conservation planning. Divers. Distrib. 20: 1120–1135.

Harris, L., R. Nel, S. Holness and D. Schoeman. 2015. Quantifying cumulative threats to sandy beach ecosystems: a tool to guide ecosystem-based management beyond coastal reserves. Ocean Coastal Manage. 110: 12–24.

Herman, L.H. 1986. Revision of *Bledius*. Part IV. Classification of species groups, phylogeny, natural history, and catalogue (Coleoptera, Staphylinidae, Oxytelinae). Bull Am. Mus. Nat. Hist. 184: 1–368.

James, R.J. 2000. From beaches to beach environments: linking the ecology, human-use and management of beaches in Australia. Ocean & Coastal Manage. 43: 495–514.

Jedrzejczak, M.F. 2004. The modern tourist's perception of the beach: is the sandy beach a place of conflict between tourism and biodiversity? Coast. Rep. BaltCoast 2004 Conf. Proc. Warnemünde, Germany 109–119.

Krelling, A.P., A.T. Willians and A. Turra. 2017. Differences in perception and reaction of tourist groups to beach marine debris that can influence a loss of tourism revenue in coastal áreas. Mar. Policy. 85: 87–99.

Lana, P.C., E. Marone, R.M. Lopes and E.C. Machado. 2001. The subtropical estuarine complex of Paranaguá Bay, Brazil. Ecol. Stud. 144: 131–145.

Lucrezi, S. and T.A. Schlacher. 2014. The ecology of ghost crabs. Oceanog. Mar. Biol.: An Annual Review. 52: 201–256.

Magalhães, W.F., J.B. Lima, F. Barros and J.M.L. Dominguez. 2009. Is *Ocypode quadrata* (Fabricius, 1787) a useful tool for exposed sandy beaches management in Bahia state (Northeast Brazil)? Brazil. Braz. J. Oceanogr. 57: 149–152.

McLachlan, A. and A. Brown. 2006. The Ecology of Sandy Shores. Academic Press, New York.

McLachlan, A. and O. Defeo. 2013. Coastal Beach Ecosystems. Encycl. Biodivers. 2: 128–136.

McLachlan, A., O. Defeo, E. Jaramillo and A.D. Short. 2013. Sandy beach conservation and recreation: guidelines for optimizing management strategies for multi-purpose use. Ocean Coastal Manag. 71: 256–268.

Micalleff, A. and A.T. Williams. 2002. Theoretical strategy considerations for beach management. Ocean Coastal Manag. 45: 261–275.

Morales, M. 2012. Sciplot: scientific graphing functions for factorial designs. Available on CRAN.

Moss, D. and D.P. McPhee. 2006. The impacts of recreational four-wheel driving on the abundance of the ghost crab (*Ocypode cordimanus*) on subtropical beaches in SE Queensland. Coastal Manage. 34: 133–140.

Negreiros-Fransozo, M.L, A. Fransozo and G. Bertini. 2002. Reproductive cycle and recruitment period of *Ocypode quadrata* (Decapoda, Ocypodidae) at a sandy beach in southeastern Brazil. J. Crust. Biol. 22: 157–161.

Nel, R., E.E. Campbell, L. Harris, L. Hauser, D.S. Schoeman, A. McLachlan, et al. 2014. The status of sandy beach science: past trends, progress, and possible futures. Estuarine Coastal Shelf Sci. 150: 1–10.

Neves, F.M. and C.E. Bemvenuti. 2006. The ghost crab *Ocypode quadrata* (Fabricius, 1787) as a potential indicator of anthropic impact along the Rio Grande do Sul coast, Brazil. Biol. Conserv. 133: 43–435.

Noriega, R., T.A. Schlacher and B. Smeuninx. 2012. Reductions in ghost crab populations reflect urbanization of beaches and dunes. J. Coastal Res. 28: 123–131.

Oliveira, M.R.L. and J.L. Nicolodi. 2012. A Gestão Costeira no Brasil e os dez anos do Projeto Orla. Uma análise sob a ótica do poder público. J. Integr. Coastal Zone Manage. 12: 91–100.

Pombo, M. and A. Turra. 2013. Issues to be considered in counting burrows as a measure of Atlantic ghost crab populations, an important bioindicator of sandy beaches. PLoS One. 8: 1–7.

Quinn, G.P. and M.J. Keough. 2002. Experimental design and data analysis for biologists. Cambridge University Press, New York.

R Core Team. 2012. R: A Language and Environment for Statistical Computing. R Foundation for Statistical Computing, Vienna, Austria (http://www.R-project.org/).

Rosa, L.C. and C.A. Borzone, 2008. Spatial distribution of the *Ocypode quadrata* (Crustacea: Ocypodidae) along estuarine environments in the Paranaguá Bay Complex, southern Brazil. Rev. Bras. Zool. 25: 383–388.

Rosa, L.C., C.A. Borzone and J.V. Vieira, A.L. Gandara-Martins, A.X.M.R. Vianna, E. Caron et al. 2015. Monitoramento das populações de *Bledius* (Insecta: Coleoptera). pp. 250–257. *In*: A. Turra and M.R. Denadai [eds]. Protocolos para o monitoramento de habitats bentônicos costeiros—Rede de Monitoramento de Hábitats Bentônicos Costeiros – ReBentos [on line]. Instituto Oceanográfico da Universidade de São Paulo, São Paulo, Brazil.

Ruiz-Delgado, M.C., J.V. Vieira, V.G. Veloso, M.J. Reyes-Martínez, I.A. Sallorenzo and C.A. Borzone. 2014. The role of wrack deposits for supralittoral arthropods: an example using Atlantic sandy beaches of Brazil and Spain. Estuarine Coastal Shelf Sci. 136: 61–71.

Sandrini-Neto, L. and MG. Camargo. 2012. GAD: an R package for ANOVA designs from general principles. R package version 1.1.1. (http://CRAN.R-project.org/package=GAD).

Santana, G. 2003. Tourism Development in Coastal Areas—Brazil: Economic, Demand and Environmental Issues. J. Coastal Res. 35: 85–93.

Santos, I.R., A.C. Friedrich, M. Wallner-Kersanach and G. Fillmann. 2005. Influence of socioeconomic characteristics of beach users on litter generation. Ocean Coast. Manag. 48: 742–752.

Scherer, M., M. Sanches and D.H. Negreiros. 2009. Gestão das zonas costeiras e as políticas públicas no Brasil: um diagnóstico. pp. 291–330. *In*: J.M. Barragán-Muñoz, P.A. Granados, J.A.C. Ruiz, J.A. Onetti and J.G. Sanabria [eds]. Manejo Costero Integrado y Politica Publicaen Iberoamérica: un Diagnóstico. Necessidad de Cambio. Red IBERMAR (CYTED), Cádiz, Espanha.

Schlacher, T.A., J. Dugan, D.S. Schoeman, M. Lastra, A. Jones, F. Scapini, et al. 2007a. Sandy beaches at the brink. Divers. Distrib. 13: 556–560.

Schlacher, T.A., L. Thompson and S. Price. 2007b. Vehicles versus conservation of invertebrates on sandy beaches: mortalities inflicted by off-road vehicles on ghost crabs. Mar. Ecol. 28: 354–367.

Schlacher, T.A., D.S. Schoeman, J. Dugan, M. Lastra, A. Jones, F. Scapini, et al. 2008. Sandy beach ecosystems: key features sampling issues, management challenges and climate change impacts. Mar. Ecol. 29: 70–90.

Schlacher, T.A. and L. Thompson. 2012. Beach recreation impacts benthic invertebrates on ocean exposed sandy shores. Biol. Conserv. 147: 123–132.

Schlacher, T.A., A.R. Jones, J.E. Dugan, M.A. Weston, L. Harris, D.S. Schoeman, et al. 2014a. Open-coast sandy beaches and coastal dunes. pp. 37–99. *In:* B. Maslo and J.L. Lockwood [eds]. Coastal Conservation. Cambridge University Press, New Jersey, USA.

Schlacher, T.A., D.S. Schoeman, A.R. Jones, J.E. Dugan, D.M. Hubbard, O. Defeo, et al. 2014b. Metrics to assess ecological condition, change, and impacts in sandy beach ecosystems. J. Environ. Manage. 144: 322–335.

Schlacher T.A., S. Lucrezi, R.M. Connolly, C.H. Peterson, B.L. Gilby, B. Maslo, et al. 2016. Human threats to sandy beaches: A meta-analysis of ghost crabs illustrates global anthropogenic impacts. Estuarine Coastal Shelf. Sci. 169: 56–73.

Short, A.D. and L.D. Wright. 1983. Physical variability of sandy beaches. pp. 133–144. *In*: McLachlan, A. and T. Erasmus [eds]. Sandy Beaches as Ecosystems. Dr. W. Junk Publishers, Boston, USA.

Silva, W.T.A.F. and T.C.S. Calado. 2011. Spatial distribution of and anthropogenic impacts on ghost crab *Ocypode quadrata* (Crustacea, Ocypodidae) burrows in Maceió, Brazil. Rev. Nord. Zool. 5: 1–9.

Silva, W.T.A.F. and T.C.S. Calado. 2013. Number of ghost crab burrows does not correspond to population size. Cent. Eur. J. Biol. 8: 843–847.

Souza, J.R.B., N. Lavoie, P.H. Bonifácio and C.M.C. Rocha. 2008. Distribution of *Ocypode quadrata* (Fabricius, 1787) on sandy beaches of northeastern Brazil. Atlântica, 30: 139–145.

Souza, G.N., C.A.G. Oliveira and A.S. Tardem. 2017. Counting and measuring ghost crab burrows as a way to assess the environmental quality of beaches. Ocean Coastal Manage. 140: 1–10.

Turra, A., M.A.O. Gonçalves and M.R. Denadai. 2005. Spatial distribution of the ghost crab *Ocypode quadrata* in low-energy tide-dominated sandy beaches. J. Nat. Hist. 39: 2163–2177.

Underwood, AJ. 1997. Experiments in Ecology. Their Logical Design and Interpretation using Analysis of Variance. Cambridge University Press, Cambridge, UK.

Veloso, V.G., E.S. Silva, C.H.S. Caetano and R.S. Cardoso. 2006. Comparison between the macroinfauna of urbanized and protected beaches in Rio de Janeiro state, Brazil. Biol. Conserv. 127: 510–515.

Veloso, V.G., G. Neves, M. Lozano, A. Perez-Hurtado, C.G. Gago, F. Hortas, et al. 2008. Responses of talitrid amphipods to a gradient of recreational pressure caused by beach urbanization. Mar. Ecol. 29: 126–133.

Vieira, J.V., C.A. Borzone, L. Lorenzi and F.G. Carvalho. 2012. Human impact on the benthic macrofauna of two beach environments with different morphodynamic characteristics in southern Brazil. Braz. J. Oceanogr. 60: 137–150.

Vieira, J.V. 2015. Efeitos dos distúrbios antrópicos associados ao uso recreativo na fauna de praias: implicações para o manejo e conservação. Ph.D. Thesis, Federal University of Parana, Curitiba, Brazil.

Vieira, J.V., M.C. Ruiz-Delgado, M.J. Reyes-Martinez, C.A. Borzone, A. Asenjo, J.E. Sanchez-Moyano, et al. 2016. Assessment the short-term effects of wrack removal on supralittoral arthropods using the M-BACI design on Atlantic sandy beaches of Brazil and Spain. Mar. Environ. Res. 119: 222–237.

Vieira, J.V. and C.A. Borzone. 2017. O uso de insetos coleópteros staphylinidae na avaliação de impacto antrópico em praias arenosas: um estudo de caso no litoral do Paraná. pp. 128–146. *In*: J.C. Melo Júnior and T.M.N. Oliveira [eds]. Ciências Ambientais: Ensaios e Perspectivas. Editora Univille, Joinville, Brazil.

WHO—World Health Organization. 2003. Guidelines for safe recreational water environments. Coastal and fresh waters.

Williams, A.T. and A. Micalleff. 2009. Beach management: Principles and Practice. Earthscan, London.

Zielinski, S., C.M. Botero and A. Yanes. 2019. To clean or not to clean? A critical review of beach cleaning methods and impacts. Mar. Pollut. Bull. 139: 390–401.

Chapter 8

Conservation Shortcuts: A Promisor Approach for Impact Assessments and Management of Sandy Beaches

Leonardo Lopes Costa[1]*, Nina Aguiar Mothé[2], Ariane da Silva Oliveira[1] and Vitor Figueira Arueira[1]

[1]Universidade Estadual do Norte Fluminense Darcy Ribeiro, Laboratory of Environmental Sciences, Av. Alberto Lamego, 2000, 28013–602, Campos dos Goytacazes, Rio de Janeiro, Brazil (costa.ecomar@gmail.com, arianee_silvaa@hotmail.com, vitorfigueira9@hotmail.com).

[2]State University of Northern of Rio de Janeiro, Laboratory of Biology of Recognition, Av. Alberto Lamego, 2000, 28013-602, Campos dos Goytacazes, Rio de Janeiro, Brazil (ninaguiarm@gmail.com).

INTRODUCTION

It is a consensus that sandy beaches are under threat in many places worldwide. Consequently, the need for correct management of beaches as valuable natural resources endowed with a unique biodiversity has never been greater. Historically, beach management focusses on economic aspects such as development, tourism and infrastructure protection. Unfortunately, managers take decisions without considering the roles that a rich, but underappreciated biodiversity plays. Meanwhile, researchers around the world have tried to alert the scientific community and society on the collapse that sandy beaches have suffered by evidencing the effects

*Corresponding author: costa.ecomar@gmail.com

of human disturbances on different levels of ecological systems. Within a bulk of impact assessments performed in the last few years, those targeting optimal cost-benefit for conservation outcomes have gained particular prominence.

Ideally, detailed studies that assess all the complexity of ecosystems should precede decisions that have ramifications for conservation. However, Caro (2010) raised some factors that hamper long-term and full assessments of the complexity of nature: an intrinsic human limitation for appraising all aspects of ecosystems coupled with absence of technologies; the vast scale of biodiversity crisis in an anthropocene epoch and need for rapid political decisions; and frequent scarcity of funds for long-term ecological research in most countries.

For many years, sandy beaches were considered systems devoid of life, but this is not true. Sandy beaches are inhabited by a broad range of plants, microalgae, microbes, invertebrates and large vertebrates. Under challenging scenarios for research, beach ecologists have applied creative approaches and depicted several patterns of biodiversity in the last years, most of them with clear implications for conservation. Not surprisingly, impact assessments have focussed mainly on single species that particularly can perform even better than complex community and ecosystem descriptors in detecting impacts (Fig. 8.1). The predictability of species-specific responses to human disturbances, therefore, is an argument beyond logistic issues that supports the prioritization of target species instead of complex communities and ecosystems approaches in impact assessment.

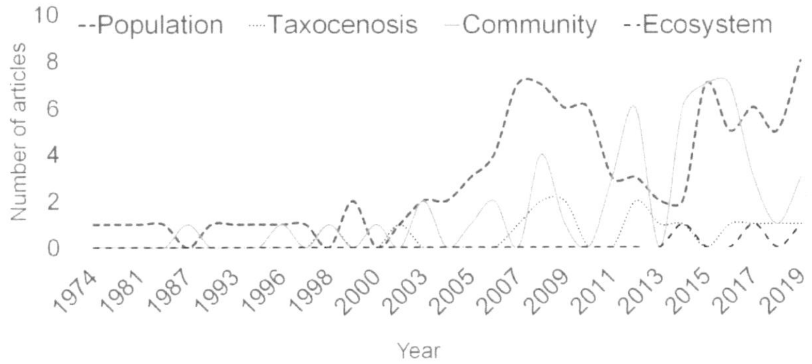

Figure 8.1 Number of impact assessments based on macroinvertebrates of sandy beaches through time (1974–2019). Dataset from the review by Costa et al. (2020a).

Target species in which one's attention is preferentially focussed to set extrapolations and to solve environmental problems are named conservation shortcuts. Many types of conservation shortcuts have emerged in scientific literature, but they are often incorrectly approached as synonymic terminologies. There are numerous ecological concepts and societal rationales behind such terminologies that can distil conservation shortcuts into 'indicators', 'keystones', 'umbrellas' and 'flagships' (Zacharias and Roff 2001). More importantly, each conservation shortcuts can play distinct, but complementary roles in conservation strategies. Thus, it is pivotal to distinguish each conservation shortcut and to apply them strategically in research, monitoring and management of an ecosystem.

Selecting conservation shortcuts requires empirical validations and robust research. Criticisms about the application of conservation shortcuts is frequently associated with the lack of proper selection methods and limitation in judging them against specific conservation objectives. Undeniably, it is pivotal at first to know about a target ecosystem in depth, including the main characteristics of its biodiversity. Later it is possible to focus on a single or on a low number of species, but provide broader ecological understanding, including those with management and conservation implications, ensuring optimal benefit-costs.

Scientific research evidencing methodological issues and novel approaches, comparing species traits over gradients of human disturbances, evaluating the effects of single species actions on co-occurring species, and showing how single-species based campaigns are useful for the ecosystem protection have been carried out on various ecosystems, including sandy beaches. There is a bulk of knowledge being produced that highlights how promising or is the application of conservation shortcuts for ecological assessments and management of sandy beaches. The objective of this chapter is to conceptualize the most common conservation shortcuts and their value for optimizing management, monitoring and conservation of sandy beaches.

SAMPLING

Feasible sampling is decisive for selecting conservation shortcuts. Therefore here one presents practicable tools for simple and effective ecological assessments based on focal species or assemblages. A traditional sampling method on sandy beaches is sediment collection, because organisms live predominantly buried. In this case, an extensive sampling effort is necessary, because invertebrates usually have a clustered distribution and a zonation driven by physical conditions along the gradient of distance from the waterline, demanding sampling of distinct beach strata. This is because different beach habitats have particular species with specific ecological tolerances. Thus, only a sampling over the entire intertidal-supralitoral interface ensures a complete representation of infaunal communities. However, some species with singular biological characteristics can be easily sampled with easier methods, acting as surrogates of community aggregate descriptors and being targets for optimal impact assessments and monitoring.

Numerous macroinvertebrates leave tracks on the sand surface that are valuable tools for ecological research. Undoubtedly, to count burrows on the sand surface demands less money and labour than sediment collection and screening. The most representative taxa in which impact assessments have relied on burrow counts is ghost crabs (Crustacea: Ocypodidae) ($n > 50$ articles). Global patterns have emerged from local studies based on burrow counts, particularly comparing burrow density and diameter among beaches with distinct human disturbances levels (Schlacher et al. 2016b).

Doubts exist regarding if burrow counts really represent the population size of ghost crabs, how physical characteristics of beach habitats or species behaviour under distinct impact levels affect burrow construction and, which conversion factors or refining methods could be applied (Gül and Griffen 2019, Pombo and Turra 2013, Schlacher et al. 2016c). However, even though burrow counts overestimate population size of ghost crabs in approximately 40% in some regions, impact assessments have not been seriously biased, meanwhile the pattern of reduced abundance and diameter of burrows on impacted beaches is maintained despite the burrow occupation rates (Costa and Zalmon 2019a, Pombo and Turra 2019). Counting ghost crabs' burrows is, indeed, a well-accepted and scientific sound method that optimizes the monitoring of human impacts on sandy beaches.

Ghost crabs are not the unique species that construct conspicuous burrows on the sand surface. As semi-terrestrial crustacean, sandhoppers (Crustacea: Talitridae) are common on beaches and they can be found burrowing within a well-defined area in retention zones and supralittoral (Williams 1988). Their burrows are smaller and have lower longevity than those of ghost crabs, as they are generally distributed downshore, where inundation rates are higher. Less attention has been given to sandhoppers burrows as tools for ecological research. Feasibility of burrow counts of sandhoppers for impact assessments is still unknown. Similarly, larvae of tiger beetles also construct burrows in the backshore and have been used for evaluating human disturbances on sandy beaches simultaneously with visual census of adults in the intertidal zone (Arndt et al. 2005). It is necessary therefore to distinguish the burrows of distinct species in dry sand (see Fig. 8.2).

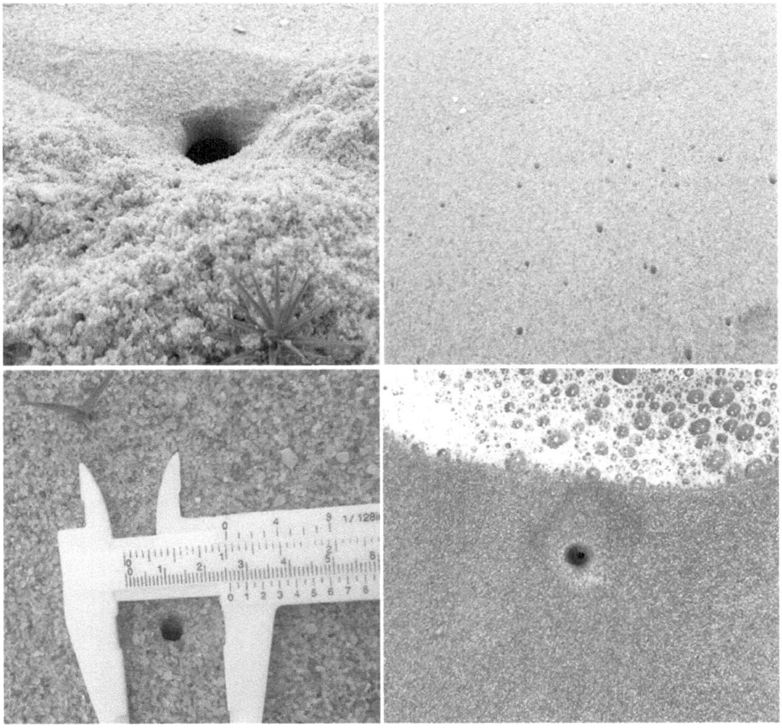

Figure 8.2 Burrows of a ghost crab (up left), sandhoppers (up right), tiger beetle larvae (down left) and ghost shrimp (down right).

If on the one hand, ghost crabs, sandhoppers and larvae of tiger beetles construct visible burrows in emerged sand of a broad range of beach types, including intermediate and reflective morpho dynamics, ghost shrimps (Crustacea: Callianassidae) are marine organisms that construct deep burrows below the upper swash limit of fine sandy and sandy-mud shores. Burrows of ghost shrimps are reasonably stable and consequently they are feasible tools for estimating populations' density mainly on dissipative beaches and tidal flats. Nevertheless, there are few studies ($n < 5$ articles) evaluating the effects of human disturbances on beaches using ghost shrimps as bioindicators.

Most beach invertebrates live buried, but some mobile species emerge from the sediment surface in specific periods of the day for feeding. This includes insects, sandhoppers, ghost crabs and spiders. Pitfalls are excellent tools for trapping mobile fauna and depending on the research objectives, it is even possible to use sea water inside traps to implement a non-destructive method. In some cases, however, it is a necessary fixation for proper taxonomic identification and counting of individuals. Disadvantages of this method are: (i) natural interference (e.g., strong winds that throw sand into the traps); (ii) sampling is limited to mobile and surface-active fauna; (iii) inundation risk that pose a difficulty of sampling species distributed closer to the swash; and (iv) time of exposure is necessary and some pitfalls are removed or destroyed by humans during this period. Nevertheless, if mobile fauna has its activity period known and the above-mentioned challenges are controlled, pitfalls can be excellent alternatives for monitoring human disturbances using target species as indicator species. The response of mobile fauna, such as sandhoppers and Tenebrionid beetles to human disturbances is similar when using pitfalls compared with results from hard sediment collections (Table 8.1).

Technology is advancing and new technological approaches appear s to be the trend for future impact assessments and monitoring on beaches. Monitoring of disturbances such as vehicle traffic and erosion and even biodiversity assessments, such as ghost crabs burrow counts, have already relied on Remote Sense in some studies, like for instance in Bycroft et al. (2019) and in Schlacher and Morrison (2008). Nevertheless, considering the need for rapid and low-cost assessments, it is reliable to select target species even if more traditional methods are implemented. There are numerous species that can be sampled even by visual census, such as birds and some insects (e.g., tiger beetles) that become complex technologies arguably dispensable. As mentioned here, burrows, pitfalls and visual census for sampling target species are very easy and low-cost tools that provide feasible proxies for assessing the complexity of biodiversity patterns on sandy beaches. Importantly, most of these species can serve as conservation shortcuts, as detailed next.

INDICATOR SPECIES

Indicator species are those that signalize either the biodiversity composition or environmental health of a particular ecosystem (Caro, 2010). The presence of a 'composition indicator' denotes the presence of other co-existing species at regional or national scales, probably by sharing some habitat requirements

Table 8.1 Mean response ratios of sandhoppers (Crustacea: Talitridae) and Tenebrionid beetles (Insecta: Tenebrionidae) to human disturbances ±95% confidence intervals using different sampling tools. Response ratios were calculated as the log of the quotient of the mean abundance value recorded for impacted sites divided by the mean abundance value at corresponding non-impacted sites from individual articles, Costa et al. 2020a

Method	Family	Species	ln (R)	Country	Reference
		Phaleria maculata	–0.415	Chile	Acuña and Jaramillo (2015)
		Phaleria maculata	–2.237	Chile	González et al. (2014)
		Phaleria bimaculata	0.588	France	Comor et al. (2008)
		Phaleria bimaculata	–1.553	Brazil/Spain	Vieira et al. (2016)
		Phaleria provinciallis	–0.223	France	Comor et al. (2008)
Sediment collection (corer)	Tenebrionidae	*Phaleria* spp	–6.392	Italy	Fanini et al. (2009)
		Phaleria testacea	–0.100	Brazil	Almeida et al. (1993)
		Phaleria testacea	–0.327	Brazil	Veloso et al. (2006)
		Phaleria testacea	0.531	Brazil	Machado et al. (2017)
		Phaleria testacea	–0.470	Brazil	Machado et al. (2017)
		Mean ± confidence interval	**–1.060 ± 1.277**	-	-
		Phaleria acuminata	–0.731	Morocco	Colombini et al. (2008)
		Phaleria bimaculata	–1.069	Morocco	Colombini et al. (2008)
Pitfall	Tenebrionidae	*Phaleria bimaculata*	–0.203	Italy	Fanini et al. (2014)
		Phaleria testacea	–3.497	Brazil	*Umpublished data*
		Mean ± confidence interval	**–1.375 ± 1.429**	-	-
		Atlantorchestoidea brasiliensis	–2.625	Brazil	Cardoso et al. (2016)
		Atlantorchestoidea brasiliensis	–0.412	Brazil	Machado et al. (2016)
		Atlantorchestoidea brasiliensis	–0.477	Brazil	Machado et al. (2017)
		Atlantorchestoidea brasiliensis	–0.972	Brazil	Suciu et al. (2018)
		Atlantorchestoidea brasiliensis	–1.127	Brazil	Veloso et al. (2008)
		Atlantorchestoidea brasiliensis	–7.267	Brazil	Veloso et al. (2009)
		Atlantorchestoidea brasiliensis	–2.443	Brazil	Vieira et al. (2012)

Table 8.1 (Contd.)

Table 8.1 (Contd.) Mean response ratios of sandhoppers (Crustacea: Talitridae) and Tenebrionid beetles (Insecta: Tenebrionidae) to human disturbances ±95% confidence intervals using different sampling tools. Response ratios were calculated as the log of the quotient of the mean abundance value recorded for impacted sites divided by the mean abundance value at corresponding non-impacted sites from individual articles, Costa et al. 2020a

Method	Family	Species	ln (R)	Country	Reference
		Orquestoidea tuberculata	–0.611	Chile	Jaramillo et al. (1996)
		Orquestoidea tuberculata	–1.529	Chile	Jaramillo et al. (2012)
		Orquestoidea tuberculata	–0.145	Chile	Rodil et al. (2016)
		Platorchestia monodi	1.113	Brazil	Borzone and Rosa (2009)
		Platorchestia monodi	–0.500	Brazil	Vieira et al. (2016)
		Talitrus saltator	–1.144	Portugal	Bessa et al. (2013)
Sediment collection (corer)	Talitridae	*Talitrus saltator*	–1.640	Portugal	Bessa et al. (2014)
		Talitrus saltator	–1.112	Spain	Junoy et al. (2005)
		Talitrus saltator	–0.916	Portugal	Nourisson et al. (2014)
		Talitrus saltator	0.000	Spain	Reyes-Martínez et al. (2015)
		Talitrus saltator	–1.865	Spain	Ruiz-Delgado et al. (2016)
		Talitrus saltator	–1.632	Spain	Veloso et al. (2008)
		Talitrus saltator	–0.968	Spain	Vieira et al. (2016)
		Talorchestia brito	–0.007	Portugal	Bessa et al. (2013)
		Talorchestia capensis	–1.535	South Africa	Harris et al. (2011)
		Talorchestia capensis	0.152	South Africa	Schoeman et al. (2000)
		Talorchestia deshayesii	0.157	Portugal	Bessa et al. (2014)
		Talorchestia tucurauna	1.140	Brazil	Borzone and Rosa (2009)
		Mean ± confidence interval	**–1.055 ± 0.628**	–	–
		Africorchestia spinifera	–1.199	Morocco	Fanini et al. (2012)
		Orquestia motangui	–3.332	Italy	Fanini et al. (2009)
		Talitrus saltator	–2.319	Morocco	Colombini et al. (2008)
		Talitrus saltator	–0.591	Italy	Fanini et al. (2007)
		Talitrus saltator	–1.954	Italy	Fanini et al. (2009)

Method	Family	Species	ln (R)	Country	Reference
		Talitrus saltator	3.892	Morocco	Fanini et al. (2012)
Pitfall	Talitridae	*Talitrus saltator*	–1.374	Greece	Fanini et al. (2014)
		Talitrus saltator	0.049	Italy	Pavesi and Matthaeis (2013)
		Talitrus saltator	–5.516	Italy	Scapini et al. (2005)
		Talitrus saltator	–3.612	Italy	Ugolini et al. (2008)
		Talorchestia brito	–3.977	Morocco	Fanini et al. (2012)
		Talorchestia deshayesii	1.012	Morocco	Colombini et al. (2008)
		Mean ± confidence interval	**–1.577 ± 1.423**	–	–

(Caro, 2010). Distinctly, a 'condition indicator' should signalize the environmental health, because changes in population or individual traits are supposed to reflect changes in the environment (Caro, 2010). Selection of composition and the condition indicator premises simple and low-cost sampling, worthwhile as it optimizes complex assessments. Consequently, high local abundance, sufficient information about biology to avoid confusing effects, predictable distribution and responses to the environment are desirable characteristics of indicator species.

Targeting hotspots of biodiversity is considered an effective strategy for protection and management in order to achieve maximum conservation with limited funds. This prioritization relies on vulnerability and irreplaceability of the biodiversity of specific areas. Species richness, endemism, rarity and presence of threatened species are common criteria for selecting biodiversity hotspots (Grenyer et al. 2006). Composition indicators can be useful for selecting hotspots, assuming that their distribution denotes the occurrence of other species, including those endangered.

The extent to which aggregate descriptors from different taxonomic groups can be congruent with each other or how they act as surrogates of common measures of biodiversity are typically applied for selecting composition indicators and to prioritize areas for conservation of terrestrial ecosystems. On sandy beaches it appears to lose importance because species richness is strongly predicted by physical factors and primary productivity (Defeo et al. 2017). Thus, selection of richer beaches could basically be on morpho dynamics factors. However, two factors make this assumption weak: first, dissipative and fine-grained beaches do not always have higher exclusive and rare species than reflective beaches (Checon et al. 2018); second, if species sampled by non-destructive, low-cost or easier methods in general such as ghost crabs, sandhoppers and tiger beetles have congruence with other species difficult to monitor, they can also be useful composition indicators.

To assess cross-taxon congruence is one strategy for selecting composition indicators. Using data from the Araçá Bay in South-eastern Brazil, Corte et al. (2017) assessed if macrofaunal (invertebrates > 0.5 mm) and meiofaunal (invertebrates from 0.04 to 0.5 mm) assemblages inhabiting from intertidal to 25 m deep in subtidal areas show concordance patterns between each other and if both groups respond to the same environmental gradients. Univariate and multivariate analysis showed strong congruent community structure patterns in all periods analyzed with both macro- and meiofauna responded similarly to environmental factors. Assuming that different monitoring teams can have specific expertise and full biodiversity assessments are almost utopic, the concordance among macrofauna and meiofauna patterns support the application of only one of the taxonomic groups for monitoring purposes at local scales according to the team's competences.

Birds that nest on the upper beach or in dunes are threatened globally and have shown some congruence degree of distribution among co-occurring species in New Jersey, USA. Using a multi-species distribution modelling approach, Maslo et al. (2016) identified overlap in spatial extent and niche characteristics among American Oystercatchers (*Haematopus palliatus*), Black Skimmers (*Rynchops niger*), Least Terns (*Sterna antillarum*) and Piping Plovers (*Charadrius melodus*) across their entire breeding range. Among them, the Piping Plover was the main composition indicator, because it co-occurs with the other bird species across most of their nesting habitat extent. The authors attributed the indicator potential of Piping Plovers at the local scale to narrow habitat preferences and sensitivity to human disturbances.

The same pattern generally occurs with tiger beetles in other ecosystems because their occurrence is positively correlated with richness of vertebrate and invertebrate communities (Pearson and Cassola 1992). Although neglected mainly in tropical regions, there are tiger beetle species on sandy beaches, most of them threatened by human disturbances (Knisley and Fenster 2005). Both tiger beetles and shorebirds are predators of beach invertebrates and their protection requires conservation of low taxonomic levels; thus, the conservation strategies for shorebirds and tiger beetles could benefit other co-occurring species, giving for these predators a potential of acting as umbrellas (more details in umbrella species).

It is important that congruence patterns be relatively stable in space and time for applying composition indicators in ecological assessments and monitoring. However, there are circumstances in which cross congruence has been shown to break down, for example under natural disturbances. Delaney and Stout (2018) assessed the cross congruence among plants, snails and water beetles in temporary ponds in coastal sand dunes. Although conceptually dune slacks are not parts of sandy beach ecosystems, they compose the Littoral Active Zone (Fanini et al. 2021), connecting beach habitats to some degree. The authors did not find evidence of concordance patterns among the three biological groups and concluded that these results have implications for conservation practice in other natural disturbed ecosystems such as wetlands and fire-dependent habitats.

The absence of an explicit cross-taxa congruence among diversity patterns of bryophytes, lichens and vascular plants was also found in secondary dunes in western Europe (Vaz et al. 2020). These findings suggest a certain limitation of applying composition indicators for sandy beaches monitoring, because the beach ecosystem is very dynamic and frequently disturbed by hydrodynamics. However, cross-taxon congruence patterns remain poorly studied in intertidal and supralittoral zones of sandy beaches and thus, there are no clear composition indicators as yet. Therefore, to identify composition indicators and selection criteria is a required study field to be explored on sandy beaches.

Condition indicators have long been suggested on sandy beaches by means of metrics such as genetic variation, age structure, behaviour and abundance (Costa et al. 2020a). However, the majority of condition indicators does not reflect proportionally the disturbance levels; metrics only signalize qualitatively environmental changes. For example, a classic conclusion of qualitative impact assessments is that a target species is more abundant on pristine beaches than on disturbed ones. In this case, the condition indicator species is a bioindicator.

Among all conservation shortcuts, the most extensive and robust scientific literature points macroinvertebrates as useful indicators of human disturbances operating at local scales. Inasmuch the majority of case studies (~160 articles) only determine qualitatively the indicator potential of species by applying a typical 'compare and contrast' approach (e.g., comparing species abundance between impacted and control beaches), sandy beaches have now many bioindicators (Costa et al. 2020a). In fact, most macroinvertebrates taxon are more abundant in low-urban beaches than in urban beaches worldwide. Among them, ghost crabs are undoubtedly the most recognized bioindicators, particularly of vehicle traffic, nourishment and synergic stressors from urban beaches (Schlacher et al. 2016b, Costa et al. 2020a).

Some manipulative experiments *in situ* and even road ecology approaches (i.e., assessments of negative impacts of roads on wildlife) evidence that ghost crabs have been crushed by vehicles into their burrows, on the sand surface or on roads surrounding beaches (Schlacher et al. 2007; Costa et al. 2020b). Mortality of ghost crabs due to vehicle traffic is a rare example of disentangled causality relationships in impact assessments based on bioindicators of sandy beaches. A recent review pointed to talitrids as indicators of trampling and mechanical cleaning because of their surface activity and detritivorous feeding habits, respectively, and donacid clams (Mollusca: Donacidae) as indicators of harvesting by means of their value for fishery (Costa et al. 2020a). Nevertheless, macrofaunal species acting as bioindicators of other specific human stressors have been difficult to reveal, because beaches are affected by multiple stressors acting simultaneously.

Some evidence points to a congruence cross-taxon regarding meiofauna and macrofauna patterns in response to human disturbances. Thus, the role of one or another as bioindicators of human disturbances have been evaluated in isolation. It is well-known that the abundance of meiofauna is usually orders of magnitude higher than macrofauna. Therefore, although difficult to identify, sampling of meiofaunal organisms requires low effort theoretically. Interstitial organisms are considered more sensitive to disturbances than macroinvertebrates, because the first exhibits a shorter response time, asynchronous reproduction, rapid turnover rate and usually a lack of larval dispersal (Sun et al. 2014). Another advantage is that meiofauna can act as early bioindicators, as it may respond to pollution before it becomes obvious visually or by its effects on the macrofauna (Raffaelli and Mason 1981).

Some of the arguments against the use of meiofauna are their small size which makes their sorting and identification difficult and time consuming, requiring a high degree of taxonomic expertise to provide reliable estimates of diversity. However, more easily identifiable taxonomic groups are targets of monitoring, particularly dominant nematodes and copepods, even though with low taxonomic resolution. Since the 1980's, studies have stressed that pollution and touristic associated disturbances predictably affects nematodes and copepods, suggesting that they can be used as early bioindicators of such disturbances (Hennig et al. 1983; Bodin 1988). Particularly for pollution ecology, the nematode/copepod ratio have been extensively examined as ecological indicators (defined here as metric of indication with an ecological sense). Copepods are expected to be more sensitive to environmental stress than nematodes, thus higher nematode/copepod ratios might be indicative of polluted beaches (Raffaelli and Mason 1981, Sun et al. 2014).

As sampling methods, effects of sediment characteristics and seasonal variability can differ among these two target groups, caution is needed for the applicability of the single nematode/copepod index as a measurement of disturbance on sandy beaches. However, if standardized methods are applied and the confounding effects are properly assessed, nematode/copepod ratio can be a very useful ecological indicator of pollution, particularly because literature about macrofauna extensively focusses in the responses of populations or communities to physical disturbances. The use of modern approaches of identification for the taxonomic assignment (e.g., DNA metabarcoding) tends to raise the efficiency of meiofauna as bioindicators of environmental changes (Martínez et al. 2020).

Among confronting arguments of which macro- or meiofauna organisms respond more predictably to human disturbances, the consensus is that sedentary characteristics of both are advantages for their application as bioindicators. On the other hand, the selection of vertebrates as bioindicators is questionable because of their high mobility and unpredictable distribution. Nevertheless, a bulk of studies show evidence that fish and birds at least avoid highly disturbed beaches as a result of food depletion or scaring (Dugan et al. 2003, Manning et al. 2013, Franco et al. 2016, Costa et al. 2017b). In addition, behaviour measurements also provide valuable early indication of how human presence affects habitat selection and energy intake of threatened shorebirds (Yasué 2006). Obviously, the local context must be always considered, but if beaches that function as temporary patches for feeding and nesting of transient vertebrates of conservation relevance are being less used than expected, it probably constitutes an indication of chronic effects of human disturbances and loss of basic ecosystem services.

Although the majority of vertebrates are facultative inhabitants of sandy beaches and reduction in abundance is related with food depletion, sea turtles obligatorily nest, and consequently depend on sandy beaches to complete their life cycle. Anthropogenic impacts on nesting include direct loss of habitat due to erosion and buildings, and indirect effects such as artificial light that disorientate hatchlings and facilitate detection of nests and hatchlings by predators such as ghost crabs (Silva et al. 2017). Sea turtles show high fidelity in the nesting site selection, whereas the nesting habitat is located within the beach where the female was born and within a few kilometres from previous nests (Miller 1997). Therefore, in some regions where nesting data have been collected for many years, sea turtles are excellent bioindicators of beach ecosystem condition, such as in Southeast Florida, USA and in Northeast Brazil (Marcovaldi and Chaloupka 2007, Marshall et al. 2014). Hatchling sex ratios also provide robust signals of thermal conditions being considered promisor ecological indicators of climate change (Monsinjon et al. 2019).

As evidenced earlier, species-specific counts are not always the response metric of a condition indicator used in impact assessments. Actually, comprehensive impact assessments must rely on simultaneous use of bioindicators and metrics, because some variables or taxa may not be as responsive as others to specific disturbances and sublethal aspects are pivotal for an early indication. Until now, scientific literature has highlighted advantages and disadvantages of selecting invertebrates and vertebrates as bioindicators. The feasibility of bioindicators and their metrics for impact assessments and monitoring would depend on funding for research, expertise of teams and aims of monitoring and assessments. Beach characteristics at local scales are also decisive for selecting bioindicators. Fortunately, there is relevant scientific literature for application of bioindicators in beach monitoring and the decision-making process supported by empirical evidences. The same is not true for quantitative condition indicators (biomonitors), in which scientific production is mainly concentrated in the Mediterranean coast.

When condition indicators denote proportionally disturbance levels in the environment, typically concentration of pollutants, they are considered biomonitors (Market et al. 2003). Numerous organisms concentrate pollutants in their tissues without efficient excretion mechanisms. Consequently, the higher the concentration of these pollutants in abiotic compartments, the higher is the uptake rates by biomonitors. For this reason, many animals or plants considered biomonitors are useful for monitoring trace elements that are usually present in the environment with low concentrations. Therefore, biomonitors must be bioaccumulators minimally resistant to pollution gradients. If most individuals of a population died because of toxicity and physiological disruption, monitoring based on biomonitors would be impossible. On sandy beaches, water renewal is typically high and grains have low sorption capacity. Thus, biomonitors could be useful to detect trace pollutants that are usually below detection limits of conventional equipment.

Macroinvertebrates have been the most suitable biomonitors of pollutants on sandy beaches (Costa et al. 2021). Nevertheless, studies have focussed predominantly on trace elements (e.g., metals and metalloids) lacking information regarding other pollutants. In addition, a consistent baseline exists only in the Mediterranean coast of Europe, in Africa and in Asia through determination of concentration of trace elements in soft tissues of donacid clams and sandhopper species (Costa et al. 2021).

One of the main assumptions of the biomonitor species is the limitation of regulating the concentration of a target pollutant in their tissues, resulting in proportional uptake according to the environmental pollution. Consequently, the concentration of a pollutant in organisms' tissues increases along with concentrations in the abiotic compartments (e.g., sediment and water). Although relatively few studies explore proportionality between environmental pollution and pollutant uptake by organisms on sandy beaches, for now at least donacid clams and sandhoppers from the Mediterranean seem to bioaccumulate trace metals in the same proportion of sediment pollution, acting as reliable biomonitors (Fig. 8.3). Despite lower sample size, this is not apparently true for hippid mole crabs; indeed, decapods usually have mechanisms of excretion that preclude their use as biomonitors of metal pollution in other ecosystems (Rainbow and White 1989).

Feeding is the main source of metals incorporation by beach macroinvertebrates. Sandhoppers are grazers, detritivores and scavengers, feeding on bacteria, stranded plant or animal organic materials of marine or terrestrial origin. Thus, in general they represent the bioavailabity of trace metals in solution and present in food (Ugolini et al. 2004). However, feeding preferences is actually species-specific, so that more studies with other species beyond *Talitrus saltator* from other parts of the world is a necessity. Distinctly, donacid clams are filter-feeders adsorbing pollutants from water and plankton. As molluscs bivalves are resistant to several environmental gradients they have been considered universal biomonitors worldwide (Tlili and Mouneyrac 2019).

The feasibility of beach species as biomonitors of non-metal pollutants is uncertain. An exception was found by Ungherese et al. (2016) who revealed a good ability of *T. saltator* to accumulate polychlorinated byphenyls, an organic persistent pollutant, at different concentrations depending on the contamination levels at the Tyrrhenian coast of Central Italy. On the other hand, sandhoppers have not signalized quantitatively the contamination of beach sediments by polybrominated diphenyl ethers (PBDEs), polyclyclic aromatic hydrocarbons (PAHs) and microplastic (Ugolini et al. 2012, Ungherese et al. 2016, Iannilli et al. 2018).

High trophic plasticity and mobility was conjectured to minimize the indicator potential of sandhoppers to pollutants such as microplastic (Iannilli et al. 2018). By quantifying the density of microplastic in sediment and in organisms, Costa et al. (2019) and Horn et al. (2019) found mainly fibres in digestive tract of ghost crabs and mole crabs, respectively. However, the frequency of individuals with these ubiquitous microplastic in their body was not related with the pollution level of sediments for both crustacean species.

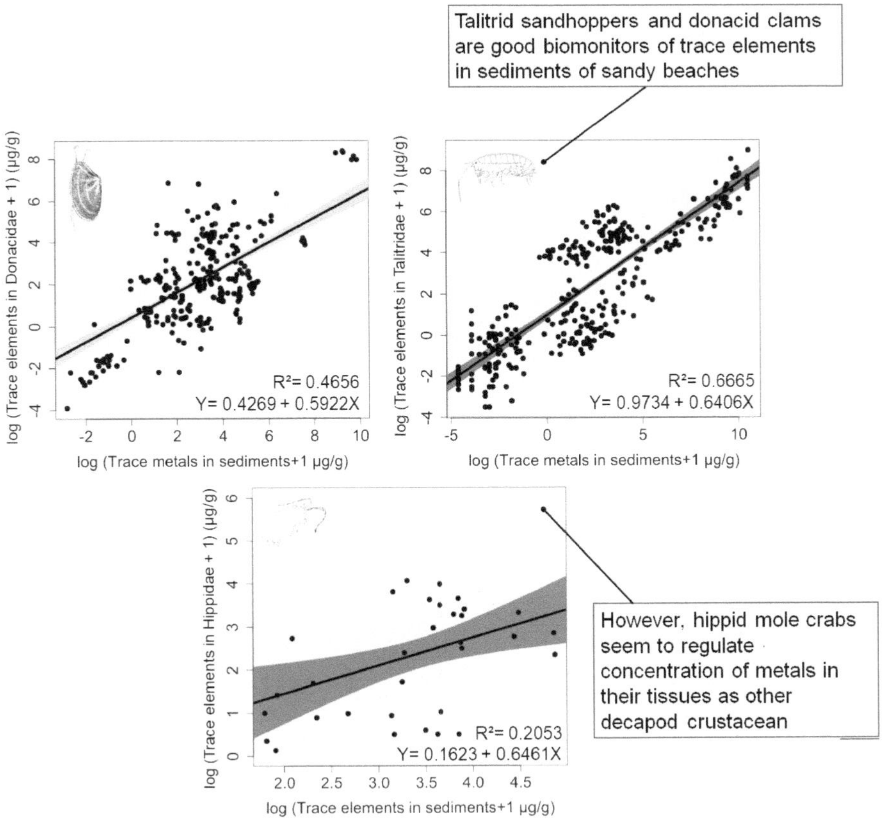

Figure 8.3 Metal concentration in the soft tissues of Donacidae, Talitridae and Hippidae in relation to their concentration in the sediment (µg/g). The black line represents the adjusted regression model, the dark shadow is the confidence interval of the model (95% confidence) and the light shadow represents the confidence interval of the prediction (95% confidence). Dataset from Costa et al. (2021).

KEYSTONES

Conceptually, keystones are those species whose 'importance' is disproportionate to their abundance, playing critical roles in ecosystems. This means that the loss of a keystone species results in broader changes over the entire community, usually expressed as trophic cascades. A keystone can be a predator regulating a strong competitive prey, releasing other species from competitive exclusion, thus, its removal could result in homogenization of communities and reduction of diversity. Similarly, the removal of certain species could induce their predator to forage in different areas or literally starve in case of rare specific predator-prey relationships, resulting in critical changes in trophic structure. It is intuitive that keystone species are interpreted as only 'important species', typically misunderstood as keystones prey or decomposers. However, consumer-resource is a fundamental ecological interaction that does not support by itself the keystone concept and application in conservation initiatives. Obviously, all organisms consume and almost all organisms are consumed, at least after to die and become detritus; therefore, it is assumed that all species are important in some degree for energy transfer and trophic functioning despite of a 'keystone' role.

Evidence of beach species acting as key prey or predators is scarce. Some studies have shown how intertidal macroinvertebrates are frequently consumed by fish, birds and crabs (Dugan et al. 2003, Tewfik et al. 2016, Costa and Zalmon 2017), but how the ecosystem functioning and the integrity of its resident species depend in part on the presence of just a single species is not clear. Applying trophic models, Costa et al. (2017b) and Reyes-Martínez et al. (2014) showed reduced energy transfer efficiency among trophic levels because of changes in some fishs' diet or decline of intertidal invertebrates on urbanized beaches, but trophic cascades were not evidenced. It is quite unlikely that trophic cascades occur in a relatively homogeneous ecosystem like a beach, usually dominated by generalist species and regulated primarily by physical factors. Yet, it is well-know that the occurrence of transitory vertebrates on beaches might be related to the availability of invertebrates (Costa et al. 2017a, Dugan et al. 2003). In some cases, even the removal of wrack by cleaning services on urban beaches has bottom-up effects fetching high trophic levels, as shorebirds (Dugan et al. 2003). Especifically in Asia and North America, horseshoe crabs (Chelicerata: Limulidae), which use beaches for mating and spawning, have been considered keystones because migratory shorebirds depend on their eggs as food resource (Botton et al. 1994).

Keystones do not necessarily affect their communities through direct trophic interactions. They can act as ecosystem engineers by modifying, creating or maintaining habitats (Caro, 2010). Typical ecosystem engineers in marine ecosystems are bioturbators. All burrowing species of sandy beaches can be considered a bioturbator. However, ghost shrimps have proved to significantly affect surrounding communities by turning over large quantities of sediment when they are in huge densities. Berkenbusch et al. (2000) showed that the number of species and abundance of co-occurring macroinvertebrates were usually lower in sites with high density of the ghost shrimp *Callianassa filholi* on an intertidal sandflat in Otago Harbour, Southeastern New Zealand. Deleterious effects of sediment turnover are prominent for tube-building and suspension-feeders organisms in particular.

Tube-building polychaetes can be an important group of marine bioengineers in soft-bottom habitats. They provide structures that potentially influence the composition of benthic communities. Santos and Aviz (2019) investigated the effects of *Diopatra cuprea* tubes on the macroinvertebrate community structure on a sandy beach in northern Brazil. They found that more than a half of the species found on the beach were associated exclusively with *D. cuprea* tubes. As large aggregations of tubes were not found in the studied area, this tube-building polychaete probably has disproportional effects in the ecosystem and its removal would induce a broader biodiversity loss. Some bivalves, sand-dollars (Echinodermata: Melitidae) and even ghost shrimps burrows are known to be inhabited by commensals, such as pea crabs (Crustacea: Pinnotheridae), and rationally, their removal could propagate for interacting species.

Most of the above-mentioned ecological interactions do not have many empirical validations for the application of keystones concept for sandy beaches management. Actually, our activities have long resulted in negative effects for beach biodiversity, from plants to large predators (Defeo et al. 2009); clearly, the absence of humans on beaches usually means different species composition, more resilient communities and proper ecosystem functioning. Conceptually, one acts as keystone allogenic engineers, changing the environment by transforming living or non-living materials, carrying litter to the beach, changing sediment properties by trampling, enhancing erosion by removing beach vegetation and installing urban infrastructure. Since sandy beaches are social-ecological systems provided with potential conflicts of use, humans might be a key target of management. One can be considered, therefore, 'key-amensalists' of sandy beaches.

UMBRELLA SPECIES

Umbrella species are broadly defined as those whose conservation confers protection to numerous co-occurring species (beneficiary species 'under the umbrella') (Caro, 2010). The percentage of co-occurring species sharing habitat requirements or interacting with each other, the same criteria that define a composition indicator, seems to conceptualize by itself the classic umbrella species. This assumes that area or configuration of areas occupied by the population of one species is able to designate where viable populations of other background species occur. However, it has been described that the rarity degree and sensitivity to disturbances must be considered for selecting umbrella species beyond the percentage of co-occurring species (Fleishman et al. 2001). Indeed, to protect habitat requirements of very ubiquitous or rare (normally generalist and specialist, respectively) or synanthropic (low sensitivity to disturbances) species hardly embrace numerous habitat requirements of co-occurring species.

It is unlikely that a unique species embraces the habitat requirements of all co-occurring species in marine systems, thus, selecting a set of focal species is possibly a more reliable approach (Lambeck 1997, Zacharias and Roff 2001). A major scientific consensus is that umbrella species must not be selected by social criteria (e.g., charisma) or assumptions that morphological traits and home ranges are enough to determine which species must be protected to embrace other co-occurring species. Similar to indicators and keystones, umbrellas must be selected by ecological criteria. Zacharias and Roff (2001) stated that umbrellas are different from keystones as communities entities will continue to exist and function in the absence of the first, but the change in the last may essentially change the community structure. Nevertheless, both shortcuts assume that protection of a single species will produce beneficial impacts for the entire community.

In as much as umbrella species are expected to cover habitat requirements of beneficiary species, percentage of co-occurring species is the most intuitive metric for selection. Similar to composition indicators, the use of umbrella species helps to maximize conservation outcomes from minimal costs by relying on assumptions that the presence of a species is a proxy of occurrence of others. This is the case of beach nesting birds in which distribution range encompassed most of other species of conservation interest on sandy beaches of central and southern New Jersey, USA (see Maslo et al. 2016). This constitutes an example of an empirical method (i.e., spatial distribution modelling) for proposing umbrellas, assuming that their area requirements are large relative to those of sympatric taxa. Thus, it is expected that if the minimum area required to support a viable population of a fairly wide-ranging bird is protected, the sufficient habitat characteristics will be available for other birds with narrow requirements.

On the other hand, it is not necessary that ubiquity equates umbrella potential. Species with extensive distribution range are often habitat generalists, which means that their habitat requirements will probably not embrace narrower habitat needs of other species. This is particularly true for marine environments, where predators are more often generalist feeders than in terrestrial environments, reducing umbrella potential at least theoretically (Zacharias and Roff 2001). In addition, it is virtually impossible to protect most of geographic ranges of ubiquitous species. Maybe this is a key issue for disconnecting composition indicators from the concepts of umbrella species regarding their utility for conservation. If assessments are part of planning for reserve design, for example, it becomes pivotal to consider other metrics beyond percentage of co-occurring species and data of distribution ranges (Fleishman et al. 2001). Protecting species sensitive to human disturbances will probably safeguard other co-occurring species equally or less negatively affected by stressors. Similarly, an intermediate degree of ubiquity is important because rare species may not be distributed across enough landscape to ensure viability of coexistent species, and ubiquitous species may be generalists, as stated earlier (Fleishman et al. 2001).

Fleishman et al. (2000) proposed an index including three main parameters for umbrella selections: degree of rarity, sensitivity to human disturbances and percentage of co-occurring species. According to Fleishman and colleagues, the umbrella potential is higher when species are neither too rare not too ubiquitous, are sensitive to human disturbances and overlap their occurrences with the maximum number of species. This index has been considered the most representative effort to assess the characteristics of effective umbrella species in terrestrial and aquatic ecosystems, to reduce the limited prospective characteristic of the umbrella concept (Caro 2010).

Until now, the unique attempt to rank umbrella species relying in Fleishman's index on sandy beaches comes from a global meta-analysis with resident macroinvertebrates (Costa and Zalmon, 2021). The authors retrieved data of communities richness and species-specific density of 726 observations, 235 beaches from almost 50 peer-reviewed articles and calculated percentage of co-occurrence and mean rarity following Fleishman et al. (2001). In addition, sensitivity of macroinvertebrates was determined empirically using response ratios (logarithmic of mean abundance values recorded for impacted sites divided by the mean values at corresponding non-impacted sites) from hundreds of impact assessments on beaches (reviewed by Costa et al. 2020a). Typical bioindicators as cirolanid isopods (Crustacea: Cirolanidae), donacid clams, hippid mole crabs (Crustacea: Hippidae), spionid polychaetes (Polychaete: Spionidae), talitrid sandhoppers and ocypodid ghost crabs were ranked as potential umbrellas along 13 marine ecoregions, extending their utility for management and conservation of beaches beyond a condition indication.

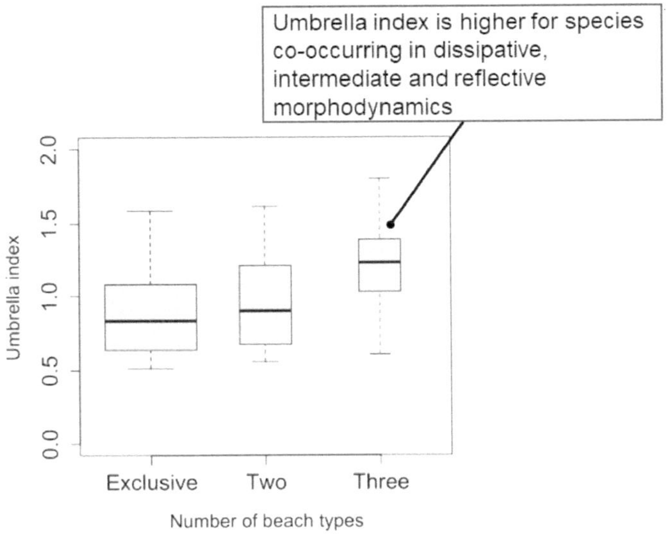

Figure 8.4 Umbrella index calculated for the pool of species according to the number of beach types (exclusive = one beach type). Horizontal lines are median values, boxes are the interquartile ranges and vertical dashed lines are the lower/upper limits of non-outliers. Wider boxes represent higher number of sampling units (species). The index was applied according to three categories: Percentage of Co-occurring Species (PCS), median rarity (RAR) according to Fleishmann et al. (2001) and sensitivity to human disturbances according to Costa et al. (2020a). Dataset from Costa and Zalmon (2021).

Because beach morpho dynamics drives the composition of macroinvertebrates communities on sandy beaches, beach type must be considered for ranking umbrella species. At first, beach ecologists could predict that species from dissipative beaches are the best umbrellas, because they co-exist with more species spatially, assuming a source-sink dynamic in meta-communities of sandy beaches. Nevertheless, Checon et al. (2018) did not support the predicted source-sink patterns from dissipative to reflective states in a set of sandy beaches in the Tropical Southwestern Atlantic ecoregion, because both dissipative and reflective beaches present similar number of exclusive species. In fact, Costa and Zalmon (2021) found species that co-occurred in different beach types and had the highest values of the umbrella index (Fig. 8.4). These findings suggest that the most tolerant and generalist species regarding natural physical dynamics tend to act as best umbrellas and could be used as targets to achieve conservation of different beach types. Checon et al. (2018) and Harris et al. (2014) stressed how it is important to set conservation targets considering varied beach types in order to protect coastal biodiversity and ecosystem services. In this context, umbrellas could be species-level targets of conservation planning, not necessarily for reserve selection because of high predictably of species richness as function of primary productivity and morpho dynamics (Defeo et al. 2017), but mainly for beach management. Assuming the feasibility of the umbrella index, the elaboration of management actions for specific macroinvertebrates as umbrellas taxa will potentialize benefits for co-occurring species, including their predators such as fish and birds (Dugan et al. 2003, Peterson et al. 2014).

FLAGSHIP SPECIES

Distinct from indicators, keystones and umbrellas, in which application for conservation actions require ecological insights, the flagship species concept has more social rationales. Flagships are species able to raise society awareness and engagement in conservation efforts. In this context, they are used as symbols in commercial sealing and environmental campaigns to raise funds and finally to achieve conservation ends. Flagship may be charismatic, economically important or threatened species, since they have enough public identification. Obviously, a major function of flagships is similar to other conservation shortcuts: to provide conservation at broader scales by means of actions based on single species. Large vertebrates such as marine turtles, seabirds and cetaceans are classical flagships in marine environments and are promising for application in marketing approaches.

Beach ecologists usually assume that beach biodiversity is neglected in management actions because it is unappreciated by the public. Indeed, if larger species are in general more charismatic than smaller ones and inspire strong positive responses in people (Berti et al. 2020), intuitively sandy beaches are in trouble, because they are dominated by inconspicuous invertebrates. According to Cardoso et al. (2011), there are public, scientific and political dilemmas impeding the effective conservation actions targeting invertebrates. The public dilemma is that the general public does not know invertebrates and their ecological services. However, it could be said that it is not necessarily true on sandy beaches.

There are some invertebrates that attract public interest and awareness, either for their harvesting importance or for being arguably charismatic. Donacid and mesodesmatid (Mollusca: Mesodesmatidae) clams, for example, support recreational, artisanal and commercial fisheries in several parts of the world (Mclachlan et al. 1996, Defeo 2003). Recreational experience provided by clams and mole crabs on ocean beaches is as valuable as the food this activity provides and could be explored by conservationists in ecotourism initiatives. Indeed, some beach species, such as ghost crabs and mole crabs, have been used for recreational purposes, but in most cases this is not an ecologically balanced activity (Kyle 1997; Gül and Griffen 2018). To reduce mere speculations, Veríssimo et al. (2014) recommended choices experiments to investigate species attributes that drive public preferences; this approach is based on marketing methods directed to specific target public. This approach is still lacking on sandy beaches, thus it could be argued that public interest and awareness with beach fauna might be context-dependent and need further investigations.

Even without systematic empirical selection, examples of successful conservation projects based on flagship species and embracing sandy beaches exist. In Brazil, *Projeto TAMAR-IBAMA*, jointly administered by the Government of Brazil and the non-governmental organization *Fundação Pró-TAMAR*, has established dozens of conservation stations which cover more than 1000 km of the Brazilian mainland coast (Marcovaldi and Dei Marcovaldi 1999). *TAMAR* was created in 1980 to investigate and implement a program for the conservation of sea turtles (Marcovaldi and Dei Marcovaldi 1999). Although sea turtles are not resident inhabitants of sandy beaches, they obligatorily nest on the supralittoral, requiring good habitat quality for that. Undoubtedly, *TAMAR* has contributed for increasing of nesting sea turtles in northwest Atlantic through environmental education, ecotourism and community-based beach patrolling for nest protection (Marcovaldi and Chaloupka 2007). However, the contribution of the project for beach conservation itself is uncertain.

Maybe it is necessary that flagships are more frequent on beaches than sea turtles to support specific actions for beach conservation. Maguire et al. (2011) showed that beach visitors have valued beaches where wildlife is present and most of them like to observe beach-nesting birds. The conservation of beach shorebirds requires the protection of their nesting habitat and maintenance of prey, mostly invertebrates (Schlacher et al. 2016a). Some proposed initiatives include adding wrack to the upper beach to increase the abundance and diversity of invertebrate prey items (Schlacher et al. 2017).

Why do birds and turtles arouse the interest of people? Where does so much charisma come from? Do ghost crabs, mole crabs, horseshoe crabs and other invertebrates not have enough strength to be symbols for raising funds for research and projects for beach conservation? It is noteworthy that, the role of beach ecologists and conservationists is to provide scientific support for decision making, not to build social capital. However, one needs to be creative in finding ways to meld conservation research with public interest (Michener 2003). Marketing and scientific outreach can be essential for the success of conservation projects of sandy beaches and higher support for research by public and private funding agencies.

MARKETING AND SCIENTIFIC OUTREACH

In a super-connected world, most people know about environmental concerns, but they are not necessarily engaged in conservation actions. When a marine conservationist goes to a beach for a survey or even for leisure, suddenly the penny drops after seeing the amount of litter on the sand. Are researchers and conservationists failing? Scientific papers are welcome, but it is not enough. In the last decades, conservation researchers and practitioners are being introduced to new tools and perspectives to achieve conservation goals. Marketing and scientific outreach are gaining prominence with specific approaches for conservation. Basically, conservation effort seeks to influence people and create an ecological behaviour, which requires multi disciplinarity and strategy. The engagement of people can couple innovative and traditional ways of communicating environmental concerns and needs.

Commercial marketing techniques to persuade a target audience and the study of consumer decision present many lessons that can be adapted to conservation ends (Fig. 8.5). In marketing, studying the needs and preferences of a target audience is pivotal to develop an assertive speech and to engage people (Smith et al. 2010). 'Conservation marketing', herein defined as the ethical application of marketing strategies to advance conservation goals, have relied on similar principles in marine and terrestrial ecosystems. Nevertheless, this is still an emerging area that needs more studies for specific conservation targets.

The new social media presents a low-cost opportunity to spread conservation messages by using an existing infrastructure. In addition, social media can help to increase credibility of conservation projects if shared by authorities and can influence more people when messages are reinforced by pairs, once behavioural change is frequently driven by awareness of the need for change and perception that people are also acting. The support of people that have a high visibility and credibility on social media has generated positive results in conservation marketing (Olmedo et al. 2020). Celebrity endorsements in conservation campaigns are pervasive indeed, but its effectiveness remains unclear because of the lack of measurable objectives, theory of change, outcome indicators and critical evaluation (Olmedo et al. 2020).

As marketing campaigns, each conservation action requires specific strategies and detailed studies about a target public, but there is no single formula for all conservation issues. Several conservation efforts rely on endangered species, disasters and other alarmist events to attract public attention. Shorebirds, marine turtles and oil spills are typically regarded for raising awareness of people and to spread, even indirectly, that sandy beaches are not ecosystems devoid of life and need attention (Escobar, 2019; Maguire et al. 2011; Marcovaldi and Dei Marcovaldi, 1999). However, there are other planned actions focussed on beaches that could fit better. Due to its scenic value, campaigns for beach conservation could use strategies based on benefits, inspiration and positive feelings that a pristine beauty of beaches provides.

Nowadays, one of the widest conservation actions on beaches are clean up days (Pasternak et al. 2019). This is so widespread that citizen science initiatives on beaches rely mainly on clean up events, contributing for research of beach pollution by litter globally (Chen et al. 2020). Indeed, this great support of people for clean-up movements comes from their relationship with the environment by itself, and not from an awareness of preserving the biodiversity or specific flagship species. However, this indirectly benefits the entire beach ecosystem and is therefore a great option for the purpose of conservation and maintenance of a harmonious relationship between people and sandy beaches. It is like clean-up events are conservation shortcuts, providing benefits beyond beach cleaning.

One of the main barriers to mobilize people in favour of beach conservation is a lack of knowledge about environmental processes, ecosystem services and reduced interaction opportunities with resident and transient species; thus, one's interest in the variety of living things is probably becoming redirected toward human artefacts (Balmford et al. 2002b). In addition, information is hidden in academics or it is often presented in a limited, simplified or biased way, posing risks for scientific credibility (Entradas et al. 2020). The fact that the ecology of sandy beaches is a relatively new and emerging discipline perhaps contributes and consequently, a low biocentric point of view of beachgoers is predominant. Studies have shown that the highest level of environmental knowledge is directly correlated to the highest level of pro-environmental behaviours (Díaz-Siefer et al. 2015).

Figure 8.5 Conservation efforts and scientific outreach usually rely on marketing strategies. On sandy beaches, litter pollution is one of the major problems raised by various organizations worldwide. Art by Camyla Freitas Viana from 'Praia Com Vida' project (www.instagram.com/praiacomvida).

Particularly, environmental education has the potential to develop intrinsic ecological behaviour of children, with the aim of promoting positive long-term effects, instead of imposing and punitive actions that tend to have only temporary outcomes (Otto and Pensini, 2017). For example, studies have speculated how the Pokémon franchise creates opportunities for didactic purposes and conservation efforts, since it replicates through cartoons and games many aspects of the real-world (Balmford et al. 2002b; Dorward et al. 2017). Why not create beach Pokémon (Fig. 8.6)?

Figure 8.6 People have interest in diversity being met by man-made variety as those created by the Pokémon Franchise (Balmford et al. 2002a). Fictional creatures inspired in sandy beach biodiversity could be inserted in games and cartoons as a promisor approach to raise awareness and engage children and Pokémon lovers in conservation initiatives. Left side: the bird of prey Carcara preying on a ghost crab; right side: the wedge clam *Donax hanleyanus*. Art by Roullien Henrique Martins Silva.

Undoubtedly, scientific outreach and popularization are pivotal to develop an ecological behaviour regarding sandy beaches. Conservation actions will have success, if the scientific information behind them is minimally understood by the public and not only for a select group of scientists who created them. Science is effectively communicated when the information reaches society. Thus, scientific outreach should not be considered an extra activity, but as an essential role of scientists, mainly because most of the countries have public funding for research (McClure et al. 2020). To recognize that scientific outreach is an assignment does not make it unviable to enjoy experiences and make a difference through engagement (Besley et al. 2018). Admittedly, communicating science to the general public is not a simple task for researchers, but it is necessary in a world where people choose to believe what reinforces their own values over empirical data. Collaborations with non-scientist groups as well as communication experts (e.g., journalists, digital influencers, marketers and producers) will certainly advance the scientific outreach.

Final Remarks

Although one has new tools to achieve conservation goals through marketing and scientific outreach, one still has a long way to go through and many questions to answer, mainly about sandy beaches. How could one increase the audience interested on threats and conservation efforts of beaches? How to bring beach science into people's daily lives? How to build a careful relationship between people and beaches? How to benefit and raise funds for a theoretically less charismatic biodiversity? Or how to increase the public appeal for the conservation of beach fauna and flora? Do beaches really not have a flagship? Strategic actions at local scales are urgent to maximize their impact with the target audience. It is important to note that these actions also directly impact government and private sector decisions about their environmental position. Creating a truly engaging and conscious society takes time, and conservation biologists can play an essential role outside the academy to help achieve it. Beach ecologists are needed to buy this idea. The challenge is to optimize information for stakeholders, but with robust science behind actions. It is believed that coupling research on conservation shortcuts, marketing and scientific outreach is a promising approach for this.

Acknowledgments

This work was supported by the *Coordenaçao de Aperfeiçoamento de Pessoal de Nível Superior—CAPES* (88882.463168/2019-01).

REFERENCES

Acuña, E.O. and E. Jaramillo. 2015. Macroinfauna en playas arenosas de la costa del Norte Grande de Chile sometidas a diferentes presiones antrópicas. Rev. Biol. Mar. Oceanogr. 50: 299–313.

Almeida, M., A. Caldas and J. de Almeida,. 1993. Variação morfométrica e demográfica em *Phaleria testacea* Say (Coleoptera, Tenebrionidae) de duas praias do Rio de Janeiro. Rev. Bras. Zool. 10: 173–178.

Arndt, E., N. Aydin. and G. Aydın. 2005. Tourism impairs tiger beetles populations—a case study in a Mediterranean beach habitat. J. Insect Conserv. Res. 9: 201–206.

Balmford, A., A. Bruner, P. Cooper, R. Costanza, S. Farber, R.E. Green, et al. 2002a. Ecology: Economic reasons for conserving wild nature. Science. 297: 950–953.

Balmford, A., L. Clegg, T.M Coulson and J. Taylor. 2002b. Why conservationists should heed Pokémon. Science. 295: 2367.

Berkenbusch, K., A.A Rowden and P.K, Probert. 2000. Temporal and spatial variation in macrofauna community composition imposed by ghost shrimp *Callianassa filholi* bioturbation. Mar. Ecol. Prog. Ser. 192: 249–257.

Berti, E., S. Monsarrat, M. Munk, S. Jarvie and J.C. Svenning. 2020. Body size is a good proxy for vertebrate charisma. Biol. Conserv. 251: 108790.

Besley, J.C., A. Dudo, S. Yuan and F. Lawrence. 2018. Understanding Scientists' Willingness to Engage. Sci. Commun. 40: 559–590.

Bessa, F., D. Cunha, S.C. Gonçalves and J.C. Marques. 2013. Sandy beach macrofaunal assemblages as indicators of anthropogenic impacts on coastal dunes. Ecol. Indic. 30: 196–204.

Bessa, F., S.C. Gonçalves, J.N.Franco, J.N. André, P.P. Cunha and J.C. Marques. 2014. Temporal changes in macrofauna as response indicator to potential human pressures on sandy beaches. Ecol. Indic. 41: 49–57.

Bodin, P. 1988. Results of ecological monitoring of three beaches polluted by the "Amoco Cadiz" oil spill: development of meiofauna from 1978 to 1984. Mar. Ecol. Prog. Ser. 42: 105–123.

Borzone, C.A and L.C. Rosa. 2009. Impact of oil spill and posterior clean-up activities on wrack-living talitrid amphipods on estuarine beaches. Brazilian J. Oceanogr. 57: 315–323.

Botton, M.L., R.E. Loveland and T.R. Jacobsen. 1994. Site selection by migratory shorebirds in Delaware Bay, and its relationship to beach characteristics and abundance of horseshoe crab (*Limulus polyphemus*) eggs. The Auk. 111(3): 605–616.

Bycroft, R., J.X. Leon and D. Schoeman. 2019. Comparing random forests and convoluted neural networks for mapping ghost crab burrows using imagery from an unmanned aerial vehicle. Estuar. Coast. Shelf Sci. 224: 84–93.

Cardoso, P., T.L. Erwin, P.A.V. Borges and T.R. New. 2011. The seven impediments in invertebrate conservation and how to overcome them. Biol. Conserv. 144: 2647–2655.

Cardoso, R.S., C.A.M. Barboza, V.B. Skinner and T.M.B. Cabrini. 2016. Crustaceans as ecological indicators of metropolitan sandy beaches health. Ecol. Indic. 62: 154–162.

Caro, T. 2010. Conservation by Proxy: Indicator, Umbrella, Keystone, Flagship, and Other Surrogates. Island Press, Washington, Covelo, London.

Checon, H.H., G.N. Corte, Y.M.L. Shah Esmaeili and A.C.Z. Amaral. 2018. Nestedness patterns and the role of morphodynamics and spatial distance on sandy beach fauna: Ecological hypotheses and conservation strategies. Sci. Rep. 8: 3759.

Chen, H., S. Wang, H. Guo, H. Lin and Y. Zhang. 2020. A nationwide assessment of litter on China's beaches using citizen science data. Environ. Pollut. 258: 113756.

Colombini, I., A. Chaouti and M. Fallaci. 2008. An assessment of sandy beach macroinvertebrates inhabiting the coastal fringe of the Oued Laou river catchment area (Northern Morocco). Trav. l'Institut Sci. 5: 81–91.

Comor, V., J. Orgeas, P. Ponel, C. Rolando and Y.R. Delettre. 2008. Impact of anthropogenic disturbances on beetle communities of French Mediterranean coastal dunes. Biodivers. Conserv. 17: 1837–1852.

Corte, G.N., H.H. Checon, G. Fonseca, D.C. Vieira, F. Gallucci, M. Di Domenico, et al. 2017. Cross-taxon congruence in benthic communities: searching for surrogates in marine sediments. Ecol. Indic. 78: 173–182.

Costa, L.L. and I.R. Zalmon. 2017. Surf zone fish diet as an indicator of environmental and anthropogenic influences. J. Sea Res. 128: 61–75.

Costa, L.L., J.G. Landmann, R. Gaelzer and I.R. Zalmon. 2017a. Does human pressure affect the community structure of surf zone fish in sandy beaches? Cont. Shelf Res. 132: 1–10.

Costa, L.L., D.C. Tavares, M.C. Suciu, D.F. Rangel and I.R. Zalmon. 2017b. Human-induced changes in the trophic functioning of sandy beaches. Ecol. Indic. 82: 304–315.

Costa, L.L. and I.R. Zalmon. 2019a. Multiple metrics of the ghost crab *Ocypode quadrata* (Fabricius, 1787) for impact assessments on sandy beaches. Estuar. Coast. Shelf Sci. 218: 237–245.

Costa, L.L. and I.R. Zalmon. 2019b. Sensitivity of macroinvertebrates to human impacts on sandy beaches: a case study with tiger beetles (Insecta, Cicindelidae). Estuar. Coast. Shelf Sci. 220: 142–151.

Costa, L.L., V.F. Arueira, M.F. da Costa, A.P.M. Di Beneditto and I.R. Zalmon. 2019. Can the Atlantic ghost crab be a potential biomonitor of microplastic pollution of sandy beaches sediment? Mar. Pollut. Bull. 145: 5–13.

Costa, L.L., L. Fanini, I.R. Zalmon and O. Defeo. 2020a. Macroinvertebrates as indicators of human disturbances: a global review. Ecol. Indic. 118: 106764.

Costa, L.L., H. Secco, V.F. Arueira and I.R. Zalmon. 2020b. Mortality of the Atlantic ghost crab *Ocypode quadrata* (Fabricius, 1787) due to vehicle traffic on sandy beaches: a road ecology approach. J. Environ. Manage. 260: 110168.

Costa, L.L. and I.R. Zalmon. 2021. Macroinvertebrates as umbrella species on sandy beaches. Biol. Conserv. 253: 108922.

Costa, L.L., M.F. da Costa and I.R. Zalmon. 2021. Macroinvertebrates as biomonitors of natural sandy beaches: overview and meta-analysis. Environ. Pollut. 275: 116629.

Defeo, O. 2003. Marine invertebrate fisheries in sandy beaches: an overview. J. Coast. Res. 35: 56–65.

Defeo, O., A. Mclachlan, D.S. Schoeman, T.A. Schlacher, J. Dugan, A. Jones, et al. 2009. Threats to sandy beach ecosystems: a review. Estuar. Coast. Shelf Sci. 81: 1–12.

Defeo, O., C.A.M. Barboza, F.R. Barboza, W.H. Aeberhard, T.M.B. Cabrini, R.S. Cardoso, et al. 2017. Aggregate patterns of macrofaunal diversity: an interocean comparison. Glob. Ecol. Biogeogr. 26(7): 823–834.

Delaney, A. and J.C. Stout. 2018. Principles of cross congruence do not apply in naturally disturbed dune slack habitats: Implications for conservation monitoring. Ecol. Indic. 93: 358–364.

Díaz-Siefer, P., A. Neaman, E. Salgado, J.L. Celis-Diez and S. Otto. 2015. Human-environment system knowledge: a correlate of pro-environmental behavior. Sustain. 7: 15510–15526.

Dorward, L.J., J.C. Mittermeier, C. Sandbrook and F. Spooner. 2017. Pokémon Go: benefits, costs, and lessons for the conservation movement. Conserv. Lett. 10: 160–165.

Dugan, J.E., D.M. Hubbard, M.D. Mccrary and M.O. Pierson. 2003. The response of macrofauna communities and shorebirds to macrophyte wrack subsidies on exposed sandy beaches of southern California. Estuar. Coast. Shelf Sci. 58: 25–40.

Entradas, M., M.W. Bauer, C. O'Muircheartaigh, F. Marcinkowski, A. Okamura, G. Pellegrini, et al. 2020. Public communication by research institutes compared across countries and sciences: building capacity for engagement or competing for visibility? PLoS One. 15: e0242950.

Escobar, H. 2019. Mystery oil spill threatens marine sanctuary in Brazil. Sci. Lett. 366: 672.

Fanini, L., G.M. Marchetti, F. Scapini and O. Defeo. 2007. Abundance and orientation responses of the sandhopper *Talitrus saltator* to beach nourishment and groynes building at San Rossore natural park, Tuscany, Italy. Mar. Biol. 152: 1169–1179.

Fanini, L., G.M. Marchetti, F. Scapini and O. Defeo. 2009. Effects of beach nourishment and groynes building on population and community descriptors of mobile arthropodofauna. Ecol. Indic. 9: 167–178.

Fanini, L., L.V. Gecchele, S. Gambineri, A. Bayed, C. Oliver and F. Scapini. 2012. Behavioural similarities in different species of sandhoppers inhabiting transient environments. J. Exp. Mar. Bio. Ecol. 420: 8–15.

Fanini, L., G. Zampicinini and E. Pafilis. 2014. Beach parties: a case study on recreational human use of the beach and its effects on mobile arthropod fauna. Ethol. Ecol. Evol. 26: 69–79.

Fanini, L., C. Piscart, E. Pranzini, C. Kerbiriou, I. Le Viol and J. Pétillon. 2021. The extended concept of littoral active zone considering soft sediment shores as social-ecological systems, and an application to Brittany (North-Western France). Estuar. Coast. Shelf Sci. 250: 107148.

Fleishman, E., D.D. Murphy and P.F. Brussard. 2000. A new method for selection of umbrella species for conservation planning. Ecol. Appl. 10: 569–579.

Fleishman, E., R.B. Blair and D.D. Murphy. 2001. Empirical validation of a method for umbrella species selection. Ecol. Appl. 11: 1489–1501.

Franco, A.C.S., M.C.N. Ramos Chaves, M.P.B. Castel-Branco and L.S. Neves. 2016. Responses of fish assemblages of sandy beaches to different anthropogenic and hydrodynamic influences. J. Fish Biol. 89: 921–938.

González, S.A., K. Yáñez-navea, and M. Muñoz. 2014. Effect of coastal urbanization on sandy beach coleoptera *Phaleria maculata* (Kulzer, 1959) in northern Chile. Mar. Pollut. Bull. 83: 265–274.

Grenyer, R., C.D.L. Orme, S.F. Jackson, G.H. Thomas, R.G. Davies, T.J. Davies, et al. 2006. Global distribution and conservation of rare and threatened vertebrates. Nature. 444: 93–96.

Gül, M.R. and B.D. Griffen. 2018. Impacts of human disturbance on ghost crab burrow morphology and distribution on sandy shores. PLoS One. 13: e0209977.

Gül, M.R. and B.D. Griffen. 2019. Burrowing behavior and burrowing energetics of a bioindicator under human disturbance. Ecol. Evol. 9(24): 14205–14216.

Harris, L., R. Nel, M. Smale and D. Schoeman. 2011. Swashed away? Storm impacts on sandy beach macrofaunal communities. Estuar. Coast. Shelf Sci. 94: 210–221.

Harris, L., R. Nel, S. Holness, K. Sink and D. Schoeman. 2014. Setting conservation targets for sandy beach ecosystems. Estuar. Coast. Shelf Sci. 150: 45–57.

Hennig, H., G. Eagle, A. Fricke, W. Gledhill, P. Greenwood and J. Orren. 1983. Ratio and population density of Psammolittoral meiofauna as a perturbation indicator of sandy beaches in South Africa. Environ. Monit. Assess. 3: 45–60.

Horn, D., M. Miller, S. Anderson and C. Steele. 2019. Microplastics are ubiquitous on California beaches and enter the coastal food web through consumption by Pacific mole crabs. Mar. Pollut. Bull. 139: 231–237.

Iannilli, V., A. Di Gennaro, F. Lecce, M. Sighicelli, M. Falconieri, L. Pietrelli, et al. 2018. Microplastics in *Talitrus saltator* (Crustacea, Amphipoda): new evidence of ingestion from natural contexts. Environ. Sci. Pollut. Res. 25: 28725–28729.

Jaramillo, E., H. Contreras and P. Quijon. 1996. Macroinfauna and human disturbance in a sandy beach of south-central Chile. Rev. Chil. História Nat. 69: 655–663.

Jaramillo, E., J.E. Dugan, D.M. Hubbard, D. Melnick, M. Manzano, C. Duarte, et al. 2012. Ecological implications of extreme events: footprints of the 2010 earthquake along the Chilean coast. PLoS One. 7(5): e35348.

Junoy, J., C. Castellanos and J.M. Vie. 2005. The macroinfauna of the Galician sandy beaches (NW Spain) affected by the Prestige oil-spill. Mar. Pollut. Bull. 50: 526–536.

Knisley, B.C. and M.S. Fenster. 2005. Apparent Extinction of the Tiger Beetle, *Cicindela hirticollis abrupta* (Coleoptera, Carabidae: Cicindelinae). Coleopt. Bull. 59: 451–458.

Kyle, R. 1997. Subsistence shellfish harvesting in the maputaland marine reserve in northern-kwazulu – natal, South Africa: sandy beach organisms. Biol. Conserv. 3207: 173–182.

Lambeck, R. 1997. Focal species: a multiple-species umbrella for nature conservation. Conserv. Biol. 11: 849–856.

Machado, P.M., L.L. Costa, M.C. Suciu, D.C. Tavares and I.R. Zalmon. 2016. Extreme storm wave influence on sandy beach macrofauna with distinct human pressures. Mar. Pollut. Bull. 107: 125–135.

Machado, P.M., M.C. Suciu, L.L Costa, D.C. Tavares and I.R. Zalmon. 2017. Tourism impacts on benthic communities of sandy beaches. Mar. Ecol. 38: e12440.

Maguire, G.S., K.K. Miller, M.A. Weston and K. Young. 2011. Being beside the seaside: beach use and preferences among coastal residents of south-eastern Australia. Ocean Coast. Manag. 54: 781–788.

Manning, L.M., C.H. Peterson and S.R. Fegley. 2013. Degradation of surf zone fish foraging habitat driven by persistent sedimentological modifications caused by beach nourishment. Bull. Mar. Sci. 89: 83–106.

Marcovaldi, M. and G.G.D. Marcovaldi. 1999. Marine turtles of Brazil: The history and structure of Projeto TAMAR-IBAMA. Biol. Conserv. 91: 35–41.

Marcovaldi, M. and M. Chaloupka. 2007. Conservation status of the loggerhead sea turtle in Brazil: An encouraging outlook. Endanger. Species Res. 3: 133–143.

Market, B.A., A.M. Breure and H.G. Zechmeister. 2003. Bioindicators and Biomonitors. Elsevier, Amsterdam, NL.

Marshall, F.E., K. Banks and G.S. Cook. 2014. Ecosystem indicators for Southeast Florida beaches. Ecol. Indic. 44: 81–91.

Martínez, A., E.M. Eckert, T. Artois, G. Careddu, M. Casu, M. Curini-Galletti, et al. 2020. Human access impacts biodiversity of microscopic animals in sandy beaches. Commun. Biol. 3: 1–9.

Maslo, B., K. Leu, C. Faillace, M.A. Weston, T. Pover and T.A. Schlacher. 2016. Selecting umbrella species for conservation: a test of habitat models and niche overlap for beach-nesting birds. Biol. Conserv. 203: 233–242.

Mclachlan, A., J.E. Dugan, O. Defeo, A.D. Ansell, D.M. Hubbard, E. Jaramillo, et al. 1996. Beach clam fisheries. Oceanogr. Mar. Biol. an Annu. Rev. 34: 163–232.

McClure, M.B., K.C. Hall, E.F. Brooks, C.T. Allen and K.S. Lyle. 2020. A pedagogical approach to science outreach. PLoS Biol. 18(4): e3000650.

Michener, W. 2003. Win-Win Ecology: How the Earth's Species Can Survive in the Midst of Human Enterprise. Oxford University Press, New York.

Miller, D.J. 1997. Reproduction in sea turtles. pp. 51–83. *In*: Lutz, P.L and J.A Musick [eds]. The Biology of Sea Turtles. CRC Press, Boca Raton, USA.

Monsinjon, J.R., J. Wyneken, K. Rusenko, M. López-Mendilaharsu, P. Lara, A. Santos, et al. 2019. The climatic debt of loggerhead sea turtle populations in a warming world. Ecol. Indic. 107: 105657.

Nourisson, D.H., F. Bessa, F. Scapini and J.C. Marques. 2014. Macrofaunal community abundance and diversity and talitrid orientation as potential indicators of ecological long-term effects of a sand-dune recovery intervention. Ecol. Indic. 36: 356–366.

Olmedo, A., E.J. Milner-Gulland, D.W.S. Challender, L. Cugnière, H.T.T Dao, L.B. Nguyen, et al. 2020. A scoping review of celebrity endorsement in environmental campaigns and evidence for its effectiveness. Conserv. Sci. Pract. 2(10): e261.

Otto, S. and P. Pensini. 2017. Nature-based environmental education of children: environmental knowledge and connectedness to nature, together, are related to ecological behaviour. Glob. Environ. Chang. 47: 88–94.

Pasternak, G., C.A. Ribic, E. Spanier, A. Ariel, B. Mayzel, S. Ohayo, et al. 2019. Nearshore survey and cleanup of benthic marine debris using citizen science divers along the Mediterranean coast of Israel. Ocean Coast. Manag. 175: 17–32.

Pavesi, L. and E. Matthaeis. 2013. Supralittoral amphipod abundances across habitats on Mediterranean temperate beaches. J. Coast. Conserv. 17: 841–849.

Pearson, D. and F. Cassola. 1992. World-wide species richness patterns of tiger-beetles (Coleoptera: Cicindelidae): Indicator taxon for biodiversity and conservation studies. Conserv. Biol. 6: 376–391.

Peterson, C.H., M.J. Bishop, L.M. D'Anna and G.A. Johnson. 2014. Multi-year persistence of beach habitat degradation from nourishment using coarse shelly sediments. Sci. Total Environ. 487: 481–492.

Pombo, M. and A. Turra. 2013. Issues to be considered in counting burrows as a measure of Atlantic Ghost Crab populations, an important bioindicator of sandy beaches. PLoS One. 8: e83792.

Pombo, M. and A. Turra. 2019. The burrow resetting method, an easy and effective approach to improve indirect ghost-crab population assessments. Ecol. Indic. 104: 422–428.

Raffaelli, D.G. and C.F. Mason. 1981. Pollution monitoring with meiofauna, using the ratio of nematodes to copepods. Mar. Pollut. Bull. 12: 158–163.

Rainbow, P.S. and S.L. White. 1989. Comparative strategies of heavy metal accumulation by crustaceans: zinc, copper and cadmium in a decapod, an amphipod and a barnacle. Hydrobiologia. 174(3): 245–262.

Reyes-Martínez, M.J., D. Lercari, M.C. Ruíz-delgado, J.E. Sánchez-moyano, A. Jiménez-rodríguez, A. Pérez-hurtado, et al. 2014. Human pressure on sandy beaches: implications for trophic functioning. Estuaries Coast. 38: 1782–1796.

Reyes-Martínez, M.J., M.C. Ruíz-Delgado, J.E. Sánchez-Moyano and F.J. García-García. 2015. Response of intertidal sandy-beach macrofauna to human trampling: an urban vs. natural beach system approach. Mar. Environ. Res. 103: 36-45.

Rodil, I.F., E. Jaramillo, E. Acuña, M. Manzano and C. Velasquez. 2016. Long-term responses of sandy beach crustaceans to the effects of coastal armouring after the 2010 Maule earthquake in South Central Chile. J. Sea Res. 108: 10–18.

Ruiz-Delgado, M.C., J.V. Vieira, M.J. Reyes-Martínez, C.A. Borzone, J.E. Sánchez-Moyano and F.J. García-García. 2016. Wrack removal as short-term disturbance for *Talitrus Saltator* density in the supratidal zone of sandy beaches: an experimental approach. Estuaries Coasts. 39: 1113–1121.

Santos, T.M.T. and D. Aviz. 2019. Macrobenthic fauna associated with *Diopatra cuprea* (Onuphidae: Polychaeta) tubes on a macrotidal sandy beach of the Brazilian Amazon Coast. J. Mar. Biolog. Assoc. UK 99: 751–759.

Scapini, F., L. Chelazzi, I. Colombini, M. Fallaci and L. Fanini. 2005. Orientation of sandhoppers at different points along a dynamic shoreline in southern Tuscany. Mar. Biol. 147, 919–926.

Schlacher, T.A., L. Thompson and S. Price. 2007. Vehicles versus conservation of invertebrates on sandy beaches: mortalities inflicted by off-road vehicles on ghost crabs. Mar. Ecol. 28: 354–367.

Schlacher, T.A. and J.M. Morrison. 2008. Beach disturbance caused by off-road vehicles (ORVs) on sandy shores: Relationship with traffic volumes and a new method to quantify impacts using image-based data acquisition and analysis. Mar. Pollut. Bull. 56: 1646–1649.

Schlacher, T.A., L.K. Carracher, N. Porch, R.M. Connolly, D. Olds, B.L. Gilby, et al. 2016a. The early shorebird will catch fewer invertebrates on trampled sandy beaches. PLoS One. 11: e0161905.

Schlacher, T.A., S. Lucrezi, R.M. Connolly, C.H. Peterson, B.L. Gilby, B. Maslo, et al. 2016b. Human threats to sandy beaches: A meta-analysis of ghost crabs illustrates global anthropogenic impacts. Estuar. Coast. Shelf Sci. 169: 56–73.

Schlacher, T.A., S. Lucrezi, C.H. Peterson, R.M. Connolly, A.D. Olds, F. Althaus, et al. 2016c. Estimating animal populations and body sizes from burrows: marine ecologists have their heads buried in the sand. J. Sea Res. 112: 55–64.

Schlacher, T.A., B.M. Hutton, B.L. Gilby, N. Porch, G.S. Maguire, B. Maslo, et al. 2017. Algal subsidies enhance invertebrate prey for threatened shorebirds: a novel conservation tool on ocean beaches? Estuar. Coast. Shelf Sci. 191: 28–38.

Schoeman, D.S., A. McLachlan and J.E. Dugan. 2000. Lessons from a disturbance experiment in the intertidal zone of an exposed sandy beach. Estuar. Coast. Shelf Sci. 50(6): 869–884.

Silva, E., A. Marco, J. da Graça, H. Pérez, E. Abella, J. Patino-Martinez, et al. 2017. Light pollution affects nesting behavior of loggerhead turtles and predation risk of nests and hatchlings. J. Photochem. Photobiol. B Biol. 173: 240–249.

Smith R.J, D. Verissimo and D.C. MacMillan. 2010. Marketing and conservation: How to lose friends and influence people. pp. 215–232. *In*: N. Leader-Williams, W.M Adams and R.J Smith [eds]. Trade-offs in Conservation: Deciding What to Save. Blackwell, Oxford, UK.

Suciu, M.C., D.C. Tavares and I.R. Zalmon. 2018. Comparative evaluation of crustaceans as bioindicators of human impact on Brazilian sandy beaches. J. Crustac. Biol. 38: 420–428.

Sun, X., H. Zhou, E. Hua, S. Xu, B. Cong and Z. Zhang. 2014. Meiofauna and its sedimentary environment as an integrated indication of anthropogenic disturbance to sandy beach ecosystems. Mar. Pollut. Bull. 88: 260–267.

Tewfik, A., S.S. Bell, K.S. Mccann and K. Morrow. 2016. Predator diet and trophic position modified with altered habitat morphology. PLoS One. 11: e0147759.

Tlili, S. and C. Mouneyrac. 2019. The wedge clam *Donax trunculus* as sentinel organism for Mediterranean coastal monitoring in a global change context. Reg. Environ. Chang. 19: 995–1007.

Ugolini, A., F. Borghini, P. Calosi, M. Bazzicalupo, G. Chelazzi and S. Focardi. 2004. Mediterranean *Talitrus saltator* (Crustacea, Amphipoda) as a biomonitor of heavy metals contamination. Mar. Pollut. Bull. 48: 526–532.

Ugolini, A., G. Ungherese, S. Somigli, G. Galanti, D. Baroni, F. Borghini, et al. 2008. The amphipod *Talitrus saltator* as a bioindicator of human trampling on sandy beaches. Mar. Environ. Res. 65: 349–357.

Ugolini, A., G. Perra and S. Focardi. 2012. Sandhopper *Talitrus saltator* (Montagu) as a Bioindicator of Contamination by Polycyclic Aromatic Hydrocarbons. Bull. Environ. Contam. Toxicol. 89: 1272–1276.

Ungherese, G., A. Cincinelli, T. Martellini and A. Ugolini. 2016. Biomonitoring of polychlorinated byphenyls contamination in the supralittoral environment using the sandhopper *Talitrus saltator* (Montagu). Chem. Ecol. 32: 301–311.

Vaz, A.S., H. Hespanhol, C. Vieira, P. Alves, J. P. Honrado and J. Marques. 2020. Different responses but complementary views: patterns of cross-taxa diversity under contrasting coastal dynamics in secondary sand dunes. Plant Biosyst. 154: 553–559.

Veloso, V.G., E.S. Silva, C.H.S. Caetano and R.S. Cardoso. 2006. Comparison between the macroinfauna of urbanized and protected beaches in Rio de Janeiro State, Brazil. Biol. Conserv. 127: 510–515.

Veloso, V.G., G. Neves, M. Lozano, C.G. Gago, F. Hortas and F.G. Garcia. 2008. Responses of talitrid amphipods to a gradient of recreational pressure caused by beach urbanization. Mar. Ecol. 4: 126–133.

Veloso, V., I. Sallorenzo, B. Ferreira and G. de Souza. 2009. *Atlantorchestoidea brasiliensis* (Crustacea: Amphipoda) as an indicator of disturbance caused by urbanization of a beach ecosystem. Brazilian J. Oceanogr. 58: 13–21.

Veríssimo, D., T. Pongiluppi, M.C.M. Santos, P.F. Develey, I. Fraser, R.J. Smith, et al. 2014. Using a systematic approach to select flagship species for bird conservation. Conserv. Biol. 28: 269–277.

Vieira, J.V., C.A. Borzone, L. Lorenzi and F.G. de Carvalho. 2012. Human impact on the benthic macrofauna of two beach environments with different morphodynamic characteristics in southern Brazil. Brazilian J. Oceanogr. 60: 135–148.

Vieira, J.V., M.C. Ruiz-delgado, M.J. Reyes-Martinez, C.A. Borzone, A. Asenjo, J.E. Sánchez-Moyano, et al. 2016. Assessment the short-term effects of wrack removal on supralittoral arthropods using the M-BACI design on Atlantic sandy beaches of Brazil and Spain. Mar. Environ. Res. 119: 222–237.

Williams, J.A. 1988. Burrow-zone distribution of the supralittoral amphipod *Talitrus saltator* on Derbyhaven Beach, Isle of Man—a possible mechanisms for regulating desiccation stress? J. Crustac. Biol. 15: 466–475.

Yasué, M. 2006. Environmental factors and spatial scale influence shorebirds' responses to human disturbance. Biol. Conserv. 8: 2–9.

Zacharias, M.A. and J.C. Roff. 2001. Use of focal species in marine conservation and management: a review and critique. Aquat. Conserv. Mar. Freshw. Ecosyst. 11: 59–76.

Chapter 9

Towards a New Integrated Ecosystem-Based Model for Beach Management: The Case of S'Abanell Beach (Catalonia-Spain: North-Western Mediterranean)

Rafael Sardá[1]*, Enric Sagristà[1] and Annelies Broekman[2]

[1]Center for Advanced Studies of Blanes (CEAB-CSIC), Road to Cala St. Francesc, 14, 17300 (Blanes), Catalonia, Spain. (sarda@ceab.csic.es, esagriso@gmail.com).

[2]CREAF, Center for Ecological Research and Forestry Applications, E08193 Bellaterra (Cerdanyola de Vallès), Catalonia, Spain. (a.broekman@creab.uab.cat).

INTRODUCTION

Delta areas and its associated beaches are one of those catchment-coastal regions in which the absence of integrated management and effective governance frameworks can have astonishing consequences. Globally, delta areas are basically retreating. Year after year, they suffer large losses of land due to human-related alteration of sediment transport and other global scale processes. This situation makes delta areas to be functioning under conditions of poor sediment availability that have deeply modified a natural process which has always existed, the process of erosion

*Corresponding author: sarda@ceab.csic.es

(Ericson et al. 2006, Valiela 2006, Nicholls et al. 2007, UNESCO-IRTCES 2011). Coastal erosion can be defined as the encroachment upon the land by the sea and is measured by averaging data over a long-time span, to eliminate the impacts of weather, storm events and local sediment dynamics (EUROSION 2004). Erosion is a generalized process worldwide that affects many coastlines producing a backward movement, that can be seen increasing dramatically on deltas and its associated beaches.

Coastal erosion is not better in the case of the Catalonian coast (North-Western Spain). A recent study carried out in its 231 kilometers of beaches (CIIRC 2010) concluded that 72% of the Catalan beaches were affected by a more or less serious erosive process, with an average annual receeding rate of 1.9 m. Another 24% of the beaches, presented processes of sedimentary accretion (gain of sediments) with an average annual increase rate of 1.5 m, while the rest of the beaches were in equilibrium. Averaging all these data, the current behavior of the Catalan beaches is that of an erosive beach with an average erosive rate of 1.03 m per year (CIIRC 2010, Jimenez et al. 2011). This erosion is even more acute due to the fact that beach management practices in the region do not allow effective management of the sedimentary balance, and responses to these problems are just reactive and slow, if anything is done at all.

At the end of 2006, the beach of S'Abanell (Banes, Girona—Catalan coast) lost a large majority of its sand. S'Abanell Beach is the area that emerges in the northern arm of the Tordera River Delta. There are no records of such a rapid erosive process in the past. Few weeks later, the Centre for Advanced Studies of Blanes (CEAB-CSIC) received a visit from the managers of the Tourism and Hospitality Association of the town of Blanes (AHB) requesting for an explanation of what might have been the causes of such rapid loss. On the other side, the town councillers of Blanes asked the present Ministry of Ecological Transition—being the responsible authority for managing this physical space—for a solution to the constant problems of the lack of sand. At the request of the AHB an investigation was started with the final objective to understand what had happened and to find a way to stabilize the beach again.

A few years before S'Abanell Beach lost its emerged sand, on 2004, the European Commission carried out the Eurosion project (EUROSION 2004). The study concluded with a set of policy recommendations meant to improve coastal erosion management in the future at the European, national, regional and local levels. To deal with erosion at whatever scale some general guidance principles were recommended; (a) increase coastal resilience by restoring the sediment balance and providing space for coastal processes, (b) internalize coastal erosion costs and risks in planning and investment decisions, (c) develop pro-active and soundly planned responses to coastal erosion, and (d) strengthen the knowledge base of coastal erosion management and planning. It was decided to follow these recommendations in order to redress the problem for S'Abanell.

After the AHB petition, a preliminary beach assessment was carried out and two initiatives started. On one side—following ordinary management procedures – nourishment activities were recommended and planned, focusing the initial response mostly on repair and restoration (Sagristà and Sardá in press). On the other side, a three-fold approach was taken, where beach regeneration was monitered after nourishment and a set of needed conditions to alleviate the general problem were described, as well as it fostered an effective governance structure in the region looking for a holistic way to facilitate the restoration of the sediment balance of the delta in the future.

In areas where natural processes can be still restored, the use of an ecosystem-based management framework could be recommended. It was believed that an important solution to deal with the erosion problem in the Tordera Delta was the creation of a governance structure able to push for a formal integrated coastal zone management program that could be running within an ecosystem-based management framework following ecosystem approach principles (CBD 1998) as it was recommended in Sardá et al. (2014, 2015). The objective of this chapter is to describe this ambitious solution. It is explained how this idea evolved and how it has been implemented until today with the hope to serve as a case study to be followed in similar processes.

S'ABANELL BEACH AND THE TORDERA DELTA

S'Abanell Beach is located in the northern arm of the Tordera Delta (41°39'07"N-2°46'43"E to 41°40'14"N-2°47'26"E, Fig. 9.1). It constitutes an example of a typical Mediterranean beach partially integrated into an urban environment that has reduced its capacity to absorb eroding forces. S'Abanell Beach can be divided in two parts. The northern and central parts of the beach (1.7 km) are located in an urban environment with a promenade in its rear front. The southern part is a suburban area where many campsites are settled (0.6 km) (Fig. 9.1). At the beginning of the last century, the width of the beach was larger than 100 m having a dune system in its rear front; this dune system was fixed anthropically by a pine plantation in the first decades of the past century, and, subsequently, the beach space was occupied by the urban development of the two coastal towns located in the delta area (Malgrat de Mar and Blanes). At that time, the beach was able to accomplish its three main functional processes, natural, protective and recreational (Sardá et al. 2013). Today, beach management is just oriented to comply with the recreational function. During the bathing season this stretch of coast is largely frequented by tourists, therefore public works to clean and smooth the beach are carried out in order to satisfy required social demand for recreational activities (Serra and Pintó 2005, Ariza et al. 2007, 2008a, b).

S'Abanell Beach is maintained by sediments coming from the fluvial course of the Tordera River. The Tordera River follows a Mediterranean regime being characterized by irregularity of flow and harsh hydrological fluctuations. Geologically, the Tordera Delta was built and developed on two other previously generated delta lobes, originated during the last glacier period about 18,000 years ago (Vila and Serra 2015). However, in line with the time scale of this work, the analysis of the last decades shows that the Tordera Delta morphology is characterized by the formation of a continuous beach on both sides (arms) of the only lobe observed in the deltaic plain, showing a typical configuration of deltas dominated by waves. For a delta to be in a growing phase, sediment contributions of the river must be greater than the transport capacity by the waves. In this case also the surrounding beaches would be stable (potentially even growing). On the contrary, as the influence of the river decreases (less sediment reaches the beach), since the swell has the same ability to take up sand, the delta starts to erode and, consequently, the surrounding beaches begin to recede. The latter is what happened to S'Abanell Beach.

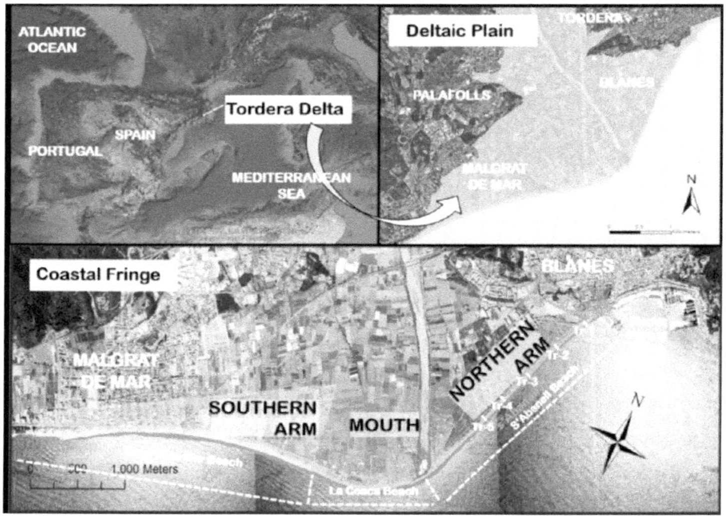

Figure 9.1 Topographic map of the study area. Top graphs) Geographical map of the Tordera Delta (Catalonia-Spain, Northwestern Mediterranean). Bottom graphs) Coastal fringe of the Tordera delta showing both delta arms and its mouth with its beach names; Tr1 to Tr5 are five sampling transects that will be described below. Source: Institut Cartogràfic de Catalunya (ICC).

The Tordera Delta front extends to the submerged area of the delta, an area where sediments carried by river dynamics normally accumulate in a growing process. The delta's submerged area protrudes into the marine environment up to about a 30 m depth, where the sediments found are not supplied by the river. The beach profile is more or less regular in its shallowest part, up to a depth of about 20 m. From this depth on, the profile varies starting from the area near the mouth of the river, where the maximum slopes are found, while up to the northern area the slope is less pronounced. The submerged platform on which the current delta front advances is located at a depth of 30 m.

DRIVERS CAUSING COASTAL EROSION IN S'ABANELL BEACH AND THEIR INTERACTIONS

During the last 60 years, the downstream part of the Tordera watershed has undergone many changes in flow dynamics and lamination conditions, due to dredging activities, land-use transformations and extractive water usages. Historically, the progressive abandonment of crops and farming areas, as well as the growth of impervious soil, such as urban areas, added more pressures to the river system which in turn influenced its coastal zone, reducing river water flow and river sediment contribution (Serra and Pintó 2005). Today, the delta area has been converted into an area of high tourist and residential potential, which exerts an enormous pressure to its natural resources (Martí 2005, Ariza et al. 2007). In order to study such transformations, the entire time span was divided considerd by the study in two different periods,

Table 9.1 An exhaustive list of human interventions carried out during the 60 years' period assessed in this work

	Historical constructive elements until 2006
1943	Construction of river banks at both sides of the river
1945–1946	Main construction of the harbor of Blanes
1950–1960	Construction of breakwaters in the central beach of Blanes
1960–1980	Dredging and extraction of sediments inside the Tordera River
1963	Construction of supply wells for Lloret de Mar in the delta
1977–1986	First enlargement of the harbor of Blanes
1980–1990	Main channeling of the low part of the Tordera River
1982	Construction of the Blanes promenade
1983–1986	Channeling of the lower reaches and main subsidiary creeks of the lower part of the Tordera River (the Vall de Burg creek)
1985	Artificial nourishment in the central beach of Blanes
1986	First artificial nourishment project in the northern part of S'Abanell Beach
1987	Enlargement and reconstruction of the Blanes promenade
1990–1995	Uncontrolled illegal extractions of sediments in the mouth of the Tordera River
1994	Large dredging Project in the submerged part of the Tordera Delta
2001	Construction of nine wells for pumping sea-water and a power generation house for the desalinization plant of Blanes at the mouth of the Tordera River
2001–2006	Construction of protective breakwaters off the campsites in the beach
	From 2006 to 2019
2007	Construction of protective breakwaters at the mouth of the river
2007	Second artificial nourishment project in the southern part (mouth) of S'Abanell Beach
2008	Third artificial nourishment project in the central part of S'Abanell Beach
2009	Fourth artificial nourishment project in the central and northern part of S'Abanell Beach
2011	Second enlargement of the harbor of Blanes
2011	Enlargement of protective jets in S'Abanell beach
2013	Construction of a new breakwater at the end of the Blanes promenade
2014	Artificial nourishment in the Malgrat de Mar beach side of the mouth of the river
2015	Dismantling of the nine wells and power generation house at the mouth; protective breakwaters were not retired
2017	New works on breakwater and other soft measures at both sides of the mouth of the Tordera River

indicating the end of 2006—when many stretches of the beach were gone—as a pivotal moment separating the two periods. In Table 9.1, a complete list of human interventions considered by the study that have contributed to beach erosion during these 60 years' period is described.

From 1956 to 2006

To assess changes in the delta region and its shoreline evolution during this period, several aerial photographs from different flights and sources were compiled (Sagristà et al. 2019). The oldest image, taken in 1956, was selected as a reference condition to estimate coastal changes, allowing to analyze the evolution of the delta system in four periods considering various and different pressures. An antropogenic sequence of events in the region was assessed using different sources of data; (a) municipal archives of the city of Blanes to date all social land-use transformations in the delta plain, (b) historical local and gray press (c) popular knowledge gained by interviewing aged people and policy actors from the city council and, (d) scientific and technical literature.

Period 1956–1977. The delta was prograding

In general, a strong advance of the coastline (around 100 m, an average of 5 m per year) was observed with a small regressive zone observed in its southern area. Altogether, the global beach surface balance of the delta beaches showed a general accretion pattern of 48,000 m^2. This period evidenced major social and economic changes that affected the dynamics of coastal demography and generated large transformations in the landscape mosaic. A progressive replacement of agricultural lands with impervious soils and enlargement of the built environment mirrored the rapid increase of tourist activities and the expansion of urban centers and related developments that demanded, in turn, new wells for water supply. Activities supporting this transformation were carried out with a large supply of gravel and sand, which were mainly extracted from the Tordera River banks and from other sites around Catalonia. Gravel extraction in the Tordera River started in 1960 and continued until 1980. Although there are no documented details, former council town managers estimated several millions of cubic meters of sand and gravel could have been extracted (pers. comm.). The extraction of sediments is believed to be responsible for the reduction of the total inputs of sedimentary materials to the beach systems in a short/medium/long-term, due to the creation of new sediment sinks in the river.

Period 1978–1986. The delta stopped its prograding tendency

A sudden southward shift of the delta coastline with initial retreats in its northern part (around 35 m), a recession in the mouth of the Delta (around 6 m per year) and a global coastline advancement in its southern part, were seen in this period. The supply of sediments by the river could not support beach maintenance where eroded surface areas (92,500 m^2) and accretion surface areas (68,000 m^2) yielded a general erosion global budget of 24,500 m^2.

During the 80s large modifications in the delta region were seen. The lower reaches of the Tordera River were channeled, changing the distribution of sedimentary inputs and reducing the mobility of the mouth of the river as well as the functionality of its secondary streams. Both coastal towns in the delta extended largely its tourist activities, with new urban developments and a series of different campsites occupying the rear front of their beaches. This was the main cause of the construction of a seafront promenade in the northern part of S'Abanell Beach (along 1.7 km of its total 2.3 km length). This promenade serves as a protection of the newly built hinterland, but induced the loss of the back of the beach, fixing its orientation and reducing its natural protective function. In addition, due to the fact that the promenade did not occupy the entire beach, beach continuity was lost where the seafront promenade was built, causing a disconnection between the part of S'Abanell Beach with promenade in its rear and the mouth of the delta. Following the construction of the Blanes promenade, an artificial nourishment of the northern part of the S'Abanell beach was carried out to rebuild the first losses that started to appear.

Period 1987–2000. The coastline was still in recession

The central areas of the Delta presented a strong erosion pattern (more than 7 m per year) while in its northern area recession was less pronounced (around 2 m per year). The situation was different in the southern part of the Delta where slight coastline advancements could be seen. The balance of erosion and accretion yielded a general erosion surface budget of 45,000 m^2.

Besides this general pattern, uncontrolled extractions of sand directly in the mouth of the Tordera River were carried out (1990–1995) to alleviate initial problems of beach retreating in the northern part of S'Abanell. These illegal extractions were not documented, but many interviewed anonymous people and local experts have confirmed this fact. These actions worsened and accelerated sediment loss in the mouth of the river and only reduced the problems of Blanes beaches during short periods of time.

The decrease in the sediment loads by the Tordera River had an influence not only on the delta area, but also on many other beaches located south in the Maresme region. The Maresme region is a regional county area that extends from Malgrat de Mar into the metropolitan area of the city of Barcelona. These beaches also started to suffer high erosive processes, and were nourished with sand extracted from sediments obtained from the submerged part of the Tordera Delta. During 1994 dredging activities in front of the Delta extracted a declared quantity of 1.5 millions of cubic meters of sand, producing a large pit in the submerged deltaic sediments. This extraction took place along both the northern and southern parts of the delta front.

Period 2001–2006. The beach was lost

The width of S'Abanell Beach was not sufficient to protect the hinterland from severe coastal storms and on many occasions caused the overtopping and destruction of the promenade with large and costly damages in local infrastructures and roads. The mouth of the delta presented a strong recession pattern but accretion was observed in its southern part. On the other side, recession in the northern part reduced although beach width was low. Altogether, even with the losses in S'Abanell Beach, a final global surface budget of accretion (19,000 m^2) was observed due to growth in the southern beaches of Malgrat de Mar.

The first desalinization plant of Catalonia was built at the end of last century in the Tordera Delta and initiated its activities in 2001. The plant was built on the river side, 500 m from the mouth of the river and required nine wells for pumping sea-water and a power generation house (black circle for geographical location in bottom graph of Fig. 9.1). These infrastructures were allocated in the rear-front of the beach inside some camping sites whose land was expropriated. These infrastructures were built at a distance of around 40 m of the coastal fringe. A brine pipeline for the extraction of the salty wastewater was placed in the seabed near the mouth. After this construction, from 2001 to 2006 a process of accelerated destabilization of the mouth of the river occurred that ended up in 2006 with all these infrastructures being surrounded by seawater. Mirroring this process, in the northern part of the Delta many problems also appeared. The beach was overtopped frequently by waves during storms, while water and sediments overwashed the promenade as well as the hinterland. As a consequence of the erosive processes seen at the mouth, breakwater constructions were built on both sides of the river to prevent flooding of the nearest campsites. This increased the channeling of the river in its mouth, with the consequent reduction in sediment distribution, especially into is northern part. Different seastorms forced to frequently reconstruct the Blanes promenade, given the protective function of the beach was compromised by the absence of sedimentary material.

From 2007 to 2019

Early in 2007, the nine wells for pumping seawater and the power generation house used for the desalinization plant were surrounded by seawater. The Catalan Water Agency carried out an emergency nourishment project (180,000 m^3) to deal with the problem. This intervention was considered as an emergengy, an exceptional condition allowing not to carry out the otherwise requiered Environmental Impact Assessment. The benefit of this nourishment project did not last, but it served to understand that, as recommended in the preliminary assessment, both wells and powerhouse should be taken out. They were disconnected and dismantled in 2015 and another system of seawater supply was installed for the desalinization plant.

After several months of considerable damages to the promenade and rear front of the city, two other nourishment projects were then carried out by the present Ministry for Ecological Transition. On April 2008, 150,000 m^3 of sand were allocated in the central part of the beach and, on August 2009, another 250,000 m^3 were deposited in its northern part. Other public works carried out during the studied period included an enlargement of the Blanes harbor north of the beach (2010–2012) with no effects on S'Abanell Beach, and another beach nourishment process carried out in 2015–2016 to restore, stabilize and build a new dune system on the beach of Malgrat de Mar, close to the mouth of the river.

As beach width is normally considered a key performance indicator for beach management processes (Sardá et al. 2015), following nourishment, S'Abanell Beach width was monitored in five transects along the beach at the northern and central part (see location in the bottom graph of Fig. 9.1) and an indirect measure was used to monitor its southern part. Weekly photographs of the entire S'Abanell beach were taken from an elevated coastal point in the 'Muntanya de Sant Joan' (Sagristà and Sardá in press). Using this data, the accomplishment of the protective function for S'Abanell beach using the Protection Partial Index (IPP) included in the Beach Quality Index (Valdemoro and Jimenez 2006; Ariza et al. 2010) were also registered. The IPP index is calculated including three different sub-measures: (i) the effective beach width (which is the distance between existing infrastructures and the shoreline); (ii) the storm reach (which is the beach width potentially eroded by a storm of a given return period); and (iii) the minimum beach width. Previous research on this index allowed one to calculate that the protective function of S'Abanell would disappear if the width of the beach gets lower than 26 m (Ariza et al. 2010). All these data were correlated with different meteorological, hydrographical and oceanographical data, mostly pluviometry, water flow, aquifer conditions and beach and swell dynamics data (Sagristà and Sardá in press).

From 2009 to 2019, S'Abanell Beach showed a different evolution according to its different areas. While the northern and central parts of the beach showed a tendency for gaining width and better functionality (Fig. 9.2), the stretch of the beach near the mouth continued its retreat trend.

Conditions for improvement in the northern and central parts were not produced in a homogeneous way. Improved beach conditions were first seen in the northern part (Transect 1: Tr-1 in Fig. 9.1) and moved south with time. Mirroring this trend, IPP improvements were also observed (Fig. 9.2). Some seasonality was detected when the width of the beach was analyzed. When a beach is functionally working well, the natural yearly evolution follows a sequence of different situations: erosion, accretion, annual oscillations and stable phases; thus, it is expected that the beach will show greater width in its summer profile than in its storm profile in winter. Therefore, the fact that the beach has recovered some seasonality related to weather conditions, together with the recovery of the beach width, would be a strong indicator of overall improvement of beach conditions. Contrary to what was detected for the rest of the beach, the southern part of the S'Abanell beach, followed a retreated pattern. From 2009 to 2018, the indirect measure taken through the photographs, indicated a reduction of the surface area by 40% (Sagristà et al. 2019).

ASSESSING THE SEDIMENTARY BALANCE IN THE REGION

To asses the sedimentary balance in the delta region, a correlation was carried out between the data related to the ecological flow of the river and its associated transport of sediment. These values were complemented with data related to the state of the aquifers of the delta and a large set of hydrological and oceanographical variables. This work allowed one to estimate the amount of sedimentary material (sand) annually supplied by the river to the sea.

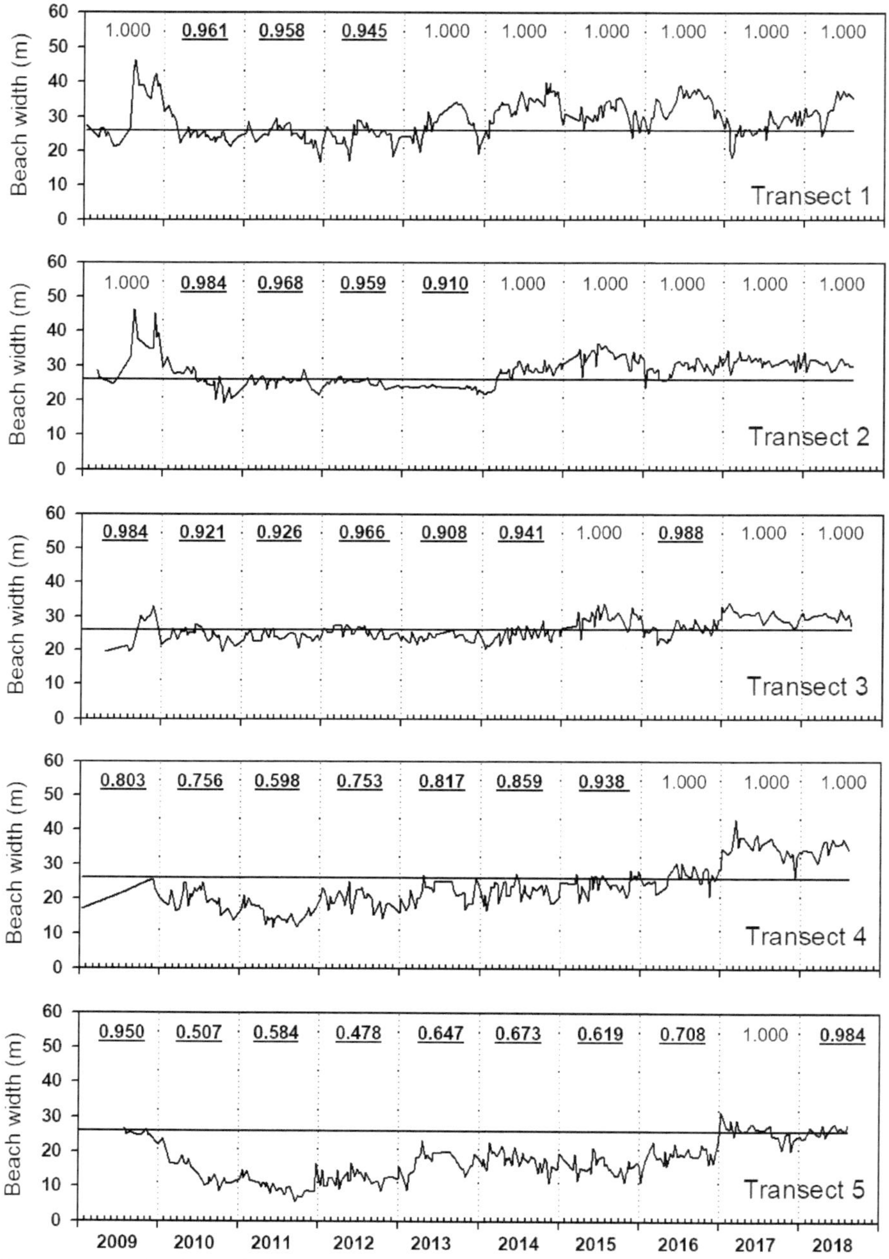

Figure 9.2 Weekly evolution of the five transects in the northern-central part of S'Abanell beach. The line at 26 m width remains the indicator for which the beach misses its protective function. The numbers on top of each graph transect show the IPP yearly average showing the tendency for improvement at entire beach.

Weekly measures of different hydraulic parameters of the Tordera River were carried out: channel width, column water height, river flow speed, river flow, the background sediment transport and the suspended sediment transport. Together with the above data, weekly observations were made to see if the sand bar that sometimes closes the river at its mouth, was open or not. A closed sand bar at the mouth means that no sediments are entering into the marine environment. In addition, with the support of the Catalan Water Agency (ACA), a weekly monitoring scheme was also set, measuring the piezometric levels of the three known aquifers of the Tordera Basin. Measurements were taken through 19 wells spread throughout the Tordera Delta area and a continuous registration—every 5 minutes—of the S-32 well by installing an autonomous probe. All these measures were correlated with other meteorological (rain, wind) and oceanographical data (waves, currents, seastorms). All the above series were obtained for a three-year hydrological cycle, from October 2015 to September 2018, in order to obtain a variation that could respond to the climatological oscillations that are seen in this Mediterranan area.

Results show that the final transport of sediments from the river into the marine environment averaged 112,657 m^3 per year (132,500 m^3 in the first hydrological cycle computed in 2015–2016; 115,500 m^3 for the 2016–2017 cycle and, 90,000 m^3 for the 2017–2018 cycle). These values were highly dependent on the seasonal conditions; in average per year, 43,349 m^3 of sediments reached the sea in fall, 21,132 m^3 in winter, 48,176 m^3 in spring and none in summer. During summer, the sand bar at the river mouth was constantly closed. The mouth of the river was closed during 147 days in the first hydrological cycle, 245 days in its second and 112 in the third one.

River flow regime in the delta region is highly connected with the state of the aquifers underlaying the delta. If these aquifers are in a bad condition and the water table is low, the flow of the river entering into the delta infiltrates into the aquifers before any of its waters can flow into the sea. During the first hydrological cycle evaluated, water in the aquifer was almost 4 m below the water table during some days in summer, 5 m in the second one and just 3 m during the last one. These reductions in the water table identified are due to high extraction rates for urban water supply to the coastal urban areas during the touristic season.

The time series analyzed found also a positive correlation between the river flow (when the sand bar at the river mouth is open) and the increase of beach width in its northern and central parts (Sagristà and Sardá, in press). This is a clear indication that the sedimentary river inputs were the main sediment source for the beach and the opening of the river mouth was the main critical factor. Therefore, in order to alleviate and to improve the erosion/accretion process and to restore the sediment balance in S'Abanell, future work should be done to increase the river flow as to maintain the sand bar open. This is necessarily linked to improved aquifer management during the summer season. As sediment inputs depend on the water flow, and the latter is highly dependent on the water table level in the Delta, integrated solutions need to be encompassed.

MAKE RESPONSES TO COASTAL EROSION PRO-ACTIVE AND PLANNED

Policy fragmentation has been recognized as an obstacle to the sustainable development of coastal resources (Cormier et al. 2010). Unfortunately, any human activity at the coast usually operates under its own regulatory framework, and best management practices designed for each intervention are applied based on its specific environmental impact assessment. This leads to fragmented and often contradictory impacts. To solve some of the problems created under such a type of development, reactive measures are promoted, which in most of the cases create consequently other environmental problems. This type of cascade mechanism needs to be adressed, especially in areas in which such type of complexity is amplified by the interactions between coastal and land-based activities. The case of the Tordera Delta is no exception. Without a global vision and integrated policy objectives aligned with ecosystem integrity, any plan that attempts to separate one driver from another is likely to be hampered by this complexity (Cormier et al. 2010). Based on this argument, it was decided to collaborate and provide support to the creation of a governance system for the Tordera Delta area with the final goal to produce a strategic and Integrated Plan for the Delta region.

During the last two decades, the Ecosystem Approach (EA) strategy, including its 12 basic principles (CBD 1998) (Table 9.2) and its direct management tool, the Ecosystem-Based Management (EBM), emerged as the dominant paradigm for managing coastal and marine ecosystems (Olsen et al. 2009, Sardá et al. 2014). In Europe, the EA is listed as a fundamental reference framework of all new European Policies listed to manage the coastal and marine environment. Besides this, beach management practices today indicate that in order to provide the best recreational attributes for beach users (clean sand for lying, clean water for bathing and landscape quality), environmental quality standards and environmental management systems are widely used and, by far, are the most important tools used in its management. Although it is clear that those frameworks improved the way in which beaches were managed, these schemes do not integrate the principles of the EA properly. A sound implementation of the EA in beach management or in territorial management would require the full integration of the set of 12 principles (Table 9.2), ensuring the inclusion of essential components such as participation, planning and decision-making, integration, promoting accountability and quality assurance, as well as the introduction of a new jargon of concepts such as social-ecological systems, ecosystem functions and services and others (see a theoretical review in Sardá et al. 2015). As many of these principles cannot be taken up in present beach managment tools, a new model of Integated Coastal Zone Management (ICZM) is needed if one wants to solve the challenges brought by today's human activities adequately.

With the final objective to make the responses to coastal erosion pro-active and planned, the ambition was to create a management approach that could run under an ecosystem-based management scheme inside an ICZM structure following Sardá et al. (2014, 2015). To carry out this idea, two preliminary pillars need to be created, a participatory pillar that allowed one to construct a governance structure and an information pillar.

Table 9.2 Ecosystem Approach principles (CBD 1998) and Integrated Coastal Zone principles

	Ecosystem Approach principles
1	The objectives of management of land, water and living resources are a matter of societal change
2	Management should be descentralized to the lowest appropriate level
3	Ecosystem managers should consider the effects (actual or potential) of their activities on adjacent and other ecosystems
4	Recognizing potential gains from management, there is usually a need to understand and manage the ecosystem in an economic context
5	Conservation of ecosystem structure and functioning, to maintain ecosystem services, should be a priority target of the Ecosystem Approach
6	Ecosystems must be managed within the limits of their functioning
7	The Ecosystem Approach should be undertaken at the appropriate spatial and temporal scales
8	Recognizing the varying temporal scales and lag-effects that characterize ecosystem processes, objectives for ecosystem management should be set for the long term
9	Management must recognize that change is inevitable
10	The Ecosystem Approach should seek the appropriate balance between, and integration of, conservation and use of biological diversity
11	The Ecosystem Approach should consider all forms of relevant information, including scientific and indigenous and local knowledge, innovations and practices
12	The Ecosystem Approach should involve all relevant sectors of society and scientific disciplines
	Integrated Coastal Zone Management principles
1	Wide ranging perspective
2	Understanding specific conditions of the area
3	Main work focused with the natural processes
4	Use of public participatory processes
5	All relevant administrative bodies should be involved
6	Use a combination of different instruments and tools
7	Consider all spatial and temporal scales

The Participatory Pillar

Olsen et al. (2009) defined coastal governance as the formal and informal agreements, institutions and behavioral norms that structure and influence: (a) how to use natural public common resources, (b) how to evaluate and analyze the problems affecting them and the opportunities derived from its uses, (c) what behaviors should be considered acceptable and/or prohibited, and (d) what rules and sanctions apply for its use. A first and necessary step to move towards coastal governance processes is to create the necessary conditions for its development. Governance implies public participation and this entails dialog, a dialog that can vary from a minimum, that is, the consultation through the provision of information to stakeholders, up to a maximum, that is to say, full participation in decision making (Sardá 2016).

In order to establish an ecosystem-based coastal management, its effectiveness must involve different actors and expertise for its design and implementation. When management strategies and activities are developed, stakeholders and society at large must be integrated in the process; their participation is important in order to address the complexity of the issues at hand and improve the acceptability of the solutions proposed. To create such a public participatory space in the delta region, we supported and promoted active participation for the creation of the so-called 'Taula del Delta i de la Baixa Tordera (TDBT)' aimed to be the governance structure dealing with the problem and adopting a long-term and strategically planned approach. The TDBT was first created on June 2017 by the city councils of the four municipalities that have part of its territory on the Tordera Delta (municipalities of Malgrat de Mar, Blanes, Palafolls and Tordera) through the development of a statutory document. Several months later, a Secretariat for the TDBT was constituted. Two projects 'ISAAC TorDelta' and 'REDAPTA', both coordinated by the CREAF with the support of the Biodiversity Foundation of the Ministry for Ecological Transition, were carried out to move and to facilitate the initial steps of the TDBT. These projects allowed setting the basis for the creation and functionality of the governance space, as well as a set of perceived challenges and opportunities for improvement of the Delta's current conditions.

The final objective of the TDBT is to recover the social and ecological balances of the Tordera Delta and to reduce its vulnerability to climate change by means of developing a strategic and integrated plan in a trasparent and participatory manner. In its first two years of activities and, after the development of a stakeholder mapping, advancement in the TDBT has been based on the compilation of different information sources, interviews, participatory seminars, focus groups and the development of a final document of objectives and recomendations. The TDBT is currently still developing a working process to define the way in which all these recommendations can be translated into a comprehensive Plan, including concrete measures to be implemented.

The Information Pillar

An essential requisite for a correct ecosystem-based management is the compilation and analysis of the best social-ecological information possible. The information pillar must provide future management with user-friendly tools to facilitate the flow of information into a decision-making process. Sardá et al. (2014, 2015) structured the information pillar as an Information Factory with two main support tools: Spatial Data Infrastructure (SDI) following standard procedures, and a platform of indicators to be used for analyzing the system at any time. Mirroring the development of the governance structure created, an environmental diagnosis of the Tordera Delta was carried out. This diagnosis followed the above recommendations, and both a Geographical Information System (GIS) and a platform of indicators were developed.

The GIS was released with the idea to describe land-use transformations in the deltaic plan throughout time. Including the submerged part of the delta, the Tordera Deltaic plain accounts for around 21 ha, 9 of them remaining in a natural state and 12 being transformed basically into urban and farming developments today. The GIS system allowed one to separate categories in a disagregated way, and referring to land-use diversity in 2018, the coastal zone—including the submerged delta—accounted for 6.32 ha (29.78% of the total amount), the farming plain accounted for 8.69 ha (40.93%), the urban environment 4.87 ha (22.93%), singular habitats 0.19 ha (0.89%) and the fluvial course 1.16 ha (5.48%).

A comprehensive characterization of the diversity of habitats in the delta region was carried out. Despite its reduced dimensions, the Tordera Delta is a natural space of high biodiversity and biogeographic features. Using the European Environment Agency CORINE land cover (Coordination of Information of the Environment) classification followed by the Generalitat of Catalonia, 148 different habitats distributed in seven different natural systems that, were identified despite clear differences, interact with each other: (a) the deltaic coastal fringe (6 habitats), (b) beach-related systems (29 habitats), (c) sandy and pine forests (7 habitats), (d) river courses (23 habitats), (e) wetlands (42 habitats), (f) the agricultural plain (31 habitats) and, (g) urban areas (10 habitats). To understand such a large variety of habitats, it is necessary to recognize that the geomorphology and hydrology of the river can be considered as unique to the Mediterranean. Its relatively small watershed is made of granitic deposits, but crowned by high mountain systems that receive large amounts of precipitation, and drained by a river and streams unregulated by reservoirs and with large seasonal variations in flow. These features have originated alluvial plains and a delta with particular characteristics that explain the presence of certain plant communities and animal species that are difficult to find elsewhere. In addition, the large anthropogenic agricultural and urban transformations recently carried out in the Delta have also contributed to enlarge the number of habitats despite losses in naturalness.

An initial assessment report was drafted to develop a common understanding of the system using the years 2016–2017 as its baseline. The assessment compiled and synthesized all the relevant information for this particular unit that becomes necessary for a correct diagnosis. The assessment also lets one learn about the pressuring factors on the system and the related stakeholders to be considered. The objective of the initial assessment was to comprehensively describe the major features of the unit that must be used sustainably, the human activities which must be managed within it and the major interactions among the unit features and the human activities observed there. The initial assessment was carried out using mostly data from the period 2016–2017 and yielded a platform of indicators under three different environmental pillars: the littoral fringe, the water vector and its fluvial course and the biodiversity. Under these three pillars, the following indicators were considered:

Littoral fringe: (a1) seastorms, (a2) beach surface, (a3) beach width, (a4) beach protection index (IPP), (a5) the Quality Beach Index (BQI), (a6) sand bar location at the river mouth, (a7) mouth sand extension, (a8) coastal artificialization, (a9) sea-bottom integrity and (a10) beach volumetry. This information was complemented with two other documents; (a) a delta bottom bathymetry and (b) a risk assessment document.

Water and Fluvial course: (b1) pluviometry, (b2) river flow, (b3) water flow quality, (b4) aquifer water level, (b5) aquifer water quality, (b6) mouth openness, (b7) water extractions, (b8) sand riverine extractions, (b9) water consumption and, (b10) sedimentary balance. This information was complemented with the hydrologic plan of watersheds in Catalonia.

Biodiversity: (c1) natural and impervious soil, (c2) natural protected area, (c3) landscape mosaic index, (c4) riverine vegetation index, (c5) group biodiversity (different groups), (c6) charismatic species and, (c7) invasive species. In this case, all this information was complemented with the earlier detailed biodiversity study commented above.

To illustrate the platform of indicators, Table 9.3 shows some data about some of the most important indicators considered in the stabilization of beaches for the delta region.

Table 9.3 Some indicators of the obtained platform in the initial assessment report

a3	Beach Width	31.9 m (2016)	To be improved
a4	Beach Protection Index	0.8 (2016)	To be improved
a5	Beach Quality Index	0.84 (2016)	To be improved
b2	River Flow	1.76 m^3 s^{-1} (2017)	Poor
b4	Aquifer Level	–1.65 m (2017)	Too low
b6	Mouth Openness	108 days (2017)	To be reduced
b10	Sedimentary Balance	115,500 m^3 (2016–17)	Poor

DISCUSSION AND CONCLUSIONS

Delta areas and its associated beaches are retreating. The historically absence of a holistic approach for its management and effective governance frameworks have produced in these catchment-coastal astonishing consequences worldwide. Although both the natural causes and human-made interventions can be causing agents, the observations described and results of this study indicate the latter normally deserves more responsibility in determining the impacts. In the Northwestern Mediterranean region, sea level increases in average 1.5 mm per year (EEA, datasets web) which correspond to around 15–20 cm retreat for beach strands on the coast following the Schwartz's Bruun rule (Schwartz 1967) and this is being aggravated by subsidence processes in many areas. However, as erosion processes were measured in Catalonian beaches averaging 103 cm per year, it is clear that most of this retreat could be explained by human interference with the natural erosion process. Human influence, particularly urbanization and other different economic activities, has turned coastal erosion from a natural phenomenon into a problem of growing intensity in the coastal zone.

In our studied case of the Tordera Delta and its associated beaches, coastal erosion resulted initially from a combination of various factors, characterized by different time and space patterns and a broad spectrum of variables of different nature (river water regulation works, illegal construction of breakwalls at the mouth of the river, dredging activities both in the river and in the submerged part of the delta, land reclamation and construction of a beach promenade, coastal artificialization and defense infrastructure, vegetation clearing or the development of hydraulic works and aquifer extractions). As a consequence, S'Abanell beach lost a large quantity of its emerged sediments by the end of 2006 (Sardá et al. 2013). By squeezing its coastal fringe, all these factors brought the beach, as well as the hinterland behind, to be more exposed and vulnerable to natural hazards such as seastorms and waves, flooding episodes, near-shore currents or sea-level rise. At the beginning of the 21st century, the explosive cocktail of successive events shown in Table 9.1 that were not correctly managed by different agencies and governmental offices, put S'Abanell Beach to the limit of collapse.

From 2007 onwards, a regional plan for the protection of the Tordera Delta and its associated beaches was elaborated in collaboration with different organizations. After a preliminary assessment to deal with the current chaotic situation, corrective measures were displayed in the area to mitigate the erosion process and restore the beach. Although some human interventions in the system were not reversible, other harmful pressures that historically happened were removed (no more dredging activities, elimination of harmful infrastructures, better sediment reworkings). All of these aspects entailed a certain degree of restoration of the Delta's natural systems, allowing one to envision the opportunity for further recovery of a correct ecological flow regime and sediment balance for the river, as well as the provision of new room for improvement for coastal dynamics.

Three nourishment interventions were carried out in the S'Abanell beach. During the next 10 years after nourishment, the northern and central part of the beach increased its width while the southern part still showed an erosive trend. The observed improvement in the northern part however could be defined as partial and incomplete. Recovery was different along the beach, being greater in the northern part than in the southern one. Recovery was also incomplete because beach width values of most transects are on the edge or below the minimum of 26 m required for achieving its protection function for a 10 year return period storm (Ariza et al. 2010) according to the Sbeach model (Larson and Kraus 1989).

Corrective measures and the nourishment process have induced that today many stretches of S'Abanell Beach show a positive trend to gain sediments and to improve its functional condition. This fact was of special relevance when in 2016 the present 'Ministry for Ecological Transition' presented the project claimed by the city council of Blanes to find a long-term solution to the problem of erosion. As a protective measure, the project was intended to introduce two or three parallel dikes off S'Abanell Beach (130 m long at a 150 m from the coastal line and projecting 70 cm above the sea) and to nourish the beach with 150,000 m^3 of sand transferred from both, the Arenys de Mar harbor (another coastal city 20 km south to Blanes) and the course of the River Tordera. According to this project, it was planned to intervene in the northern and central part of the beach, leaving the south for free evolution. In December 2016, this project was halted due, among others, to the communications sent from the muncipality which, based on the results presented in this chapter, advised against its development, based on the increased beach width trend observed during the past decade.

The identified drivers of coastal erosion in the region and their interactions increased the pressure on the Delta's system over time. Many of the past measures implemented to manage coastal erosion in S'Abanell beach were just reactive and designed from a strongly local and fragmented perspective; they ignored the influence of non-local driving agents and disregarded the sediment transport processes within the river and coastal system at large. Sediment transport in the river was calculated to be in the order of 112,657 m^3 per year.

Historical measures of natural suspended sediment transport provided by the Tordera watershed system to the sea ranged from 20,000–45,000 m^3 y^{-1} (DGPC 1986) or 60,000 m^3 y^{-1} (Copeiro 1982) up to a potential maximum of 200,000 m^3 y^{-1} (Serra and Montorí 2003). These sediments are responsible for around 90% of the total contribution of sand to the delta and its beaches. Other secondary streams could represent a remaining residual 10% contribution to the system. This study indicates interventions oriented at recovering a sedimentary balance in the order of 150,000–200,000 m^3 per year could be enough to stabilize delta beaches in the region.

Besides nourishment practices and the alleviation of antropogenic pressures on the beach system, a strong effort was also initiated to create the conditions to manage the system in an integrated way following Ecosystem Approach guidelines within an Integrated Coastal Zone Management (ICZM) approach (Sardá et al. 2013, 2014, 2015, Billé and Rochette 2015). ICZM frameworks are normally called to deal with these situations, but despite the attempts to display guidelines for many regions and to develop governance conditions for these processes, the reality shows that nowadays these guidelines are still being underused and few working conditions for ICZM frameworks are seen in practice. As the River Tordera is still a non-regulated flowing river, this case would be ideal to describe the possibility to apply an ICZM of second generation (ICZM 2.0) based on the application of the full set of principles of the ecosystem-based management framework. Although greatly advised, this is not simple.

Spain is particularly susceptible to the challenge of policy fragmentation given a significant number of overlapping jurisdictions and lack of supporting legislation for integrated coastal zone management. Thus, it became necessary to develop a public-private partnership scheme to revert the causes of the problem and to facilitate the recovery of ecosystem functioning by the development of an effective governance structure in the region. A few months later, the TDBT governance structure was created and some objectives and recommendations outlined. The TDBT is thought to be recognized as the governance mechanism to evaluate options for recovery of the sedimentary balance, to assess a possible implementation of these options identifying responsible key stakeholders, to deal with conflicts and benefits, as well as to issue the optimal timing for implementing the measures identified. This governance mechanism under development can constitute a key instrument to deal with the unsolved land planning problems.

Thirteen years ago, many stretches of S'Abanell Beach were gone. Today, the beach shows certain new signs of functionality and all related stakeholders work on mantaining and strengthening this pace of recovery, with the aim of achieving the highest possible degree of renaturalization. Still, no miracle solution can eliminate the fact that erosion is driving the region, but if all the parties involved could work together to get a sedimentary balance of 150,000–200,000 m^3 per year and the TDBT governance framework works well, the functional condition of the beach could be improved.

Acknowledgements

This chapter was carried out within the framework of the ECOPLAYA project (CGL2013-49061) of the National Research Plan of Spain in R+D+i. The authors gratefully acknowledge the ISAAC-TorDelta and the REDAPTA projects coordinated by CREAF with the support from the Biodiversity Foundation of the Ministry for Ecological Transition of Spain that contributed to the development and implementation of the called 'Taula del Delta i de la Baixa Tordera (TDBT)' for which we also thank many regional stakeholders involved in the process. We are also grateful to those actors involved that supported the process started and now continue working in the governance structure created.

REFERENCES

Ariza, E., R. Sardá and J.A. Jiménez. 2007. Beyond performance assessment measurements for beach management: application to Spanish Mediterranean beaches. Coast. Manage. 36: 47–66.

Ariza, E., J.A. Jimenez and R. Sardá. 2008a. Seasnal evolution of beach waste and litter during the bathing season on the Catalan coast. Waste Manage. 28: 2604–2613.

Ariza, E., J.A Jiménez and R. Sardá. 2008b. A critical assessment of beach management on the Catalan coast. Ocean Coast. Manage. 51: 141–160.

Ariza, E., J.A. Jimenez, R. Sardá, M. Villares, J. Pinto, R. Fraguell, et al. 2010. Proposal for an integral quality index for urban and urbanized beaches. Environ. Manage. 45: 998–1013.

Billé, R. and J. Rochette. 2015. The mediterranean ICZM protocol: paper treaty or wind of change. Ocean. Coast. Manage. 105: 84–91.

Convention on Biological Diversity (CBD). 1998. Report of the workshop on Ecosystem Approach. Lilongwe, Malawi, 26–28 January 1998, UNEP/CBD/COP/4/Inf.9, p. 15

Copeiro, E. 1982. Playas y obras costeras en España. Revista de Obras Públicas. 82: 125–130.

Cormier, R., A. Kannen, I. Davies, R. Sardá and A. Diedrich. 2010. Policy fragmentation implications in ecosystem-based management in practice. Working Paper. Proceedings of the ICES Annual Science Conference, ICES Annual Science Conference, Nantes, France.

Centre Internacional d'Investigació dels Recursos Costaners (CIIRC). 2010. Estat de la Zona Costanera a Catalunya. Departament Política Territorial i Obres Públiques. Barcelona.

Dirección General Protección Costa (DGPC). 1986. Investigación tecnológica de las acciones a tomar para la estabilidad del tramo de costa de los términos municipales de Malgrat y Sta. Susana. DGPC-Generalitat de Catalunya, 2: 318.

EEA, datasets web www.eea.europa.eu/data-and-maps/data

Ericson, J.P., C.J. Vörösmarty, S.L. Dingman, L.G. Ward and M. Meybeck. 2006. Effective sea-level rise and deltas: Causes of change and human dimension implications. Global Planet. Change, 50: 63–82.

EUROSION. 2004. Living with Coastal Erosion in Europe: Sediment and Space for Sustainability. Service Contract B4-3301/2001/1329175/MAR/B3. Directorate General Environment, European Comission.

Jimenez, J.A., V. Gracia, H.I. Valdemoro, E.T. Mendoza and A. Sánchez-Arcilla. 2011. Managing erosion-induced problems in NW Mediterranean urban beaches. Ocean Coast. Manage. 54: 907–918.

Larson, M. and N. Kraus. 1989. SBEACH: numerical model for simulating storm-induced beach change. Report 1. Empirical foundation and model development. Tech Report CERC-89, 9. US Army Corps of Engineers, Vicksburg (USA).

Martí, C. 2005. La transformació del paisatge litoral de la Costa Brava: Anàlisi de l'evolució (1956–2003), diagnosi de l'estat actual i prognosi de future. Ph.D. Thesis. Univeristy of Girona. Girona, Spain.

Nicholls, R.J., P.P. Wong, V. Burkett, J. Codignotto, J. Hay, R. McLean, et al. 2007. Coastal systems and low-lying areas. pp. 315–357. *In*: M.L. Parry, O.F. Canziani, J.P. Palutikof, P. van der Linden and C.E. Hanson [eds]. Climate Change 2007: Impacts, Adaptation and Vulnerability. Contribution of Working Group II to the Fourth Assessment Report of the Intergovermental Panel on Climate Change. Cambridge University Press. Cambridge, UK.

Olsen, S.B., G.G. Page and E. Ochoa. 2009. The Analysis of Governance to Ecosystem Change: a handbock for assembling a baseline. LOICZ reports and studies, n° 34. GKSS Research Center, Geesthacht. 87 p.

Sagristà, E., R. Sardá and J. Serra. 2019. Consecuéncias a Largo Plazo de la Gestión Desintegrada en Zonas Costeras: el Caso del Delta de la Tordera (Cataluña, España). Costas, 1: 1–22.

Sagristà, E. and R. Sardá. Decadal shift in the northern arm of the Tordera delta (north-eastern Spain) as a consequence of shoreline remediation practices and societal engagement. Reg. Environ. Change. (in press)

Sardá, R., J. Pintó and J.F. Valls. 2013. Hacia un nuevo modelo integral de playas. Documenta. Publ. Girona.

Sardá, R., T.G. O'Higgins, R. Cormier, A. Diedrich and J. Tintoré. 2014. A proposed ecosystem-based management system for marine waters: linking the theory of environmental policy to the practice of environmental management. Ecol. Soc. 19: 51. http://dx.doi.org/10.5751/ES-07055-190451.

Sardá, R., J.P. Valls, J. Pintó, E. Ariza, J.P. Lozoya, R. Fraguell, et al. 2015. Towards a new integrated beach management system: the ecosystem-based management system for beaches. Ocean Coast. Manage. 118: 167–177.

Sardá, R. 2016. La Gestió Integrada del Litoral i els models de Governança. pp. 33–47. *In*: F. Romagosa [ed.]. La Governança i la Gestió Integrada del Litoral a Catalunya.— Kitbook Servicios Editoriales. Barcelona, Spain.

Serra, J. and C. Montori. 2003. Tordera River delta relict lobes and Holocene Sea level rise (Maresme Coast, Spain, NW Mediterranean) Proc. IGCP 437 Conference on Coastal Sediments. 485–940.

Serra, I. and J. Pintó. 2005. La transformació del paisatge del delta de la Tordera en els darrers cent cinquanta anys. Una anàlisi per mitjà dels canvis en els usos i les cobertes del sòl. Documents d'Analisi Geografica. 46: 81–102.

Schwartz, M. 1967. The Bruun theory of sea-level rise as a cause of shore erosion. J. Geol. 75: 76-92.

UNESCO-IRTCES. 2011. Sediment Issues and Sediment Management in Large River Basins Interim Case Study Synthesis Report. International Sediment Initiative. Technical Documents in Hydrology. UNESCO Office in Beijing & IRTCES.

Valiela, I. 2006. Global Coastal Change. Blackwell Publ. USA.

Valdemoro, H.I. and J.A. Jiménez, 2006. The Influence of Shoreline Dynamics on the Use and Exploitation of Mediterranean Tourist Beaches. Coast. Manage. 34: 405–423.

Vila, I. and J. Serra. 2015. Tordera River Delta system build up (NE Iberian Peninsula): sedimentary sequences and offshore correlation. Sci. Mar., 79: 305–317.

Chapter 10

The Beach Ecology Coalition: Enhancing Ecosystem Conservation and Beach Management to Balance Natural Resource Protection and Recreational Use

Karen L.M. Martin[1]*, Dennis R. Reed[2], Dennis J. Simmons[3], Julianne E. Steers[4] and Melissa Studer[5]

[1]Department of Biology, Pepperdine University, 24255 Pacific Coast Highway, Malibu, CA 90263, United States of America (karen.martin@pepperdine.edu).

[2]City of San Clemente, Beach Ecology Coalition, 24255 Pacific Coast Highway, Malibu, CA 90263, United States of America (info@beachecologycoalition.org [retired]).

[3]City of San Diego, Beach Ecology Coalition, 24255 Pacific Coast Highway, Malibu, CA 90263, United States of America (info@beachecologycoalition.org [retired]).

[4]Ocean Institute of Dana Point, Beach Ecology Coalition, 24255 Pacific Coast Highway, Malibu, CA 90263, United States of America (info@beachecologycoalition.org).

[5]Grunion Greeters, Beach Ecology Coalition, 24255 Pacific Coast Highway, Malibu, CA 90263, United States of America (melissastuder@beachecologycoalition.org).

*Corresponding author: karen.martin@pepperdine.edu

INTRODUCTION

Perception of Beach Management by Municipalities and Scientists

Historically, beaches have been perceived in terms of hazards and recreational opportunities for humans (Bird 1996). However, recent studies have clarified the ecological roles and services that beaches provide to marine and terrestrial ecosystems (McLachlan and Brown, 2006, Schlacher et al. 2007, Dugan et al. 2010). A goal of ecologically sensitive beach management should be to strive to balance the needs and wishes of human beachgoers with the needs and benefits of the natural resources that comprise the sandy beach ecosystem (Defeo et al. 2009, Pilkey and Cooper 2014, Harris et al. 2014).

Beaches in urban settings are often managed more for human recreation than for ecological values (Griggs et al. 2005, Nelsen 2012, King et al. 2018). Municipalities and local businesses may consider the ideal beach to be a sterile, uniform environment, with daily raking to pick up trash and remove natural wrack along the shore. Beach raking (or beach grooming) began in the 1960s and continues to this day along many public and private beaches of the southern California coastline (Orme et al. 2011). Recently, scientists and environmentalists have pointed out the importance of beach wrack as a nutrient subsidy to the sandy shores for numerous animals and plants (Llewellyn and Shackley 1996, Dugan et al. 2003, 2011, Dugan and Hubbard 2010, Suursaar et al. 2014). The importance of beaches as nursery areas for marine mammals, sea turtles, shore birds and even some fishes requires that eggs, nests and the young have protection from mechanical disturbance of the sand and removal of nutrients (Robbins 2006, Martin et al. 2006, Schlacher et al. 2014a).

Conflicts (Perceived or Real) between Recreation and Natural Resource Conservation

Changes to beach management and maintenance protocols are sometimes discouraged by the perception that inherent conflicts exist between management for people versus management for natural resources. Concerns about public access, recreational opportunities, safety and the possibility of unwieldy constraints from regulations may cause managers to hesitate to make changes in management of public open spaces (King et al. 2018).

One approach to changes in policy is through public pressure. An informed and active public can stimulate change more quickly, more broadly and less controversially through a grass-roots effort than one enlightened manager can do alone. Educating people about the ecological role of beaches is critical for formulating new management strategies to implement changes in practice (Martin et al. 2007, 2011, MPAMAP 2018). When the public expects a response to concerns, managers generally oblige.

Gaining Community Involvement and Support

While community involvement and support are necessary, it is equally important to explain the reasons for policy changes to the people responsible for implementing or enforcing them. In the case of beaches, these workers include maintenance employees, lifeguards, vendors that drive on beaches, park rangers, game wardens, resource managers and public safety officers. These groups are quite diverse and may not interact on a regular basis with one another. For this reason, it was important to form a new community based on shared interests and goals.

Origins: The Beach Ecology Coalition

Scientists and volunteers working with city government

The organization that was to become the Beach Ecology Coalition began with a concern raised by a local resident in San Diego in the year 2001. Pat Gallagher watched the city's daily beach grooming activity, which was to rake the entire beach near her home, all the way down to the water line. One day, she noticed flocks of birds following in the tracks of the tractor pulling the rake. On closer inspection, she saw hundreds of small orange eggs turning up on the furrows left behind. She realized that the birds were feeding on the exposed eggs of a beach-spawning fish, the California grunion *Leuresthes tenuis* Ayres, 1860.

California grunions are famous for their dazzling display during spawning runs. Fish surf out of water onto sandy beaches, late at night, after a full or new moon, on the highest tides and bury their eggs a few inches deep in the sand (Sandrozinski 2013). Because of their placement at semilunar high tides, these eggs remain out of water throughout incubation, only hatching when washed back to sea by the subsequent semilunar high tides of the next new or full moon (Griem and Martin 2000, Martin et al. 2009).

This concerned citizen spoke with the operator of the vehicle, who did not understand her dismay. Gallagher took her complaint to her City Councilman Scott Peters, telling him that beach grooming was killing all the grunion. He asked a local environmental organization, Project Pacific, to examine this issue. They agreed and President Melissa Studer began putting together a blue-ribbon panel of scientists and city workers to assess the situation.

Blue-Ribbon panel, composition and direction

The blue-ribbon panel included the supervisor of beach maintenance and other city workers, scientists from resource agencies (such as the U.S. Fish and Wildlife Service and California Department of Fish and Wildlife), plus ichthyologists from the nearby Scripps Institution of Oceanography (Jeffrey Graham and Richard Rosenblatt). The group decided that a study was needed to determine the status of the California grunion on San Diego beaches, namely whether the fish were present, as well as to examine the effects that grooming would have on their eggs buried in the sand. In the meantime, they demanded the beach maintenance workers to stop grooming the beaches, which lead to an outcry from beachfront businesses suggesting that tourism would drop.

Local network of volunteers

Guided by M. Studer, a brief pilot program was conducted in summer of 2001, in order to test the concept that volunteers could be recruited to monitor local beaches for grunion runs. However, work started very late in that summer and few fish were seen on any of the beaches. Still, relationships between Project Pacific and the Birch Aquarium at Scripps were established. More than 30 members of the public volunteered. With the guidance of J. Graham and R. Rosenblatt, M. Studer secured a grant from the U.S. Fish and Wildlife Federation for a formal study in the city of San Diego. In the following year, K. Martin joined the team. The monitoring dates were changed, new methods were developed for the types of data to be taken and the trained volunteers were now called 'Grunion Greeters'. In 2002, over 200 members of the public participated in training workshops at the Birch Aquarium, lured by promises of 'behind the scenes' tours with the Director J. Graham, as much as by their desire to volunteer, along with a fondness for the California grunion.

Cooperative relationship between scientists and city workers

The study required volunteers to monitor grunion runs, which could be forecasted by tide charts and were always late at night (Sandrozinski 2013). Volunteers signed up for specific nights and locations on city beaches. They reported their data via an online web portal, so that the scientific team had instantaneous access to results of the observations (Martin et al. 2007, 2011). With the revised dates and methodology changes, grunion spawning runs were observed at every city beach location monitored in 2002.

Following the reports, the scientific team assessed the presence of eggs on the shore to check the veracity of the observations. In some of the areas identified with eggs, the study called for experimentally grooming over the eggs to observe its effects. Other areas with eggs were not raked, as a control. This study would not have been possible without the cooperation of the city beach maintenance workers, supervised at that time by D. Simmons. Beach maintenance personnel worked, side by side with the scientific team, to carry out the experiments at multiple locations along the shoreline. The maintenance staff quickly learned to recognize the spawning sites and find the buried eggs.

The survey showed that directly raking over the spawning zone destroyed the eggs (Martin et al. 2006). These results were presented to the City Council, along with a written policy for beach maintenance developed by the staff. The new Grunion Grooming Protocol recommend that raking could be carried out at the upper beach, but absolutely not in areas below the highest high tide line, where grunion eggs could be hidden. This tideline was visible by examining the wrack washed in during the nights of the highest semilunar tides, the same nights as grunion runs were expected. The results and written policy were endorsed by the City Council and adopted unanimously for future grunion seasons.

First meeting of the group, establishment of network

The experience of collegial interactions and a constructive outcome of the grunion study was carried forward in several ways. For one thing, the Grunion Greeters program expanded to other areas along the California coast, with workshops and volunteer engagement patterned after the initial meetings (Martin et al. 2007, 2011). Over 5000 citizen scientists have volunteered as Grunion Greeters to the present day.

Separately, the collaboration between the city beach maintenance organization, the scientists and environmental organizations, that started in the Blue Ribbon Panel, continued. Spring training sessions for the San Diego City beach workers were held in 2003, to ensure the new practices for Grunion Grooming Protocols were understood. D. Simmons proposed gathering other beach managers from the southern California coast to discuss making this relatively simple (but effective) change to their beach grooming practices during the grunion season. He had several informal contacts, but beach managers did not have any professional organization. Therefore, beach managers only spoke with one another when they had specific questions or problems that they hoped to solve with the benefit of another's experience, through ad hoc personal recommendations. Then, Simmons suggested that there were surely other important issues that could be addressed by an organization, namely for ecologically sensitive beach maintenance, with the input of a scientific perspective.

Establishment of semi-annual meetings, alternating field sites and classroom settings

The first meeting of this beach management working group was held in a conference room at Pepperdine University in January 2004, being attended by 14 people. Although small, it included beach managers from the cities of San Diego, Solana Beach, Newport Beach, Long Beach, Santa Barbara, plus Orange and Los Angeles Counties, along with the staff scientist from Heal the Bay—an environmental organization from the Los Angeles area (Fig. 10.1A). The positive interactions encouraged the group to try meeting regularly. Over time, half-yearly meetings were organized. The winter meetings were held indoors in an auditorium or meeting room. The spring meetings were typically held outdoors, at a beach, for demonstrations and site visits. By spring 2005, over 60 participants joined the meeting at Doheny State Beach, in Orange County, which included state park rangers, staff of the California Coastal Commission and California Department of Fish and Wildlife, coastal ecologists, surfers, lifeguards and representatives of many non-governmental environmental organizations (Fig. 10.1B). Since then, the Coalition has met twice a year at alternating locations, including aquariums, nature centers, a variety of beaches, lifeguard headquarters, universities and youth centers. Attendance is usually between 60 and 75 participants at each meeting, from multiple organizations and affiliations (Table 10.1).

(A)

(B)

Figure 10.1 (A) The first meeting of the Beach Ecology Coalition at Pepperdine University, in Malibu, had 14 participants, including beach managers from the cities of San Diego, Solana Beach, Newport Beach, Long Beach and Santa Barbara, plus Orange and Los Angeles Counties, along with scientists from Heal the Bay—an environmental organization in California. Co-founders K. Martin and D. Simmons are on the right in the second row. (B) Within two years, the meeting at Doheny State Beach had expanded to 60 participants, including state park rangers, ecologists, lifeguards, surfers and representatives of governmental and non-governmental environmental organizations.

Table 10.1 Cooperating organizations from the Beach Ecology Coalition included government and non-government agencies, educational institutions and industry. Organizations paid the expenses of the representatives attending the semi-annual meetings

Federal organizations	
• Camp Pendleton US Marine Base • Channel Islands National Marine Sanctuary • Cordell Banks National Marine Sanctuary • Coronado Island US Naval Amphibious Base • Los Alamitos US Naval Base • National Marine Fisheries Service: Southwest Region, Habitat Conservation Division	• National Oceanic and Atmospheric Administration • Jet Propulsion Laboratory, NASA • US Fish and Wildlife Service • National Parks Service, Golden Gate National Recreation Area • Pt. Reyes National Seashore • US Army Corps of Engineers
Local and state organizations	
• California Coastal Coalition • California Coastal Commission • California Coastal Conservancy • California Department of Fish and Wildlife • California Marine Protected Areas • California Oil Spill Prevention and Response • California State Parks • California State Beaches: San Elijo State Beach, Bolsa Chica State Beach, Doheny State Beach, Topanga Beach, Malibu Lagoon 'Surfrider' State Beach, Crystal Cove State Beach. • City of Encinitas • City of Huntington Beach • City of Imperial Beach • City of Long Beach • City of Malibu	• City of Monterey • City of Newport Beach • City of Oceanside • City of Oxnard • City of Santa Barbara • City of San Clemente • City of San Diego • City of Santa Monica • City of Solana Beach • City of Ventura • County of Los Angeles • County of Orange • County of Santa Barbara • County of Ventura • East Bay Regional Parks District • Los Angeles County Beaches and Harbors • Resource Conservation District, Santa Monica Mountains
Nongovernmental organizations (NGOs)	
• Activist San Diego • Audubon Society (Buena Vista, Los Angeles, and Santa Monica Bay chapters • Aquarium of the Pacific in Long Beach • Birch Aquarium at Scripps Institution of Oceanography • Cabrillo Marine Aquarium, San Pedro • Heal the Bay, Santa Monica • L.A. Works • Monterey Bay Aquarium Research Institute • Moss Landing Marine Laboratories • Muth Interpretive Center in Newport Beach • National Wildlife Federation, National Wildlife Watch Week	• Natural History Museum of Los Angeles County • Ocean Institute at Dana Point • Ocean Protection Council • Pacific Grove Museum • Roundhouse Aquarium, Manhattan Beach • Santa Barbara Channel Keepers • Santa Monica Bay Restoration Commission • Santa Monica Pier Aquarium • Sea Center, Santa Barbara Natural History Museum • Surfrider Foundation, National and local chapters • The Bay Foundation

Industry	
• Beach Tech • Bonterra Consulting • Chambers Group Consulting • ECorp Consulting • Entrix Consulting • Lawson's Landing, Tomales Bay	• Littoral Ecological & Environmental Services • MBC Consulting • Rincon Consultants • San Marino Environmental Associates • Southern California Edison • Port of Oakland
Academic Institutions	
• California State University, Channel Islands • California State University, Fullerton • California State University, Long Beach • California State University, Northridge • California State University, Stanislaus • Loyola Marymount University • Mira Costa College • Oxnard Harbor College • Palomar College	• Pepperdine University • Pomona College • Scripps Institution of Oceanography • University of California, Davis • University of California, Los Angeles • University of California, San Diego • University of California, Santa Barbara • University of California, Santa Cruz • University of Southern California, Wrigley Science Center

Incorporation, board of directors, mission statement

As the beach management organization grew and became more active, the need for formal recognition was clear. In 2007, the group formally incorporated as a 501(c)3 non-profit public benefit corporation in California, with assistance from a legal counsel at Pepperdine University. With bylaws, officers, a Board of Directors and an Advisory Board, the organization was now eligible to apply for external funds from foundations and to accept tax-deductible donations.

During a lively organizational meeting at the Ocean Institute in Dana Point, it was decided that the Ecologically Sensitive Beach Management Working Group would thereafter be known as the Beach Ecology Coalition, with the intent to be inclusive and collaborative among all interested parties, from scientists to equipment operators. The charter calls for educational outreach and scientific research and forbids lobbying or political activism.

The mission statement is intended to be inclusive of all participants and clear on the purpose: "to enhance ecosystem conservation with beach management, in order to balance natural resource protection and recreational use" on sandy beaches in California. The objectives emphasize the educational purpose of the organization:

- to develop and identify best practices for beach management, combining conservation and restoration of natural resources with human recreational opportunities;
- to provide educational outreach for beach workers, managers, scientists and the public on management methods, for greater professionalization, broader consensus and more consistent implementation of best management practices;
- to promote research, information and expertise exchange among beach workers, managers, scientists and the public, toward better understanding of natural resource management on sandy beaches.

E-mail and Web-based Communications, 'Reach' (Local and Global) of Organization

The Grunion Greeters always emphasized outreach by web connections for communication, education and data acquisition (www.Grunion.org). In turn, its offshoot, the Beach Ecology Coalition has several e-mail distribution groups, with hundreds of members, for announcements of important dates or meetings, plus news of interest to beach managers and workers. A web site for the Beach Ecology Coalition (www.BeachEcologyCoalition.org) allows public outreach global in scope. Beach managers in US Great Lakes, US Atlantic coast, Hawaii, Canada, Europe and India have sent queries about management issues.

CURRENT STATUS OF BEACH ECOLOGY COALITION

Meeting Agendas, Attendees

The Beach Ecology Coalition has already had important impacts for habitat monitoring on sandy beaches and for the involvement of a variety of stakeholders on management practices in California. The major outreach of the organization occurs at the semi-annual meetings. Agendas are distributed well in advance, listing the speakers, topics and locations. In addition to updates on the California grunion, animals ranging from beach-nesting birds to marine mammals, to surf fish, to great white sharks and sea turtles, are discussed for their ecological relationships to beaches and beachgoers. Native and invasive beach and dune plants are identified, with recommendations for protection of the natives and removal of the invasives. The ecological roles of beach wrack and benthic invertebrates are shared as part of the food web for the Pacific Flyway, resident and migrating shorebirds.

Presenters include ecologists, coastal geologists, environmental scientists, water quality managers, lifeguards, professors, beach managers, aquarists, surfers, park rangers, other Non-Governmental Organizations (NGOs) and informal educators. Each meeting includes a variety of subjects and no presentation lasts longer than 45 minutes. See Table 10.2 for examples of previous speakers.

Attendees include beach managers, resource managers from local, state and federal agencies, public works professionals, US Army Corps of Engineers, environmental consultants, lifeguards, beach maintenance staff, park rangers, scientists, informal educators, surfers, aquarists, botanists and other wildlife professionals, as well as some members of the general public.

Best Management Practices (BMPs)

Prior to the Beach Ecology Coalition formation, there was no cohesive structure for communication of best management practices for sandy beaches to those people who would implement them. Each municipality or agency that managed beaches took an independent approach, typically based on past experience at that location, or translated from management of terrestrial parks. Even though similar concerns and issues arise at different places along the coast, the fragmented nature of managing coastal areas prevents easy communication between people working at different locations, regarding routine maintenance or management activities. Since beach maintenance entails work by cities, states, counties, districts, private contractors and even the federal government, often on the same beach, few forums exist to coordinate all interested parties and stakeholders.

Table 10.2 Selected examples of speakers at Beach Ecology Coalition meetings. Each meeting typically has about five presentations highlighting scientific studies, restoration activities, maintenance issues, human safety issues and other beach activities

Speaker	Affiliation	Topic
Rosi Dagit	Marine Ecologist, Resource Conservation District, Santa Monica Mountains	'Fish, Fire, and Flows: Response to Wildfire'
Dr. Chris Lowe	Professor, California State University, Long Beach	'Recovery of white sharks off California and what that means to beach communities'
Lana Nguyen	Biologist, California State Parks	'Least Terns and Western Snowy Plovers at Huntington State Beach'
Susan Brodeur	Coastal Engineer, Orange County Parks	'Preparations for El Nino on Orange County Beaches'
Scott John	US Army Corps of Engineers	'Oceanside Harbor Dredge Operations and Sediment Placement'
Karina Johnston, Melodie Grubbs	The Bay Foundation	'Beach Restoration at Santa Monica'
Diane Alps, Bernardo Alps	Wildlife biologists, American Cetacean Society and Cabrillo Marine Aquarium	'An Amazing Year for (all sorts of) Whales in Southern California'
Dr. Bill Patzert	Oceanographer, JPL—NASA	'Sea Level Rise and California's Beaches: The Future Ain't What it Used to Be'
Susan MacLeod	Artist, Painter	'The Beauty of Beach Wrack: Seashore Wrack Art'
Dana Murray	Staff Scientist, Heal the Bay	'What the new Southern California Marine Protected Areas mean to beach managers and visitors'
Tedd Woodford	Owner, Coastal COMs	'Coastal Cameras and Advanced Software: Grunion by Night and People by Day'
Dr. Kimball Garrett	Curator of Ornithology, Natural History Museum of Los Angeles County	'Seabirds and Shorebirds and a Walk on the Beach'
Dennis Reed	Beaches and Parks Maintenance Manager, City of San Clemente	'Driving on the Beach: Developing Best Management Practices'

Table 10.2 Contd.

Table 10.2 (Contd.) Selected examples of speakers at Beach Ecology Coalition meetings. Each meeting typically has about five presentations highlighting scientific studies, restoration activities, maintenance issues, human safety issues and other beach activities

Speaker	Affiliation	Topic
Dave Revell	Coastal Geomorphologist, Revell Consulting	'Coastal Process, Sea Level Rise, and Adaptation'
Dr. Karen Martin	Professor, Pepperdine University	'Screening of new documentary film: *Surf, Sand and Silversides: The California Grunion*'
Michael Schaadt	Director, Cabrillo Marine Aquarium	'An Inside and Outside Look at Cabrillo Coastal Park and Cabrillo Marine Aquarium'
Jim Turner, Joe Delgado	Batallion Chief, Marine Operations, and Beach Maintenance Supervisor, City of Newport Beach	'An Overview of Beach Management in Newport Beach'
Dan Harding	Professional Photographer	'Tidepool and Underwater Photography with Macro Lenses'
Nick Steers	Captain, Ocean Lifeguards, Los Angeles County (retired)	'How Lifeguards and Beaches Have Changed in 30 Years'
Lester Thompson	Beach Maintenance Supervisor, City of Long Beach	'A Unique Educational Outreach Program for Local School Children'
Dr. Jenny Dugan	Director, Marine Science Institute, UC Santa Barbara	Studies of Beach Wrack, Shorebirds, and Invertebrates at Oceano Dunes'
Dr. Craig Shuman	Marine Biologist, California Department of Fish and Wildlife	'How California's Marine Life Protection Act May Affect Beach Management'
Bob Grove	Marine Biologist, Southern California Edison (electric power company)	'Grunion in Del Mar, and the Harbor Dredging Process'
Julianne Steers	Director of Husbandry, Ocean Institute	'Tour of new education facilities: dockside, shipboard, and classrooms'
Melissa Studer	Program Director, Grunion Greeters	'Grunion Greeting in San Diego and Beyond'
Tony Hoffman	Filming Coordinator, California State Parks and National Park District	'On Location on the Beach'

Once the Beach Ecology Coalition began, communication across and between entities vastly increased. Questions could be asked to the entire group and conversations started at a meeting could be continued later. One important result was the development of several Best Management Practices (BMP) documents. These can be used across agencies for supporting changes and improvements in different aspects of coastal management. Each BMP is peer-reviewed and vetted both by a scientist and by a beach manager for accuracy, clarity and applicability. Once accepted, the BMPs are posted on the Beach Ecology Coalition's web site and freely available to all for download. The first BMP was the Grunion Grooming

Protocol resulting from the grunion study described above. Later BMPs for diverse management issues, including beach bonfire rings, plant management and beach driving, were developed. These are brief and to the point, providing guidance without proscriptive messaging. Additional BMPs are in development.

SUCCESSES OF THE BEACH ECOLOGY COALITION

Inclusion of Stakeholders

Among the successes of this unique coalition is the ability to accommodate people from multiple organizations, backgrounds, experiences, expertise and viewpoints. The meetings provide a warm, collegial atmosphere, with peaceful, calm discussions and opportunities for questioning experts. This is a rare, non-confrontational situation for parties on different sides to reach out and share concerns on some difficult, controversial issues.

The semi-annual meetings and e-mails have allowed the Beach Ecology Coalition to address multiple issues with recognized experts. These include dune restoration projects, beach driving, oil spills, habitat preservation, sea level rise, sand replenishment methods, sea star wasting syndrome, wildfires and their effects on coastal habitats, marine mammals, harbor dredging impacts on beaches, the potential for drone surveys or web cameras on beaches and many other topics. Protection of beach animals and plants is addressed in ways that respect public access (Table 10.2).

The presence of professionals across hundreds of kilometers of coastline, with different areas of expertise, encourages networking. New ideas are shared at the meetings, solutions to specific problems are communicated and group discussion promotes creative approaches to new concerns. Discussions of funding struggles for projects inevitably inspires an attendee to offer insightful suggestions. Or if an intriguing topic arises, which may interest the group at a future meeting, an attendee may have someone appropriate in mind to contact for a presentation about it. Letters of support can be requested and promised from a variety of agencies in a short time. Collaborations can be initiated with direct connections to resources and contacts. All this conversation means that the actual agenda items for the meeting are only a portion of the information that is disseminated among the group. Attendance and interest for these meetings, plus the information provided, are increasing over time amongst the beach community.

Effects on Beach Management Practices

The purpose of the Beach Ecology Coalition's outreach and communication is to make beach management more consistent and more responsive to scientific information. This is a form of translational science, making changes in management, based on the most current, actionable knowledge. This mechanism provides a path for the community to share solutions, share expertise and adjust protocols quickly, supported by strong rationales.

Over the decade and a half of this organization, beach grooming has changed across southern California. State Parks no longer groom any of its beaches. Virtually all the public beaches in central and southern California employ the Grunion Grooming Protocol during a grunion season, a time that coincides with the heaviest beach use. Many agencies now routinely groom only above the high tide line, to let the lower beach face develop naturally, providing year-round habitat for beach wrack invertebrate communities that feed shorebirds and surf fishes, while also decreasing erosion. In spite of dire predictions to the contrary, the presence of natural wrack has not negatively impacted tourism or beach-going by the public. As the equipment operators understand the purpose of leaving areas to nature, they are able to respond directly and knowledgably to questions, unlike what the unfortunate driver faced with the initial grunion egg crisis back in 2001. In addition, the Grunion Greeter data were used for a stock assessment of this endemic species, data that could have been obtained no other way (Martin et al. 2020). Currently, increased protection is being evaluated for this species.

Due to increased communication, beach restoration projects have been re-defined. Instead of simple sand replenishment, pre- and post-monitoring for ecological features encourage protection of native plants, with avoidance of vehicle and foot traffic in particularly sensitive areas. Beaches in state parks show increased recognition of the conservation aspect of management in coastal, as well as terrestrial habitats. Docents and informal educators have greater understanding of the ecosystem functions and food webs on sandy beaches. Therefore, the public is more supportive of these efforts.

The increased communication and collaboration between government agencies, municipalities, scientists and NGOs have improved protection of living natural resources and enhanced the ecological knowledge of beach workers in general. Better training for drivers and heavy equipment operators on beaches boosts protection of both humans and natural habitats. The assessment of the effects of the Cosco Busan fuel spill in San Francisco Bay and the Refugio oil pipeline spill in Santa Barbara County included assistance and protocols developed by members of the Beach Ecology Coalition.

The effects of the Beach Ecology Coalition on management have generated clearer, more consistent guidelines for protecting living natural resources across agencies, such as the California Coastal Commission, the California Department of Fish and Wildlife, the National Marine Fisheries Service and the US Fish and Wildlife Service (King et al. 2018). These include changes to permitting requirements for coastal construction and other projects, increased attention to conservation when planning recreational events, or events likely to draw large crowds. It also brought more agreement between agencies on appropriate measures. The Beach Ecology Coalition has developed a large network of experts for information and advice on all kinds of topics. These frequent interactions create personal relationships that improve negotiations when difficult decisions must be made. The city of San Clemente, a major tourist destination in southern California, adopted an environmentally based policy for beach maintenance, which served as a template for one of the Beach Ecology Coalition's BMPs. A dune restoration project in the city of Santa Monica inspired similar efforts in the cities of Manhattan Beach and Malibu.

Connections made during the Beach Ecology Coalition meetings have supported funding for various projects, internships, restorations and research. Most importantly, there has been great expansion and amplification of the information provided, – e.g., increased outreach to the public via aquariums, environmental organizations (such as the Surfrider Foundation and the Audubon Society), plus increased connections between professionals and citizen scientists. Outreach materials in English and Spanish for the general public and a beautiful, laminated field guide for beach ecology, which meets the educational standards for K-12 teaching (Ericson et al. 2017) are available. This field guide was used for junior lifeguard training in Los Angeles County and in the summer Teachers' Institute of the Monterey Bay Aquarium, as well as by many docents along the coast.

Lessons Learned and Recommendations for the Future

As the Beach Ecology Coalition has emphasized an informal approach, as opposed to a prescriptive legislative approach, changes may be implemented quickly and smoothly. The peer-reviewed BMPs inform policies, allowing managers to explain their actions to the public and to their own governing bodies. However, when there is resistance, the limitations of this approach become evident, as best practices may not be implemented or enforced consistently. In addition, the high turnover of staff in positions directly working on beaches means that maintaining contact with active managers and equipment operators is challenging. While some of the recommendations of the organization have been incorporated into the permitting process by the California Coastal Commission, other recommendations are applied less consistently. Increased public pressure by individuals and NGOs may be important in the future for significant, permanent policies.

The up-to-date, open access to vetted information and BMPs, which have been reviewed by experienced managers, are the key elements to the credibility of this organization. Up to the present time, small sources of funds have provided open access to the web site and meetings at no cost to participants. This is ideal from an information–dissemination viewpoint, but may not be sustainable in the long run. In the future, the Beach Ecology Coalition will need to wrestle with the issues of reliable funding. Perhaps, it will take the form of charging for meetings and/or memberships or finding a sponsoring foundation, willing to commit support. This may mean that the Beach Ecology Coalition will need to affiliate under the umbrella of a larger organization, in order to survive.

The organizational structure will also have to be reviewed. The five founding members of the Board of Directors are still active, willing and able to serve, but efforts to recruit additional Directors have been ineffective. There is an Advisory Board that oversees the actions of the Directors, but its membership is mainly engaged via e-mail and telephone communications, or in conversation during the general meetings. Board members serve as volunteers and there is no funding for separate meetings of the Advisory Board, although this type of more focused meeting would be desirable for planning and problem solving.

Looking to the future, next steps for the Beach Ecology Coalition ideally will include increased outreach and training activities for beach ecology. The intent is to promote greater beach ecology awareness. One way to do this is to have monitoring of beaches by citizen scientists, similar to the efforts of the Grunion Greeters that started this organization. A pilot project was developed in the past five years with external funds, but funding to formalize implementation of the program is lacking. Parts of that program have been used at some universities and state parks, but there is neither consistent implementation, nor a central database for the results.

Alternatively, beach workers themselves could keep track of some basic ecological information, similar to the monitoring efforts developed for and by the citizen scientists. These observations could be used to develop baseline data for each beach or coastal space, along the California coast. Staff, who are on the beaches every day, could become part of a frontline network, to recognize seasonal and long-term changes in a beach's physical and biological components. Part of the difficulty in managing sandy beaches is the sheer number of threats facing them (Zhang et al. 2004, Defeo et al. 2009) and the other part is the highly mutable nature of beaches themselves (McLachlan and Brown 2006, Schlacher et al. 2014b, Vitousek et al. 2017). In any case, the need for monitoring during this time of climate upheaval is clear, particularly along the coast where great loss of habitats is predicted (Vitousek et al. 2017).

In an ideal world, improved training for beach managers and beach workers would move towards certification, to ensure a basic level of knowledge about how beaches evolve seasonally and over time (Yates et al. 2009, Revell et al. 2011, Schooler et al. 2017), to understand all the ecological goods and services they provide. The development of a curriculum for this training course, and possible video delivery, are under consideration, along with the development of a published training manual.

Issues that demand future attention include sediment management, particularly sand replenishment and winter berm building. Both are traditional methods to address shoreline erosion, but both have unwanted ecological impacts on the sandy beach food web (Leonard et al. 1990, Peterson et al. 2006) and coastal fishes (Lawrenz-Miller 1991, Manning et al. 2013). Ecological effects of shoreline armoring (Fletcher et al. 1997, Dugan et al. 2008, 2018) present strong arguments in favor of soft solutions to coastal erosion (Gittman et al. 2015), but some difficult choices will have to be made (Griggs et al. 2005; Schoeman et al. 2014). Of course, climate change and sea level rise will lead to new concerns (Feagin et al. 2005, Flick and Ewing 2009) and habitat shifts (Martin et al. 2013, Martin 2015).

This chapter is a reminder that beach management is a global concern. Many of the issues addressed are important on any sandy beach, albeit different species and cultural approaches. The many requests for information received by the Beach Ecology Coalition from other countries, and the efforts of scientists around the world, suggest that a global network of coastal organizations could be affiliated with the Beach Ecology Coalition. Clearly this would require a serious effort and substantial funding to initiate, but the benefits in coastal resilience would be immediate and would justify the work.

Finally, it is important to recognize the importance of individuals to the functioning of organizations. The initial collaborative spirit that started the Beach Ecology Coalition has continued to the present day. Each of the people from the founding group was essential to the organization and to its ability to attract participation of such a wide range of people that work or play on beaches. The presence of respected scientists from Scripps Institution of Oceanography, marine biologists from state and federal resource agencies, coupled with the support of the city government of San Diego and the cooperative relationship between environmentalists and beach maintenance staff, brought credibility and gravitas from the very start. The frequent and accessible web communications, plus e-mail messages, provide a sense of community outside of the meetings. The involvement of aquariums, NGOs and other informal educators amplifies and publicizes the information, far beyond the scope of those directly involved in the Beach Ecology Coalition. The ability of this organization to survive and thrive on a shoestring budget is testament both to the dedication of the volunteers on the Board and to the desire for this targeted information among all those involved. It is hoped that this can provide a model for other locations, to develop similar programs, involving local stakeholders and scientists in guiding coastal planning and management.

Acknowledgements

We are thankful for funding from Coastal America Foundation, US Fish & Wildlife Service, 'Connecting People with Nature,' California Coastal Commission Whale Tail Program WT-13-22, National Science Foundation DBI 1062721, National Science Foundation, REU-1560352, USC Sea Grant College—Urban Oceans Program NOAA —NA14OAR4170089/Subaward 6094463, National Marine Fisheries Service, Southwest Region, Habitat Conservation Division Contract 8-819, National Geographic Society CRE 8105-07, Pepperdine University, BeachTech, Edison International and many private donors. We are grateful to thousands of Grunion Greeters for their long walks on moonlit beaches.

REFERENCES

Bird, E.C.F. 1996. Beach Management. John Wiley and Sons, New York.

Defeo, O., A. McLachlan, D.S. Schoeman, T.A. Schlacher, J. Dugan, A. Jones, et al. 2009. Threats to sandy beach ecosystems: a review. Estuar. Coast. Shelf Sci. Estuar. Coast. Shelf Sci. 81: 1–12.

Dugan, J.E., D.M. Hubbard, M. McCrary and M. Pierson. 2003. The response of macrofauna communities and shorebirds to macrophyte wrack subsidies on exposed sandy beaches of southern California. Estuar. Coast. Shelf Sci. 58S: 25–40.

Dugan, J.E., D.M. Hubbard, I.F. Rodil, D.L. Revell and S. Schroeter. 2008. Ecological effects of coastal armoring on sandy beaches. Mar. Ecol. 29: 160–170.

Dugan J.E. and D.M. Hubbard. 2010. Loss of coastal strand habitat in southern California: the role of beach grooming. Estuaries Coast. 33: 67–77.

Dugan, J.E., O. Defeo, E. Jaramillo, A.R. Jones, M. Lastra, R. Nel, et al. 2010. Give beach ecosystems their day in the sun. Science 329: 1146.

Dugan, J.E., D.M. Hubbard, H.M. Page and J. Schimel. 2011. Marine macrophyte wrack inputs and dissolved nutrients in beach sands. Estuaries Coast. 34: 839–850.

Dugan, J.E., K. Emery, M. Alber, C. Alexander, J. Beyers, A. Gehman, et al. 2018. Generalizing ecological effects of shoreline armoring across soft sediment environments. Estuaries Coast. 41: 180–196.

Ericson, D.N., K. Martin and L.T. Groves. 2017. Pacific Coast Sea Shores. Nature Unfolding, Manta Publications, Ventura, California, USA.

Feagin, R.A., D.J. Sherman and W.E. Grant. 2005. Coastal erosion, global sea-level rise, and the loss of sand dune plant habitats. Front. Ecol. Environ. 3(7): 359–364.

Fletcher, C.H., R.A. Mullane and B.M. Richmond. 1997. Beach loss along armored shorelines in Oahu, Hawaiian Islands. J. Coast. Res. 13: 209–215.

Flick, R.E. and L.C. Ewing. 2009. Sand volume needs of southern California beaches as a function of future sea-level rise rates. Shore Beach 7: 36–45.

Gittman, R.K., F.J. Fodrie, A.M. Popowich, D.A. Keller, J.F. Bruno, C.A. Currin, et al. 2015. Engineering away our natural defenses: an analysis of shoreline hardening in the US. Front. Ecol. Environ. 13(6): 301–307.

Griem J.N. and K.L.M. Martin. 2000. Wave action: the environmental trigger for hatching in the California grunion, *Leuresthes tenuis* (Teleostei: Atherinopsidae). Mar. Biol. 137: 177–181.

Griggs, G.B., K.B. Patsch and L.E. Savoy. 2005. Living with the Changing California Coast. University of California Press, Berkeley, California, USA.

Harris, L., E.E. Campbell, R. Nel and D. Schoeman. 2014. Rich diversity, strong endemism, but poor protection: addressing the neglect of sandy beach ecosystems in coastal conservation planning. Divers. Distrib. 2014: 1–16.

King, P.G., C. Nelsen, J.E. Dugan, D.M. Hubbard and K.L. Martin. 2018. Valuing beach ecosystems in an age of retreat. Shore Beach 86: 1–15.

Lawrenz-Miller, S. 1991. Grunion spawning versus beach nourishment: nursery or burial ground? Coast. Zone 1991(3): 2197–2208.

Leonard, L., K. Dixon and O. Pilkey. 1990. A comparison of beach replenishment on the U.S. Atlantic, Pacific and Gulf coasts. J. Coast. Res. 6: 127–140.

Llewellyn, P.J. and S.E. Shackley. 1996. The effects of mechanical beach-cleaning on invertebrate populations. Br. Wildl. 7: 147–155.

Manning L.M., C.H Peterson and S.R. Fegley. 2013. Degradation of surf-fish foraging habitat driven by persistent sedimentological modifications caused by beach nourishment. Bull. Mar. Sci. 89: 83–106.

Martin K., T. Speer-Blank, R. Pommerening, J. Flannery and K. Carpenter. 2006. Does beach grooming harm grunion eggs? Shore Beach 74: 17–22.

Martin K., A. Staines, M. Studer, C. Stivers, C. Moravek, P. Johnson, et al. 2007. Grunion Greeters in California: Beach-spawning fish, coastal stewardship, beach management and ecotourism. pp. 73–86. *In:* M. Lück, A. Gräupl, J. Auyong, M.L. Miller and M.B. Orams [eds]. Proceedings of the 5th International Coastal & Marine Tourism Congress: Balancing Marine Tourism, Development and Sustainability. New Zealand Tourism Research Institute, Auckland, New Zealand.

Martin K.L.M., C.L. Moravek and J.A. Flannery. 2009. Embryonic staging series for the beach spawning, terrestrially incubating California grunion *Leuresthes tenuis* with comparisons to other Atherinomorpha. J. Fish Biol. 75: 17–38.

Martin K.L.M, C.L. Moravek, A.D. Martin and R.D. Martin. 2011. Community based monitoring improves management of essential fish habitat for beach-spawning California Grunion. pp. 65–72. *In:* A. Bayed [ed.]. Sandy Beaches and Coastal Zone Management—Proceedings of the Fifth International Symposium on Sandy Beaches. Travaux de l' Institut Scientifique, Série Générale 6, Rabat, Morocco.

Martin K.L.M., K.A. Hieb, and D.A. Roberts. 2013. A southern California icon surfs north: local ecotype of California grunion *Leuresthes tenuis* (Atherinopsidae) revealed by multiple approaches during temporary habitat expansion into San Francisco Bay. Copeia 2013(4): 729–739.

Martin K.L.M. 2015. Beach-Spawning Fishes: Reproduction in an Endangered Ecosystem. Taylor & Francis Group, CRC Press, Oxford, UK.

Martin, K.L.M., E.A. Pierce, V.V. Quach and M. Studer. 2020. Population trends of California's beach-spawning grunion *Leuresthes tenuis* monitored by citizen scientists. ICES J. Mar. Sci. 77(6): 2226–2233.

McLachlan, A. and A.C. Brown. 2006. The Ecology of Sandy Shores, 2nd Ed. Academic Press, San Diego, California, USA.

[MPAMAP] Marine Protected Area Monitoring Action Plan. 2018. California Department of Fish and Wildlife and California Ocean Protection Council, Sacramento, California, USA.

Nelsen, C. 2012. Collecting and using economic information to guide the management of coastal recreational resources in California. Ph.D. Dissertation, University of California, Los Angeles, California, USA.

Orme, A.R., G.B. Griggs, D.L. Revell, J.G. Zoulas, C.C. Grandy and H. Koo. 2011. Beach changes along the southern California coast during the 20th century: a comparison of natural and human forcing factors. Shore Beach 79: 38–50.

Peterson, C.H., M.J Bishop, G.A. Johnson, L.M. D'Anna and L. Manning. 2006. Exploiting beach filling as an unaffordable experiment: benthic intertidal impacts propagating up to shorebirds. J. Exp. Mar. Biol. Ecol. 338: 205–221.

Pilkey, Jr., O.H. and J.A.G. Cooper. 2014. The Last Beach. Duke University Press, Durham, North Carolina, USA.

Revell, D.L., J.E. Dugan and D.M. Hubbard. 2011. Physical and ecological responses of sandy beaches to the 1997–98 El Niño. J. Coast. Res. 27: 718–730.

Robbins E. 2006. Essential Fish Habitat in Santa Monica Bay, San Pedro Bay, and San Diego Bay: A Reference Guide for Managers. MS Thesis, Duke University, Durham, North Carolina, USA.

Sandrozinski, A. 2013. California grunion. Status of the fisheries report, an update through 2011. California Department of Fish & Wildlife, Sacramento, California, USA.

Schlacher, T.A., J. Dugan, D.S. Schoeman, M. Lastra, A. Jones, F. Scapini, et al. 2007. Sandy beaches at the brink. Divers. Distrib. 13(5): 556–560.

Schlacher, T.S., A.R. Jones, J.E. Dugan, M. Weston, L. Harris, D.S. Schoeman, et al. 2014a. Open-coast sandy beaches and coastal dunes. pp. 37–94. *In:* B. Maslo and J. Lockwood [eds]. Coastal Conservation. Series in Conservation Biology. Cambridge University Press, Cambridge, UK.

Schlacher, T.S., D.S. Schoeman, A.R. Jones, J.E. Dugan, D.M. Hubbard, O. Defeo, et al. 2014b. Metrics to assess ecological condition, change, and impacts in sandy beach ecosystems. J. Environ. Manage. 144: 322–335.

Schoeman D.S., T.A. Schlacher and O. Defeo. 2014. Climate-change impacts on sandy-beach biota: crossing a line in the sand. Glob. Chang. Biol. 20: 2383–2392.

Schooler, N.K., J.E. Dugan, D.M. Hubbard and D. Straughan. 2017. Local scale processes drive long-term change in biodiversity of sandy beach ecosystems. Ecol. Evol. 7: 4822–4834.

Suursaar, Ü., K. Torn, G. Martin, K. Herkül and T. Kullas. 2014. Formation and species composition of stormcast beach wrack in the Gulf of Riga, Baltic Sea. Oceanologia. 56: 673–695. 10.5697/oc.56-4.673.

Vitousek. S., P.L. Barnard, P. Limber, L. Erikson and B. Cole. 2017. A model integrating longshore and cross-shore processes for predicting long-term shoreline response to climate change. J. Geophys. Res. Earth Surf. 122: 782–806.

Yates, M.L., R.T. Guza, W.C. O'Reilly and R.J. Seymour. 2009. Overview of seasonal sand level changes on southern California beaches. Shore Beach 77: 39–46.

Zhang, K., B.C. Douglas and S.P. Leatherman. 2004. Global warming and coastal erosion. Clim. Change 64: 41–58.

Chapter 11

Sandy Beach Management and Conservation: The Integration of Economic, Social and Ecological Values

Iván F. Rodil[1*], Linda R. Harris[2], Serena Lucrezi[3] and Carlo Cerrano[4]

[1]Tvärminne Zoological Station, University of Helsinki, Finland. *and* Departamento de Biología, Instituto Universitario de Investigación Marina (INMAR), University of Cádiz, Spain. (ivan.rodil@helsinki.fi, ivan.franco@uca.es).

[2]Department of Zoology, Institute for Coastal and Marine Research, Nelson Mandela University, Port Elizabeth, 6001, South Africa (harris.linda.r@gmail.com).

[3]TREES—Tourism Research in Economics, Environs and Society, North-West University, Potchefstroom, South Africa (23952997@nwu.ac.za).

[4]Department of Life and Environmental Sciences (DiSVA), Polytechnic University of Marche, UO CoNISMa, Ancona, Italy (c.cerrano@staff.univpm.it).

INTRODUCTION

The Value of Intact Sandy Shores

Sandy beaches and dunes are iconic environments that dominate the world's open-ocean shorelines. They attract higher densities of human settlement compared

*Corresponding author: ivan.franco@uca.es

All authors contributed equally to this work

to areas further inland (Small and Nicholls 2003) because the attractive natural environment facilitates activities such as recreation and tourism, subsistence harvesting, fishing and transportation (see Schlacher et al. 2008). Therefore, beaches have been particularly linked to human history and activities due to the provision of ecosystem services from which people obtain many benefits. As dynamic natural systems, beaches are home to a rich diversity of organisms that perform numerous supporting and regulating services, e.g., water filtration and nutrient cycling (McLachlan 1982, Lastra et al. 2008, Dugan et al. 2011, Rodil et al. 2019); as social systems, they provide aesthetic and recreational assets associated with enhancing human health and wellbeing, and underpinning cultural or personal identities (Taylor 2007, Ashbullby et al. 2013, Bell et al. 2015, Voyer et al. 2015); and as economic systems, they provide income-generating services to people that can strengthen local and national economies (Stronge 2005, Houston 2013).

Figure 11.1 A natural beach environment (left panel, Byron Bay, Australia) versus a heavily urbanized beach after a typical Atlantic storm event (right panel, Porto, Portugal). Photo credits: IFR.

Importantly, delivery of beach ecosystem services is largely related to the ecological condition of this environment. Despite their recognized social relevance, the value of some of these services is sometimes difficult to estimate in economic terms, especially where the benefits are intangible. In addition, beaches have been recognized as ecosystems only relatively recently (McLachlan and Erasmus 1983), such that their management has been historically built on a poor understanding of their ecological processes and value (e.g., La Cock and Burkinshaw 1996). In consequence, management responses are frequently inappropriate, focussing on limited aspects of beaches and often neglecting the ecological value of beaches and critical ecological processes that connect systems across the ecotone, eventually causing cumulative impacts to the shore. As human population density and economic activity along the sandy shoreline increases, cumulative pressures also increase (Harris et al. 2015), on which climate-change stressors are superimposed, e.g., sea-level rise. This results in coastal squeeze, inevitably exacerbated by extreme weather and wave events, and poorly managed human use of beaches (Fig. 11.1).

In this chapter, the environmental, social and economic values of sandy shores, and how these values can be best integrated into conservation and management strategies for sustainable development that consider the social-ecological system as a whole are explored.

Sandy Beaches as Ecosystems

To the occasional visitor, beaches may simply appear as coastal deserts empty of life. However, sandy beaches support a great variety of mostly small (< 1 cm) living organisms that have high rates of endemism (Harris et al. 2014a), and are usually found inconspicuously buried in the sand. Inhabitants of beaches include representatives from all major trophic groups, such as decomposers (bacteria, fungi, meiofauna), primary producers (phytoplankton and microphytobenthos), and primary and secondary consumers, including filter feeders (e.g., clams), scavengers (e.g., ghost crabs), and predators (e.g., fish and birds) (see McLachlan and Defeo 2018). The ecological role of these organisms is important for the functioning of the beach ecosystem, and for connectivity with other marine and terrestrial habitats (e.g., Le Gouvello et al. 2017a, b). In addition to the resident species, some migratory species rely on beaches for only a small part of their lives, often to fulfil important life-history stages. For example, sea turtles and horseshoe crabs need beaches to lay their eggs in sandy nests, and some marine mammals (e.g., sea lions) use beaches to rest and during mating seasons (Brockmann 1990, van Buskirk and Crowder 1994, Augé et al. 2012). Shorebirds, such as terns and plovers, require beaches and dunes to nest and lay eggs, and migratory birds use beaches as roosting points, gathering annually in large numbers and providing attractive sites for birdwatching (Steven et al. 2014). All these animals are among the most iconic on sandy shores, many of which are also threatened species (Fig. 11.2).

Figure 11.2 Species with a high public profile (from top to bottom and clockwise: sea turtles, sea lions, seagulls and chinstrap penguins) use beaches for many different purposes (e.g., breeding, feeding or resting) during sometimes short but important life-history stages. Photo credits: LRH (top left) and IFR.

As unique ecotones, sandy beaches influence ecological processes and food webs across the contiguous terrestrial, estuarine and marine realms, well beyond the immediate land-sea interface (McLachlan and Defeo 2018). Notably, they link vegetated dunes with the truly marine surf zone through a constant interchange of nutrients and sand. In terms of nutrients, the beach sediment body plays a fundamental role as a filtering system for coastal water that mineralizes organic matter and recycles nutrients back to the sea (e.g., McLachlan 1982, Dugan et al. 2011). Groundwater also flows from the land through beach sediment into the sea, fuelling phytoplankton communities in the surf (Campbell and Bate 1991) that are important for primary productivity on beaches, particularly by accumulations of surf diatoms (e.g., Campbell and Bate 1988). In turn, the rich biodiversity that thrives on beaches worldwide is fuelled mainly by inputs from the ocean. The sea delivers plankton to the beaches' consumers; washes wrack (beach-cast macroalgae) and carrion (dead organisms) ashore, which are recycled by microorganisms and eaten by scavengers (Koop and Griffiths 1982, Koop et al. 1982, Schlacher et al. 2013). Marine species also provide nutrients to beaches, e.g., nutrients from horseshoe crab and sea turtle eggs that are exported to both terrestrial and marine food webs (Botton and Loveland 2011, Le Gouvello et al. 2017a, b). Some beaches naturally accumulate considerable amounts of wrack on the upper shore near the dunes (Fig. 11.3). This material may be unsightly, and even a nuisance to the casual beachgoer, but such organic material is a vital food source and habitat to several species (Dugan et al. 2003; Lastra et al. 2008), and is important for carbon cycling (Coupland et al. 2007; Rodil et al. 2019). Sandy beaches are therefore marine metabolic hotspots with exports to both land and sea. They consequently play an important role in some coastal recreational, subsistence and small-scale fisheries; for example, by providing important nursery and recruitment areas for fish in the surf zone (Able et al. 2013), and providing rich supplies of nutrients and prey items in the form of resident macrofauna (Lasiak 1983), some of which species form the basis for fisheries themselves (McLachlan and Defeo 2018).

Sand moves between the dunes, beach and sand bars in the surf zone (collectively called the littoral active zone; Tinley 1985) through dynamic exchanges from erosion and accretion, especially during storm-calm cycles. Their natural fluctuations in morphology in response to wind, waves, tides and storms protect the hinterland against flooding and erosion as they absorb the impacts from heightened-wave energy (Nordstrom 2000, Feagin et al. 2005). Under periods of prolonged erosion (e.g., in response to sea-level rise), sandy beaches and dunes respond by migrating landward *sensu* Bruun's rule (Bruun 1962). By maintaining form and function, they continue to reduce impacts from storms by acting like a buffer along the coastal edge and absorbing and dissipating the energy of breaking waves, either seaward on sand bars or on the beach itself. In this way, sandy beaches and dunes have a key role to play in climate change adaptation (i.e., through coastal protection services). The wider a beach or dune system is, and the more space between the sea and any developed or populated areas, the more effective the system will be at reducing the impacts of coastal hazards. Sandy beaches and coastal dunes are also natural climate regulators, and recent research emphasizes the important role of sandy beaches and vegetated dunes for climate-change mitigation given their capacity to sequester and store carbon (Beaumont et al. 2014, Drius et al. 2016).

Figure 11.3 From top to down: Accumulations of beach-cast macroalgae (i.e., wrack) are a key food source (top panel) for cryptic sandhoppers (middle panel) that aggregate around and consume the algal blades, burrowing under and around the wrack until it dries out (bottom panel). Photo credits: F. Barreiro (top), LRH (middle) and IFR (bottom).

The Social and Cultural Values of Sandy Beaches

Sandy beaches and dunes have been inextricably linked to the evolution, survival and wellbeing of people in many ways (Erlandson and Fitzpatrick 2006, Everard et al. 2010, Barbier et al. 2011). Primarily, they are highly valued habitats for human settlements and protect urban developments along coastlines (Schlacher et al. 2008). They also provide raw materials (e.g., minerals, timber), water and food (Barbier et al. 2011). Beaches and dunes ultimately offer a variety of socio-cultural services and benefits to humankind (Table 11.1). Beach-based tourism and recreation are among the top material services offered by beaches, constituting a billion-dollar industry in several countries (Rolfe and Gregg 2012, Houston 2013). Beach-based recreation embodies a lifestyle to many coastal residents, and contributes to human wellbeing through aesthetic qualities, contact with nature, and physical or restful activities, from sightseeing and meditation to swimming and surfing (Everard et al. 2010, Voyer et al. 2015, Domínguez-Tejo et al. 2018).

Table 11.1 Summary of socio-cultural values and benefits ascribed to sandy beaches and dunes

Values	Underlying themes
Recreation and tourism	Beach-based and water-based activities Sports Physical and restful activities Contact with nature Encounters with wildlife
Cultural heritage	Enriching culture through traditional events Maintaining cultural traditions Passing cultural traditions to others Cultural identity Unique coastal cultures Surfing subculture
Historical heritage	Archaeological remains Events in history (e.g., battles)
Aesthetic	Beauty and attractiveness of beachscapes Contact with nature Moving, emotional coastal sceneries
Spiritual and religious	Religious practices by indigenous peoples Praying and meditation Ancestral connections Sacred places Surfing as a religion
Wellbeing	Physical and mental health, fitness Personal challenges Personal fulfilment Escapism Solitude The beach as a place for meditation and reflection The beach as a mood changer The beach as a place offering a desirable lifestyle

Values	**Underlying themes**
Sense of place and identity	Belonging (people belonging to the beach, the beach belonging to people) Emotional attachment The beach as a home The beach as a familiar place The beach as a national symbol
Artistic	Inspiration for the arts (e.g., painting, poetry, film, photography), architecture and folklore
Personal and social relations	A place of gathering for hobbyists, nature lovers and sportspeople Strengthening friendship and family bonds Personal growth—the rite of passage Reconnecting with one's own past (the beach as a place of memories) Community development
Education and research (i.e., learning and ingenuity)	Using the beach for schooling Providing examples for teaching concepts like the 'succession theory' Providing information on the ocean and other ecosystems (e.g., through beach cast) Learning lessons in life Acquiring wisdom Acquiring and refining skills
Legacy (i.e., bequest) and stewardship	The beach for present and future generations The beach as a place to help others (e.g., lifesaving) The beach as a place for all (public use rights) The beach as a place to help the environment (e.g., litter clean-ups, monitoring, activism)

Sources: Tunstall and Penning-Roswell 1998, La Mar 2006, Taylor 2007, Thompson 2007, West-Newman 2008, Ford 2009, Collins and Kearns 2010, Everard et al. 2010, Barbier et al. 2011, Kearns and Collins 2012, Ashbullby et al. 2013, Collins and Kearns 2013, Wakita et al. 2014, Lawson 2016, Vivian and Schlacher 2015, Voyer et al. 2015, Lucrezi and Van der Walt 2016, Domínguez-Tejo et al. 2018, Rodríguez-Revelo et al. 2018, Liu et al. 2019, Lucrezi et al. 2019.

People ascribe a variety of values to beaches (Taylor 2007, Everard et al. 2010, Lawson 2016). Some examples include: aesthetic and artistic values expressed through the representations of beaches in photography and painting (Ford 2009, Everard et al. 2010); social values represented by the use of beaches as theatres for the development of social relations (e.g., amongst hobbyists and family members) (Taylor 2007, Everard et al. 2010); and spiritual and cultural values, expressed in the maintenance of traditional bonds between indigenous people and the beach (Collins and Kearns 2010, 2013). Beaches can be perceived as environments that, through their complexity, stimulate curiosity, learning and ingenuity (Voyer et al. 2015, Domínguez-Tejo et al. 2018; Rodríguez-Revelo et al. 2018). Studies have ultimately identified the bequest and stewardship values attributed to sandy beaches as environments that ought to be preserved for the intrinsic qualities they possess and for the sake of future generations (Voyer et al. 2015, Lucrezi et al. 2019).

Threats and Stressors to Sandy Shores

Given their aesthetic appeal, numerous benefits and diverse values for people, beaches are consequently highly used ecosystems. As a result, sandy beaches and dunes are undergoing striking anthropogenic modifications, with habitat destruction and fragmentation increasing at an unprecedented rate around the world (Schlacher et al. 2008, Dugan et al. 2010). Key pressures that are escalating and posing a major risk for the shore and coastal sand dunes include direct impacts from coastal development and some recreational activities (e.g., Nordstrom 2000, Feagin et al. 2005, Peterson and Bishop 2005, Dugan et al. 2008, Schlacher et al. 2008). Other important anthropogenic pressures occurring along shorelines worldwide relate to beach grooming, mining, fishing, pollution, and biological invasions (see Schlacher et al. 2008 and Defeo et al. 2009 for detailed reviews).

Coastal development is a particular cause of concern for beach ecosystems because it is largely an irreversible stressor (Harris et al. 2014b). The charm of beachfront properties raises real estate values, making them attractive options for investors, resulting in increasing urbanization and ribbon development that can invade, and in some cases destroy entire beach and dune systems. Once a coastal area has become highly developed, and the width of the beaches has become considerably narrower, there is a greater threat from coastal hazards because the protective capacity of the natural system is reduced (Fig. 11.1). In addition, the capacity of coastal habitats (including sandy beaches and vegetated dunes) to sequester and store carbon is significantly reduced in areas where they are lost to development, with consequent ecological and economic losses (Beaumont et al. 2014). Approximately one-third of the world's population resides within 100 km of the coast, and almost 40% live in coastal cities; further, coastal agglomeration and increasing urbanization rates are expected to continue in the near future, generating changes in land use never seen before (Barragán and de Andrés 2015). Therefore, strategic coastal planning combined with plans to manage ecosystems and biodiversity (Barragán and de Andrés 2015) will be indispensable for sustainably developing coastal habitats such as sandy beaches and dunes.

Keeping the littoral active zone intact is important for maintaining the ecological condition of beaches in the future, especially given the observed and predicted impacts from climate change (IPCC 2014). Sea-level rise and storm surges have increased the levels of coastal erosion and will continue to do so in the near future (Bender et al. 2010). Thus, beaches that have been developed right up to the shore cannot respond as they normally would, and instead of migrating inland, they are trapped in a coastal squeeze between the development and rising seas where they, and their associated biodiversity, will ultimately disappear (e.g., Dugan et al. 2008, Hubbard et al. 2014). Episodes associated with climate change, such as increasing hurricane landfalls (Bender et al. 2010) and catastrophic episodic events such as earthquakes and associated tsunamis (Jaramillo et al. 2012) exacerbate the impacts to beaches (Fig. 11.4). Intact sandy shores have a natural capacity to adapt and to recover from such extreme events, while still preserving their structure and function over geological timescales (Berry et al. 2014). However, engineering solutions to defend coastal infrastructure heighten impacts to beaches, resulting in higher long-term costs (i.e., to address coastal squeeze) compared to more ecologically friendly solutions such as scientifically determined setback zones that proactively avoid impacts and damage (Dugan et al. 2008, Jaramillo et al. 2012, Rodil et al. 2015).

Figure 11.4 Shorelines around the world are routinely affected by winds and storms, and occasionally by extreme events such as hurricanes or earthquakes (e.g., tsunami warning in the Biobío region, Chile in the left panel) that typically result in engineering solutions (e.g., seawall at Arroyo Quemado, California in the right panel). Photo credits: IFR.

How humans manage coastal systems can therefore alter the natural trajectory of both beach and dune integrity and their ecological condition, and the concomitant delivery of services and benefits for people (Peterson and Bishop 2005, Schlacher et al. 2008, Rodil et al. 2015; see also below). Anthropogenic pressures and their multiple interactions pose a long-term threat to beaches and the restoration efforts that are likely to follow (e.g., nourishment, filling and armouring), with significant impacts to the ecological value of the beach. Consequently, there have been increasing calls for implementing ecosystem-based approaches to beach management and coastal planning. If the biodiversity, attributes and benefits that are valued today are to persist into the future, coastal management and conservation need to integrate economic, social and ecological values better, and imperatively, facilitate maintenance of core ecological processes across the ecotone.

A CALL TO CHANGE THE CURRENT APPROACH TO BEACH MANAGEMENT AND CONSERVATION

Convenient Coastal Jurisdictions are Problematic for Beach Management and Conservation

Contemporary coastal governance is generally such that each realm (terrestrial, estuarine and marine) at the land-sea interface is managed or conserved under separate legal instruments and by different organs of state. For example, the jurisdiction of local authorities often extends to the high-water mark, with their mandate to manage all land-based activities; whereas national government is often responsible for all sea-based activities below the high-water mark. Respectively, different tools might be used for land-use planning and Marine Spatial Planning (MSP), usually undertaken as mutually exclusive planning processes. As much as Integrated Coastal Zone Management (ICZM) intends to bring these processes together, the relationship between ICZM, land-use planning and MSP is often unclear. The result is that ICZM tends to relate more to managing local user-user or user-environment conflicts in the short term rather than providing spatial prioritization and planning at ecologically relevant scales to guide conservation and management for the long term (Harris 2012). The typical focus of ICZM is sediment (mostly erosion) management and property defence, and the recreational use of beaches, but has largely ignored the ecological values of beaches (Schlacher et al. 2008). Consequently, many beach-management practices have often resulted in ecologically harmful interventions (see next).

For some coastal ecosystems, the discrete terrestrial, estuarine and marine jurisdictions are appropriate, e.g., for conserving and managing dunes, estuaries and rocky shores, respectively. However, beach management and conservation rely strongly on considering the entire tripartite ecotonal interface comprising the linked dune, shore and surf (i.e., littoral active zone) as a single geomorphic unit. Furthermore, estuaries play a key role in supplying sediment and nutrients to the coast, in turn, maintaining beaches (and other river-influenced marine ecosystem types) and fuelling coastal food webs. One of the most important roles beaches play is as transitional systems, linking terrestrial, estuarine and marine realms through connecting ecological processes, as discussed above. Consequently, the greatest hindrance to beach conservation and management is that the ecotonal interface between land and sea is split by our convenient divides of ocean, estuaries and land. This artificial split in coastal governance (usually at the high-water mark) and lack of alignment in management priorities across the ecotone breaks down the capacity of beaches to fulfil their fundamental connecting role. It also reduces their adaptive capacity to stressors from climate change. In turn, this enhances our vulnerability to impacts from sea-level rise and the increase of frequency and intensity of extreme storms, as well as concomitant accelerated erosion degrading beaches that are key assets for recreation and tourism.

A Legacy of Mismanagement

The current state of beach management practices (including inappropriate jurisdictions) has come about largely because, as noted earlier, beaches were only recently recognized as ecosystems (McLachlan and Erasmus 1983) and despite several decades of research since (Nel et al. 2014), are still often overlooked (e.g., Dugan et al. 2010). There are several examples of interventions where managers have made decisions without fully considering the affected ecological processes, which in turn have negatively impacted sandy shores. These decisions are made in the absence of information because either they lacked it historically or because of the poor public profile beaches carry such that the available information is overlooked. For example, a common management action currently implemented on many beaches relates to removing wrack from beaches that are popular with tourists. Typical mechanical beach grooming actions have significant consequences to wrack-associated consumers (i.e., invertebrates), that rapidly propagate up through the trophic levels, also affecting ecological processes related to the marine-terrestrial transfer of nutrients and energy (e.g., Dugan et al. 2003) and the marine-atmospheric transfer of carbon dioxide (Lastra et al. 2018, Rodil et al. 2019). Another example is large-scale dune stabilization with invasive alien trees that has had significant effects on beaches, reducing flows of sand to the beach, accelerating erosion, enhancing coastal squeeze and increasing the risk of failure of inappropriate development (McLachlan and Burns 1992, La Cock and Burkinshaw 1996), as well as potentially affecting sea turtle nesting (de Vos et al. 2019).

The breakdown in ecological processes, loss of resilience and declines in service delivery as a result of this legacy of poor management indicates the consequences of inappropriate actions. Concomitantly, it also provides a key motivation to revolutionize the way one perceives, manages and conserves beaches so that one can maintain the many benefits they confer, not only for oneself but also for future generations. For the most part, countries still have sufficient proportions of intact sandy shores where prudent and judicious decision-making can proactively contribute to protecting beach biodiversity assets and associated services that provide benefits for sustainable economic development, buffering impacts from climate change, and enhancing one's health and wellbeing. The question that follows is: how does one achieve this?

The Need to Integrate Economic, Social and Ecological Values

Beaches have a range of stakeholders and beneficiaries that require access to and benefit from sandy shores (Schlacher et al. 2008, Maguire et al. 2011). At the same time, recent research has started to identify broader environmental values that need to be upheld, and elements that are required to sustain the ecological processes and services provided by beaches (e.g., Schlacher et al. 2008, Harris 2012, Harris et al. 2014a, b, c, Lucrezi et al. 2016). Developing and implementing appropriate beach management strategies thus requires considering the complexity of the relationships among user groups and with the environment, including their socio-cultural and economic dimensions, so that the interests of both people and nature are included (Fig. 11.5). Different backgrounds (including cultures, demographics and access to knowledge) will influence the attitudes, values and preferences of people, and eventually their behaviour and priorities, including activities undertaken, sense of stewardship and management actions. In turn, these behaviours and priorities determine the relative strength and dominant direction of interactions among the three elements of social-ecological systems: the environment, people and the economy. Achieving sustainability requires a balance among these three elements. Ultimately, this means that beach management needs to redirect the traditional scope of maximizing recreation and coastal development, and adopt the concept of beaches as complex social-ecological systems rather than recreational playgrounds (James 2000). This requires a more deliberate recognition of the environmental, socio-cultural and economic values and potential trade-offs if one dimension is favoured over others, such that management actions aim to strike a balance among the three dimensions (Fig. 11.5).

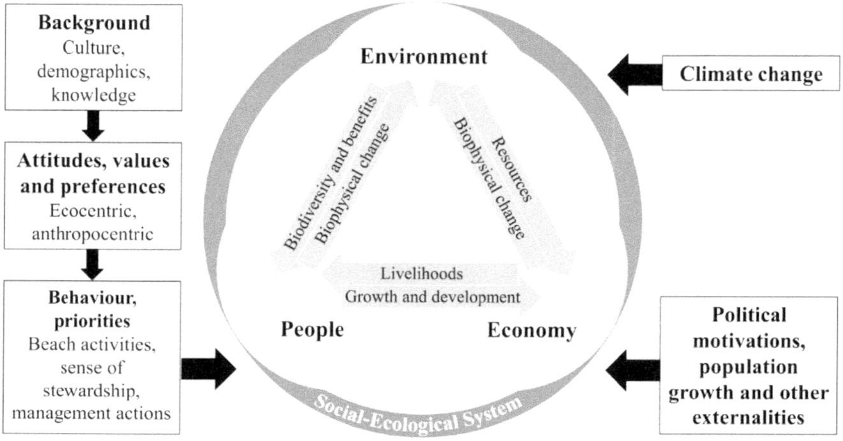

Figure 11.5 Conceptual model illustrating links between components of the beach social-ecological system. Environment is placed at the top because sustainable development requires an ecosystem-based approach: if harnessing benefits and extracting resources causes too much biophysical change, it undermines the provisioning of resources and benefits, causing negative feedback loops and ultimately a collapse of the system. Effects from climate change, political motivations and population growth are superimposed on the system, providing additional drivers that affect the balance among the three elements.

The nature of user-user and user-environment relationships depend largely on people's backgrounds, geographical access to beaches, coastal knowledge, values and attitudes toward the environment, and nature of interactions with coastal ecosystems (James 2000, Wolch and Zhang 2004, Maguire et al. 2011, Lucrezi and van der Walt 2016, Domínguez-Tejo et al. 2018). Many of these attributes also influence people's attitudes towards beach management, such that information on community values and preferred beach activities is recognized as being critical to fostering new trends in beach management (James 2000, Lucrezi et al. 2016, Domínguez-Tejo et al. 2018). Adopting the new management paradigm of social-ecological systems over the historically utilitarian view may then also require supplementary actions to help people transition from predominantly anthropocentric perspectives on beaches to those that include more ecological and ecocentric values.

A Two-Part Action Plan

In recent times, scientists have been calling for two main streams of conservation strategies for sandy beaches and dunes that are highly complementary. The first stream primarily involves decision-makers, managers and scientists in formal actions ranging from filling research gaps to active interventions on threats, zoning activities, systematic conservation planning, delineating setback lines, and specifically representing beaches and dunes in policy (e.g., Schlacher et al. 2008, Harris 2012, Harris et al. 2014a, c, Schlacher et al. 2014a). The second stream entails the engagement of stakeholder groups in understanding the ecological value, functions and vulnerability of beaches and dunes, understanding the role of humans in changing the coast, accepting and promoting sustainable uses of beaches and dunes, participating actively in more informal conservation, and stewardship (Schlacher et al. 2008, Devictor et al. 2010, Maguire et al. 2013, Schlacher et al. 2014a, Burger et al. 2017, Lucrezi et al. 2019). Actions belonging to the second stream are expected to use the current suite of social and cultural values ascribed to beaches (Table 11.1) as an opportunity to endorse a fundamental culture of stewardship of beach and dune ecosystems across stakeholder types globally. This two-part action plan is thus proposed as a new approach to beach management and conservation that operates from the level of regions and countries (spatial prioritization informing formal management and conservation) down to the level of the individual (informal actions of beach stewardship).

ACTION PLAN PART 1: FORMAL BEACH MANAGEMENT AND CONSERVATION STRATEGIES

Towards Sustainable Development

The first step in any planning process is to address the question: what is the goal? The contemporary answer to that question is invariably 'sustainable development', in alignment with countries' commitment to achieving the 2030 Agenda for Sustainable Development (United Nations 2015) and related Sustainable Development Goals, and the Aichi Biodiversity Targets (Convention on Biological Diversity 2011). Coasts provide people with a variety of benefits, especially from sandy shores where recreational activities are concentrated (Harris et al. 2019a). It is important for people to be able to access these benefits, but not beyond their sustainable limits. Therefore, truly achieving sustainable development requires an ecosystem-based approach, where a sufficient proportion of the underlying biodiversity assets are secured in a way that they can continue to deliver benefits in perpetuity (Fig. 11.5). One of the most robust ways in which to identify the core set of biodiversity priority areas, while also taking into account existing uses and values to maximize conflict avoidance, is through systematic conservation planning (Margules and Pressey 2000; see also SANBI and UNEP-WCMC 2016).

In this chapter, we distinguish between management and conservation practices. Management "involves dealing with or controlling human activities in an environmental context, particularly where activities overlap and/or have negative impacts on ecosystems"; conservation "explicitly concerns protecting natural systems, the species they contain, and the processes that maintain their functioning" (Harris 2012). These are not necessarily mutually exclusive; however, there is a subtle difference in their respective primary objectives. Sustainable development requires a combination of both: conserving some areas in a natural or near-natural state, and managing other areas as multi-use spaces where biodiversity and processes are modified by human uses but are maintained in at least a functional state (see SANBI and UNEP-WCMC 2016). Given the legacy of beach mismanagement, some areas may be irreversibly impacted, or much degraded and in need of restoration. This consequently gives three primary actions to safeguard biodiversity assets: protection, avoidance of further habitat loss/degradation and restoration (SANBI 2013).

The scale at which management and conservation actions are planned and take place is also a key consideration. Lessons from real-world systematic conservation planning for sustainable development in South Africa have shown that planning is done increasingly to administrative rather than ecological boundaries, or at a national scale (Botts et al. 2019). This is so that plans can be easily taken up into other government planning and regulatory processes (Botts et al. 2019). Although this is true, it does provide a genuine challenge to spatial prioritization and planning for beaches because, as discussed above, administrative boundaries tend to bisect the littoral active zone. The recommendation, therefore, is to plan at ecologically relevant scales and to ecologically relevant boundaries, and if necessary, subset the outputs to administrative boundaries to improve implementation (Harris 2012, Harris et al. 2019b). The planning domain must, at the very least, span the entire littoral active zone and ideally should be undertaken in conjunction with terrestrial and marine planning processes so that the coastal prioritization is not just about aligning priorities across the land-sea interface, but rather about deliberate planning to secure biodiversity assets in this critical ecotone. Integrating coastal priorities in respective land-use and marine spatial plans then ensures that, even if beaches and dunes are unavoidably managed separately, management actions are contributing to the same objective for each sandy shore (see Harris et al. 2019b). If this is the case, it is recommended that working groups are established that comprise decision-makers from all realms and relevant authorities to formulate a joint management plan to formalize the overarching goals, objectives, monitoring and reporting for specific sandy shores.

Protecting, Avoiding Further Loss and Restoring Beaches

Formal management and conservation practices need to account for ecological processes and connectivity across the ecotone explicitly and move away from piecemeal planning and decision-making. This requires integration and place-based solutions. The key advantage of taking a systematic conservation planning approach is that it is a tool to support decision-making that deliberately integrates economic, social and ecological information in a single, over-arching, coherent prioritization. It is spatially explicit, which means that it can identify sites of importance for conservation, as well as assist in zoning activities to reduce user-environment conflicts (e.g., by excluding those activities that are incompatible with the underlying biodiversity features) and user-user conflicts (e.g., by separating conflicting activities into different use zones). It can also identify priority areas for restoration by analyzing ecological condition and ecosystem service delivery. It is data-driven, robust and transparent; allows existing priorities to be built into future planning; and allows for scenario planning and evaluating trade-offs among management options (e.g., Linke et al. 2011, Harris et al. 2014b). It is thus one of the best available tools to address many of the shortcomings in current beach management and conservation.

The most critical foundational dataset to develop to represent beaches accurately, and to align priorities across realms, is a map of ecosystem types that is seamless across the ecotone (see Harris et al. 2011, 2019b for guidelines and details on how to do this). A systematic conservation plan also requires spatial mapping of other biodiversity features, e.g., species distributions (Harris et al. 2014a); human activities and uses, including income-generating activities that are important for livelihoods and economies (Harris et al. 2015); existing conservation areas; and social values (Table 11.1), including people's connections to areas that confer therapeutic benefits. The latter are often overlooked perspectives in beach conservation and management, but these connections, values and intangible benefits are important to include in decision-making (Bell et al. 2015); the values prioritized in the plan should ideally reflect (or at the very least, accommodate) those held by local communities and regular users of the planning domain. The planning units (small areas of evaluation, analogous to pixels) must be developed in a way that allows the prioritization to be easily deconstructed to administrative boundaries; they also need to be of high enough resolution to allow for appropriate prioritization and planning for sandy shores that are very narrow relative to adjacent terrestrial and marine ecosystem types (see also Harris et al. 2014b, c, and SANBI and UNEP-WCMC 2016 for more technical details on the planning process).

Areas with high selection frequency in the systematic conservation plan should be prioritized for conservation, where the objective is to keep the sandy shore in a natural or near-natural state. Protected areas that span the ecotone are an ideal way to achieve this objective; however, where this is not possible, Other Effective Conservation Measures (OECM) can be equally effective if applied correctly, such as restrictions or prohibitions on resource extraction, and very conservative setback lines to preclude development on the dunes. Other areas may be multi-use areas, where careful management is required to minimize user-user and especially user-environment conflicts, also through OECM. As described above, connectivity across the ecotone is one of the core ecological roles fulfilled by sandy shores, with the added benefit of intact littoral active zones conferring coastal protection services in the face of sea-level rise and increased storminess. Consequently, the key imperative to maintain sandy shores in a functional state is ensuring land-sea connectivity within the littoral active zone for ecological processes to remain intact (McLachlan and Defeo 2018, Harris et al. 2019b). For the most part, this means keeping development behind scientifically determined setback lines. Secondarily, management actions to ensure beaches remain functional relate to sustainable use, which will depend on local priorities, e.g., sustainable limits for recreational or small-scale fisheries (see McLachlan and Defeo 2018 for a review),

Off-Road Vehicle (ORV) driving or other recreational uses (that may call for beach grooming, greater access and related amenities).

Restoration is perhaps one of the most needed actions for sandy beaches globally, and in the immediate future, is neatly aligned with the UN Decade on Ecosystem Restoration 2021–2030 (UN General Assembly 2019). This restoration could be of degraded dunes, where actions such as formalizing access points to minimize dune trampling, restoring natural sand flows (where possible), limiting ORV traffic, and removing alien invasive species can improve the ecological condition of the dunes. Similarly, some fisheries may require additional restrictions (e.g., in gear, size limits, catch and/or effort) in space and/or time or closures in extreme cases, to help recover depleted stocks. Beach grooming and kelp harvesting may need to be temporarily suspended until natural populations of invertebrates recover in order to provide more prey for predators such as shorebirds. Solid waste and wastewater may also require improved management to reduce pollution impacts, improve water quality and reduce anoxia in sediments. Returns on investment in beach restoration are in the form of enhanced delivery of ecosystem services and more benefits for people, e.g., improved coastal protection from more resilient dunes, more fishing opportunities from recovered stocks and better recreational experiences in the form of birdwatching from increased population abundances and safe bathing from improved water quality.

Monitoring Success

For all three actions above, the management objective relates to a desired ecological state: protection requires beaches to be in a natural or near-natural state; avoidance of further habitat loss/degradation requires maintenance of the ecological condition without further decline; and restoration requires improvement in the ecological condition. Progress toward achieving the management objective can be tracked using a suite of ecological indicators in a monitoring programme. Recommended indicators include: (1) traits of bird populations and assemblages; (2) breeding or reproductive success of birds, turtles, invertebrates and/or plants; (3) population parameters and distributions of beach- or dune-associated vertebrates (e.g., birds, turtles and mammals); and (4) population- or assemblage-level measures of the abundance, cover and/or biomass of beach and dune biota, e.g., plants, invertebrates, and/or vertebrates (see Schlacher et al. 2014a for further details). These indicators lend themselves to include citizen scientists in the monitoring programmes, e.g., involving local birdwatching groups in bird surveys or school groups in estimating vegetation cover or assisting with invertebrate sampling. This is one way in which formal conservation (part one of the action plan) can encourage informal conservation action (part two of the action plan) in the complementary approach to managing and conserving sandy shores as social-ecological systems for sustainable development.

ACTION PLAN PART 2: TRANSFORMING VALUES INTO THE STEWARDSHIP OF BEACHES THROUGH EDUCATION AND CITIZEN SCIENCE

Opportunities for Conservation-Oriented Stewardship of Sandy Beaches

The social and cultural values characterizing sandy beaches highlight a strong sense of co-dependence, where people and the beach are inextricably linked together, and the wellbeing of one will be affected by the other and vice versa (West-Newman 2008, Collins and Kearns 2010, Wakita et al. 2014, Lawson 2016, Voyer et al. 2015). Examples of stakeholder groups likely to face this sentiment include, among others, resident communities of the coast, indigenous people who have traditional bonds with the coast, sub-cultures (e.g., surfers), frequent users of the coast, and hobbyists (e.g., bird watchers) (Taylor 2007, Collins and Kearns 2010, Everard et al. 2010, Voyer et al. 2015). A strong sense of co-dependence between people and the beach can translate into active stewardship of the beach, and the agency in beach conservation (Voyer et al. 2015). Coastal communities and regular beach users may engage in various acts including informally monitoring changes such as erosion and weed invasion; reporting illegal acts; supporting regulations; lobbying against development; offering to pay to preserve beaches; signalling extraordinary and unusual events; organizing and participating in beach litter clean-ups; formally participating in volunteering for restoration and wildlife rehabilitation; and participating in scientific research through citizen science (Storrier and McGlashan 2006, Maguire et al. 2011, Ferreira et al. 2012, Kearns and Collins 2012, Collins and Kearns 2013, Voyer et al. 2015, Liu et al. 2019, Lucrezi et al. 2019, Rodella et al. 2019).

Despite the above and paradoxically, the valorization of sandy beaches and dunes as social and cultural hubs still fails to fully translate into an appreciation of their ecological features, resulting in a low public profile when compared with other marine ecosystems such as coral reefs (Schlacher et al. 2008, Nel et al. 2014). Indeed, the massive economic value of beaches (for tourism, resource extraction and urban development) counts against their effective protection and in favour of meeting societal demands to maintain utilitarian uses of beach and dune environments (Schlacher et al. 2014b). Additionally, efforts by various user groups to preserve the integrity of sandy beach and dune ecosystems sometimes struggle to keep up with the growing pressures affecting coastlines, including urban development, pollution, recreation, biological invasions by alien species, and resource extraction (Schlacher et al. 2016), discussed above. These pressures are exacerbated by climate change, a legacy of historical mismanagement (e.g., McLachlan and Burns 1992, La Cock and Burkinshaw 1996), and the limited appropriate inclusion of sandy beaches in contemporary management and conservation plans (Harris et al. 2014a, c). Notwithstanding, the values that people hold for beaches can be used as an opportunity to grow a culture of beach stewardship at the level of individuals and communities, with environmental education and citizen science proposed to be the tools by which this could be achieved.

Education and Citizen Science as Tools to Support Stewardship of Sandy Beaches and Dunes

Education and participatory processes have the potential to achieve several goals across the two main streams of conservation strategies identified above. There are two prime examples of the applicability of education and participatory processes in support of sandy beaches and dunes. One is the introduction of beach and dune ecology in the formal education system, including curricular and extra-curricular school-based activities. The other is the establishment of citizen science networks for monitoring and assessing beaches and dunes. If applied over large spatial and temporal scales, these tools can increase the public profile of beaches, create a culture of coastal stewardship, and generate valuable human resources for science, management and conservation (Fanini et al. 2019, Lucrezi et al. 2019).

Beach and dune ecosystems in the formal education system

The introduction of basic education on sandy beaches and dunes, their functionality and role as ecosystems, can be embedded within the broader concept of ocean literacy: the understanding of the influence of the ocean on people and of people on the ocean (Cava et al. 2005). Ocean literacy aligns with environmental education objectives defined by UNESCO, which include awareness of the global environment and its problems, positive attitudes towards the environment, an agency in conservation, skills development and participation in environmental problem-solving (Fauville et al. 2019).

There have been global efforts to endorse ocean literacy as part of standard education due to a discovered lack of public understanding of marine phenomena, marine issues and the ocean in general (Cava et al. 2005). A discourse of ocean literacy in schools is valuable in stimulating interest in marine issues among school pupils, but also in enhancing their overall scientific literacy, providing a more solid background in preparation for higher education experiences, especially in Science, Technology, Engineering and Mathematics (Lambert 2006). Ocean literacy would represent a unique opportunity for school children to develop environmental awareness and ecological sensibility, which are indispensable prerequisites for future responsible actions (Castle et al. 2010, Guest et al. 2015).

Due to the abstract nature of the ocean, transitionary ecosystems that are more familiar, such as sandy beaches and dunes, can be considered a stepping stone in the acquisition of knowledge underpinning ocean literacy, particularly among communities in countries with extensive coastlines. Sandy beaches and dunes are considered excellent theatres for implementing education activities revolving around the sea and the coast for various reasons (Fanini et al. 2019). Apart from their evident accessibility compared with other marine ecosystems, sandy beaches and dunes possess the characteristics typical of unplanned landscapes favoured by children in educational processes (Hart 2013); they are dynamic and living spaces affording the observation of physical processes (sand movement, waves, tides) and ecological processes (e.g., animal activities) at play; they provide simultaneous images of changing ecosystems through dune successions and beach zonation; and provide information on adjacent terrestrial and marine ecosystems, e.g., by analyzing beach-cast wrack (Schlacher et al. 2008). These characteristics make sandy beaches and dunes suitable places to apply scientific thinking and interdisciplinary learning (Cambers and Diamond 2010; Ferreira et al. 2012).

Several examples of short- and long-term educational initiatives, both in and out of school and with the partial or whole focus on sandy beaches and dunes, have been provided over the last decade (Table 11.2). The education interventions described range from standalone events to long-term initiatives and even formal inclusion of beaches and dunes in school curricula and school-based assessments. These interventions tended to yield positive results, including increased knowledge of beach and dune ecosystems, awareness of and concern regarding various anthropogenic impacts, positive attitudes towards conservation, motivation for self-driven learning and interest in a career in the natural sciences.

Based on the lessons learnt from these initiatives, it is possible to synthesize 10 key success factors and issues to take into account for effective inclusion of sandy beach and dune ecology in school curricula.

1. Stakeholders' contribution: Successful education would first require the participation of key stakeholders to assist and facilitate the work of schools and teachers. Stakeholders can include the families of school pupils, trained volunteers, local scientists, beach ecologists, marine scientists and institutions like aquaria and research stations. For example, beach ecologists could visit schools and engage in question-and-answer sessions with school pupils, and parents could accompany children during field trips to support learning.

2. Education balancing knowledge and fun: Educational initiatives with an element of fun can stimulate the curiosity of school pupils while administering scientific knowledge. For example, field trips to the beach, games, photography, drawings and simulations of the natural environment can make for positive learning experiences.

3. Interdisciplinary approach: Education should apply an interdisciplinary approach, both theoretical and practical, through the cooperation of different subjects and teachers. This is made possible by the multidisciplinary nature and practical appeal of the marine sciences.

4. Time-intensive education: Time-demanding activities, such as measurements and long-lasting initiatives, such as week-long field trips, are encouraged. These types of education interventions can enhance learning and create a sense of ownership of acquired knowledge and skills among school pupils.

5. Familiar environments: Initiatives should consider local and familiar environments for theory as well as outdoor experiences. These environments can relate to local traditional knowledge, sense of place and stewardship among school pupils. However, an understanding of processes at play in these environments may be possible only by comparisons with other, either more or less impacted environments.

6. Issues of local and global concern: Initiatives are best to focus on a mixture of issues of both global concern, such as pollution and local concern, such as habitat

Table 11.2 Summary of literature describing initiatives of environmental education, including elements of sandy beach and dune ecology, as well as beach- and dune-based citizen science targeting schools and youth.

Source	Location	Stakeholders involved	Initiative description	Main findings/discussion	Issues and challenges
EDUCATION					
Fanini et al. 2007, Scapini and Fanini 2009	Nefza and Zouaraa, Tunisia.	Local school, local teachers, local children aged 8–11 years, local researchers, international researchers in the framework of a research project (MEDCORE).	Single scientific dissemination field trip to the sandy beach with 64 school children. Activities included: a pre-field trip and post-field trip questionnaire (prepared by researchers in collaboration with local school teachers) on beach and dune characteristics; a game at the beach; and knowledge exchange between children and researchers at the beach.	Children were more aware of the damages of pollution to the beach, but not aware of the origins of dunes. There was an increase in knowledge of beach-dune systems among children after the field trip. The study highlighted the importance of education at the local level, considering local traditional knowledge and familiar environments to foster local coastal stewardship.	Importance of paying attention to education revolving around local environments. Gaps between researchers and local people, who may possess important knowledge. Need to consider the efficacy of different information flows in environmental education.
Piwowarczyk 2010	Gdynia, Poland.	High school students, biology teachers, external marine scientists.	Three educational programmmes run from 2000 to 2009, including theory (biological problem and research techniques); time-demanding measurements of various biological (macro-detritus, macrofauna, nekton, and fauna in the algal mats) and physical data (water transparency, sea surface temperature, pH, salinity	The initiative highlighted the multidisciplinary nature and practical appeal of marine science.	It is necessary to formulate educational programmes that will involve integrative approaches (and the cooperation of different subjects and teachers), and also engage younger children. Importance of measuring whether these programmes

Source	Location	Stakeholders involved	Initiative description	Main findings/discussion	Issues and challenges
			and meteorological data) of selected sandy beaches; planning and execution of an experiment; sandhopper monitoring campaigns; data processing and analysis.		shape personal attitudes towards science and increase scientific culture.
Hartley et al. 2015	UK.	Schools, school children aged 8–13 years, university, national aquarium.	176 children from nine schools participated in an intervention to raise awareness, change attitudes and increase self-reported litter-reducing behaviour. The intervention, based at an aquarium, was designed to highlight the types, sources (including the beach) and impacts of marine litter, particularly plastics. The intervention was set within a larger event, including sea kayaking, beach conservation, and a tour of the aquarium.	The intervention increased awareness of the problems associated with litter and increased litter-reducing behaviour. Children became more concerned and better understood the sources of litter and the ecological consequences of litter on ecosystems.	Only schools located near the coast were involved.
Bryant 2018	Shannon Point, Washington, USA.	Local school children, marine centre, undergraduate volunteers, primary investigator.	A five-day-long summer programme (Beach Investigators) teaching about the beach and how to carry out an in-depth exploration of the local beach ecosystem. The programme included:	A total of 17 students participated. Participation increased ecological knowledge and motivation for continued self-driven learning. The programme can be	The short span of the initiative. Lack of interest in the classes by some pupils. Importance of working with local ecosystems. Coordination issues.

Table 11.2 (Contd.)

Table 11.2 (Contd.) Summary of literature describing initiatives of environmental education, including elements of sandy beach and dune ecology, as well as beach- and dune-based citizen science targeting schools and youth.

Source	Location	Stakeholders involved	Initiative description	Main findings/discussion	Issues and challenges
			lessons on nature journalism, photography and observation; free explorations of the beach; and an evaluation of aggregate student learning.	a powerful foundation for environmental and outdoor education.	
Baker and Readman 2019	Plymouth, UK.	Beach rangers, local families, local primary schools.	A programme run during school holidays aimed at encouraging people to visit and use local blue spaces sustainably. The programme combines cutting-edge science and fun for local communities.	The programme has successfully engaged 7000 local people. The programme has begun delivering marine-themed assemblies in primary schools to promote awareness and engagement with natural blue spaces. The programme has also begun a 'guest scientist' scheme to design games and activities (mud to music, paint to plankton, and crafts to crabs).	The project is unable to establish a national reach due to local focus and limited resources.
Riedinger and Taylor 2019	Wallops Island, Virginia, USA.	Parent chaperones, school children, marine science field station.	Participation in the four-day Coastal Ecology field trip programme at the Chincoteague Bay Field Station. The programme included the dunes field experience to a barrier	Learning among youth was enhanced by positive interactions with parent chaperones (i.e., supporting learning, managing learners). Chaperones enforced proper conduct on sandy beach	Parent chaperones can still interfere with learning.

Source	Location	Stakeholders involved	Initiative description	Main findings/discussion	Issues and challenges
			island focussing on coastal geology, ocean circulation, and processes such as weather, erosion, waves, currents tides and deposition. The youth examined beach profiles and discussed the geological processes that shape the formation of dunes and barrier island systems. The youth also looked for evidence of human interference with these processes (e.g., seawalls, jetties, rip rap) and observed the effect.	and dune ecosystems (e.g., no taking of shells).	
EDUCATION AND CITIZEN SCIENCE					
La Mar 2006	Victoria, Australia.	Primary and secondary school students, community groups, industry groups, marine centre, local councils and committees.	A project on dune restoration active since 1985. The project includes dissemination and learning activities (coastal formation and dunes, the importance of vegetation in protecting the coast from erosion); dune stabilization using brush matting; and dune revegetation.	In 2004 alone, 1179 participants completed 4617 hours of restoration work. Participation increased knowledge and awareness of coastal dunes and their role. Participants gained a sense of ownership of coastal dunes.	Importance of working in local areas the stakeholders are familiar with and can gain ownership of.

Table 11.2 (Contd.)

Table 11.2 (Contd.) Summary of literature describing initiatives of environmental education, including elements of sandy beach and dune ecology, as well as beach- and dune-based citizen science targeting schools and youth.

Source	Location	Stakeholders involved	Initiative description	Main findings/discussion	Issues and challenges
Ballard et al. 2017, Freiwald et al. 2018	San Francisco Bay Area, California, USA.	Local schools, local youth including middle school and college students, local teachers, local researchers in the framework of the LiMPETS programme.	A one-year programme involving 2500 youth and teachers in 40 classrooms and out-of-school groups, in training and data collection using sandy beach monitoring protocols. The programme also involved data processing, analysis and dissemination of findings.	Data collected contributed to site and species management. The programme was also used to monitor marine protected areas. Youth displayed evidence of environmental science and conservation knowledge and agency during participation.	Extended time working on the same project with the same group was essential. The importance of pre-existing relationship to the places they study in citizen science. Perception of the project as contributing to 'real' science, working with data and dissemination are essential.
Lucrezi et al. 2019	Ponta do Ouro, Mozambique.	Local school, local teachers, local children aged 9–12 years, tourism operators, marine park rangers, marine park manager, international researchers in the framework of a research project (Green Bubbles).	A two-day module on ocean literacy and marine science for school children, delivered by trained local stakeholders (rangers, teachers, tourism sector). The module included principles of ocean literacy, a description of various local marine ecosystems (beach, dune, rocky shore, reefs) and their characteristics, a laboratory session with microscopes, and a field	Local stakeholders delivering the module increased their knowledge of sandy beach and dune ecosystems, ecosystems' connectivity and function. The stakeholders grew an appreciation for conservation as a way to guarantee sustainable livelihoods in a marine reserve. The bequest value of sandy beaches and dunes was deemed important. Data collected through citizen science can help to monitor	The effects of the module on knowledge among children were not measured. Uncertainty of the temporal continuity of the initiative. Difficulties in keeping children focussed on tasks at the beach. Difficulties in ensuring effective citizen science. Spatial issues (small scale of the initiative

Source	Location	Stakeholders involved	Initiative description	Main findings/discussion	Issues and challenges
			session using citizen science protocols to collect data at the sandy beach and rocky shore (Reef Check and ghost crab burrow count protocols).	and manage the intertidal zone in the marine reserve.	compared with the large extent of the marine reserve). Limited funding to support citizen science activities.
Weiss and Chi 2019	California, USA.	High school students from groups underrepresented in the sciences, undergraduate mentors, marine research station, working research laboratory in the framework of the project Youth & the Ocean!.	A programme running from 2010 to 2012, including participation in intertidal monitoring during the school year (monitoring surveys of the distribution and abundance of Pacific mole crabs (*Emerita analoga*)), and the development of research questions to be explored during a summer residence at a marine research station. The monitoring followed protocols of the LiMPETS citizen science programme.	The programme deepened participants' understanding of the nature and practices of science. The programme expanded access to practising scientists and academia. The programme maintained or deepened interest in pursuing careers in the natural sciences. The program influenced commitment to ocean conservation and further learning about the ocean.	—
CITIZEN SCIENCE					
Bravo et al. 2009	Chile.	Social organizations, citizen groups, schools, school children aged 10–16 years, coordinators, villages of the Chilean coast.	Beach litter surveys involving 1531 high school students during 2008, as part of a project introducing the scientific method to school children.	The schools were supported by 46 different institutions, including local city governments, corporations, private foundations and other nongovernmental organizations.	Importance of collaborations between scientists and schools (teachers and students), to answer questions from all branches of natural sciences (biology,

Table 11.2 (Contd.)

Table 11.2 (Contd.) Summary of literature describing initiatives of environmental education, including elements of sandy beach and dune ecology, as well as beach- and dune-based citizen science targeting schools and youth

Source	Location	Stakeholders involved	Initiative description	Main findings/discussion	Issues and challenges
			The programme included: booklets with an educational story informing about the fate of plastic bags in the marine environment, to motivate and inform participants; and beach surveys of litter using a standard protocol. Surveys were done at local accessible beaches.	The students discussed possible origins of litter and ways of mitigating the problem (increased sensitivity towards their local beaches). Enthusiasm for the surveys.	environmental ecology, geography, sociology).
Cambers and Diamond 2010	Africa, Asia, Europe, Caribbean, Pacific and Indian Ocean.	Primary school children, secondary school children, youth, teachers, adults in the framework of the Sandwatch project.	Sandwatch aims to contribute to an understanding of how climate change affects beach systems and to build ecosystem resilience and adaptation to climate change. The project is intended for integration in school curricula and interdisciplinary learning. It uses a protocol of standardized methods to measure beach changes including erosion, accretion, wave and current action, water quality, plants, animals (e.g., sea turtle monitoring) and human impacts on the beach.	The project has been active for 16 years. Some countries have formally included Sandwatch into the school curriculum and have also used it for school-based assessments (school examinations).	—

Source	Location	Stakeholders involved	Initiative description	Main findings/discussion	Issues and challenges
			Students follow a manual, carry out fieldwork activities from monitoring to restoration, organize projects, discuss findings of their projects in class, and disseminate findings (events, fairs, international symposia).		
Ferreira et al. 2012	Portugal (including Azores).	Junior school and high school teachers and students, municipal officials, university students, members of nongovernmental organizations, environmental educators. This initiative was part of the Coastwatch programme.	Workshops between 2003 and 2007 with 140 stakeholders. The workshops included instruction and practice on beach profiling, as well as data processing.	Teachers considered the method a useful tool, with scientific and pedagogic (interdisciplinary) potential. Junior school teachers reported their students' enthusiasm for being outdoors and doing 'scientific work'. Data could support management decisions (valid data). Students could apply theoretical concepts learned at school and learn to relate concepts from different disciplines. 'First-hand learning through intent participation' constituted an extremely valuable and memorable learning experience.	Time remains a constraint for some people. Challenges in obtaining feedback post participation.

Table 11.2 (Contd.)

Table 11.2 (Contd.) Summary of literature describing initiatives of environmental education, including elements of sandy beach and dune ecology, as well as beach- and dune-based citizen science targeting schools and youth.

Source	Location	Stakeholders involved	Initiative description	Main findings/discussion	Issues and challenges
Hidalgo-Ruz and Thiel 2013	Chile.	Schools, school children, teachers, marine scientists in the framework of the citizen science projects National Sampling of Small Plastic Debris and Litter Scientists.	39 schools and 1000 students from continental Chile and Easter Island participated in documenting distribution and abundance of small plastic debris. Activities included: class reading of a 28-page children story created to motivate the children; familiarizing with the citizen science protocol and materials in class; counting and sorting of small plastic debris on the beach, using quadrats; data entry and reflection in the class; a post-participation survey.	The students were able to follow the instructions and generate reliable data. Participation increased willingness to be involved in citizen science activities in the future.	—
Eastman et al. 2014	Chile.	School children <18 years, school teachers, marine biologists.	A programme involving 30–48 schools and including: education materials (storybooks introducing the problem of pollution); discussions on marine litter in the class; quantitative surveys (using plot counts and sieving) of marine litter at the beach; assessment of beach users' perceptions	Participation increased knowledge of microplastics and willingness to participate in citizen science. Participation promoted curiosity and independent inquiry. Data were used for scientific publications.	Communication of results throughout educational establishments, local communities, and via scientific publications is a critical process to promote pride, validation and continued participation.

Source	Location	Stakeholders involved	Initiative description	Main findings/discussion	Issues and challenges
			and behaviour concerning littering; data entry and processing (graphs); discussion of the data collected and reflections in the classroom; and a post-participation survey with the children.		
Honorato-Zimmer et al. 2019	Chile and Germany.	Schools, school children aged 5–12 years, school teachers, scientific advisors, coordinators at local universities.	Nationwide beach litter surveys as part of a semester-long project, in which children acquired insights into the scientific method, the marine ecosystem, and the marine litter problem, guided by their teachers. Each student received a copy of a workbook designed for the activity. Teachers were trained to learn the sampling protocol and how to apply it in practice with the children. Before the beach surveys, all students were introduced to the marine litter problem, and they learned about the scientific method, through theoretical and experimental activities and tasks contained in the workbook.	The data informed management and policy actions concerning litter problem. The project was successful thanks to the constant communication between teachers and coordinators.	Children need better understanding of natural debris.

degradation and the reduction of species abundance. In so doing, children are enabled to reflect upon problems affecting their local beach and dune environments and acquire a sense of ownership, while also understanding the magnitude of bigger problems caused by human impacts on coastal ecosystems.

7. Balanced representation of topics: Educational content ought to balance descriptions of beach and dune ecosystems' structure and function (that tend to receive little attention in informal education narratives) and the exposition of issues of concern (that tend to receive more attention and therefore dominate the child's awareness).

8. Education monitoring: It is important to measure whether educational programmes shape personal attitudes towards science and increase scientific culture. This can be done, for example, through pre- and post-participation surveys and discussions.

9. Education beyond school: It is important to consider the efficacy of different information flows in environmental education. Transfer of knowledge, for example between school pupils and their families, should be taken into account in the design and delivery of content on sandy beaches and dunes to maximize educational benefits beyond the school.

10. Citizen science: Many educational interventions with a focus on sandy beaches and dunes have been paired with active participation in citizen science because the two tend to be closely linked. Coupling education with citizen science would enable school pupils to put acquired knowledge into practice, catalyzing scientific thinking and science-based problem-solving.

One excellent example of how environmental education can inspire action for change is through the recent campaigns around ocean pollution, especially plastic pollution. The scientific and media coverage has brought attention to the coastal zone, and particularly to sandy beaches as sinks for litter originating from both land and sea (Hind-Ozan et al. 2017). Widely publicized international commitments (such as those announced during the annual Our Ocean Conference), both monetary and managerial, towards prevention and elimination of litter have led to a mobilization of crowds in support of these commitments in various forms and across different scales (Thiel et al. 2017). It could be argued that this mobilization is somewhat facilitating the establishment of deeper connections between people and the beach ecosystem.

Beach and Dune Citizen Science

Citizen science, or the participation of non-specialized citizens in the production of scientific research (Bonney et al. 2009, Silvertown 2009, Thiel et al. 2014), is increasingly regarded as a fundamental way to create and strengthen the relationship between society and science (Bear 2016, Vann-Sander et al. 2016, Martin 2017). Primarily, citizen science facilitates the scientific process through the collection of data by volunteers over extended spatial and temporal scales (Bonney et al. 2009, Shirk et al. 2012, Martin 2017). Citizen science also has positive effects on society, including better attitudes towards science; improved scientific literacy; scientific thinking; awareness of various issues; the ability to use science to solve daily problems; capacity building; personal growth; social learning; and pride, stewardship and proactivity (Foster-Smith and Evans 2003, Brossard et al. 2005). Citizen science can promote formal or informal partnerships between scientists, citizens, economic actors and decision-makers; it can guide and inform management and endorse social democratization through public participation in policy processes (Dickinson et al. 2012, Vann-Sander et al. 2016). The data collected by citizen scientists, including school pupils, are increasingly recognized as valid and meeting or even exceeding quality standards (Earp and Liconti 2020). As a result, citizen science is supported by governments and scientific organizations globally.

Marine citizen science is still less prevalent than terrestrial citizen science (Thiel et al. 2014). This calls for social marketing highlighting the versatility and cross-cutting benefits of citizen science for marine conservation, including beach and dune ecosystems. For example, citizen science projects can take many forms and address different needs, from promoting action and participation in local issues to educating, investigating a specific research question, and creating a connection between inland communities and the marine environment (Earp and Liconti 2020). Ultimately, marine citizen science can contribute to achieving several conservation outcomes, including those for policy, education, capacity building, management of sites and species and research (Cigliano et al. 2015).

Although intertidal ecosystems, including sandy beaches, are easily accessible, and therefore characterize a great proportion of marine and coastal citizen science projects (Thiel et al. 2014, Earp and Liconti 2020), the expansion of citizen science projects revolving around sandy beaches and dunes is still required, considering the vastness of these ecosystems globally and their relevance in terms of functions and services to humankind. There are numerous citizen science projects that either partly or fully revolve around sandy beach and dune ecosystems and that engage different stakeholders (Table 11.3). Projects may be locally, nationally or internationally based and have different focusses, e.g., collecting and counting beach litter; monitoring stranded animals; monitoring species' distributions, abundance and activity (e.g., nesting); testing seawater quality; monitoring geophysical and climatic changes; monitoring phenomena such as algal blooms and oil spills; and monitoring dune rehabilitation. Effective legislation protecting sandy beaches and dunes needs to be supported by large data sets (Cigliano et al. 2015, Hyder et al. 2015); survey-based citizen science can generate such data sets over large spatial and temporal scales (Ward et al. 2015). Therefore, future avenues of developing beach-based citizen science could include expanding or increasing the number of networks addressing specific issues, such as: erosion, sea-level rise and extraordinary climatic events; changes in species distributions, possibly as a response to climate change; biological invasions by alien and pest species; and human impacts including pollution, engineering and recreation. Beach and dune citizen science could also see a steady increase in the number of projects and networks addressing all four areas of marine policy (Hyder et al. 2015), namely biodiversity, the physical environment, pollution and resource management.

Table 11.3 A selection of citizen science projects revolving, either partly or fully, around sandy beaches and dunes

Project name	Project web page
Adopt-a-Beach	https://www.naturesvalleytrust.co.za/programmes/conservation-education/adopt-a-beach/
A Rocha	https://www.arocha.org/en/projects/?filters[habitat]=marine
A Tide Turns	http://www.vikalpsangam.org/article/a-tide-turns-coastal-communities-of-climate-change/#.XbLu1ugzaUk
Audubon Western Everglades	https://audubonwe.org/our-work/shorebird-stewardship/
BEACH	https://ecology.wa.gov/Water-Shorelines/Water-quality/Saltwater/BEACH-program
BeachCare	https://www.griffith.edu.au/cities-research-institute/griffith-centre-coastal-management/community-engagement/beachcare
Beachwatch	https://beachwatch.farallones.org/
Beach COMBERS	https://montereybay.noaa.gov/getinvolved/volunteer/bchmon.html
Beach-nesting Birds	http://birdlife.org.au/projects/beach-nesting-birds
Beach Observer	http://www.coastalandoceans.com/Core-Services/Education/
British Columbia Beached Bird Survey	https://www.bsc-eoc.org/volunteer/bcbeachbird/
California King Tides Project	https://www.coastal.ca.gov/kingtides/
Clean Sea Life	http://cleansealife.it/
COASST	https://coasst.org/
Coastal Research Volunteers	https://seagrant.unh.edu/crv
CoastSavers	https://www.coastsavers.org/
CoAST SB	http://explorebeaches.msi.ucsb.edu/education
CoastSnap	https://www.environment.nsw.gov.au/research-and-publications/your-research/citizen-science/digital-projects/coastsnap
Coastwatch	http://coastwatch.org/europe/
DuneWatch	https://www.griffith.edu.au/cities-research-institute/griffith-centre-coastal-management/community-engagement/dunewatch
eBird	https://ebird.org/home
Grunion Greeters	http://www.grunion.org/
iNaturalist	http://www.inaturalist.org/
iSpot	https://www.ispotnature.org/
Jellywatch	http://www.jellywatch.org/
LiMPETS	https://limpets.org/
Litter Scientists	http://www.cientificosdelabasura.cl/es/
Maine Sea Grant Beach Profile Monitoring	https://seagrant.umaine.edu/extension/southern-maine-volunteer-beach-profile-monitoring-program/about-volunteer-beach-profile-monitoring/
Marine Conservation Society Beachwatch	https://www.mcsuk.org/get-active/beachcleans

Project name	Project web page
Marine Debris Action Team	https://seaturtles.org/
Marine Debris Monitoring	https://olympiccoast.noaa.gov/science/citizenscience/marinedebris.html
Ocean Conservancy International Coastal Cleanup	https://oceanconservancy.org/trash-free-seas/international-coastal-cleanup/
Operation Posidonia	https://www.operationposidonia.com/storm-squad-1
OSPAR Beach Litter	https://www.ospar.org/work-areas/eiha/marine-litter/beach-litter
Reef Check Med 'Mac Emerso'	https://www.reefcheckmed.org/italiano/reef-check-med/mac-emerso/
Sandwatch	http://www.sandwatchfoundation.org/
Save Coastal Wildlife	https://www.savecoastalwildlife.org/
Shorecombers	https://www.shorecombers.org/
Shoresearch	https://www.wildlifetrusts.org/get-involved/other-ways-get-involved/shoresearch
SpeSeas	https://speseas.org/projects/sea-turtle-citizen-science/
Surfrider Foundation (water testing)	https://www.surfrider.org/
TrashBash	https://www.aquarium.co.za/blog/entry/trash-bash-invitation-two-oceans-aquarium-beach-cleanup-cape-town
TurtleCare	https://www.cooloolacoastcare.org.au/projects/turtlecare
Turtle THiS	https://uwf.edu/centers/gulf-islands-research-and-education-center/citizen-science/
TurtleWatch Maldives	https://www.mrc.gov.mv/en/programmes-and-collaborations/new-progmarine-turtle-research-programmeramme-page/

Youth-based citizen science

In the context of the above, youth-based citizen science, especially within formal education and extracurricular settings, is valuable in many ways. Cigliano et al. (2015) and Ballard et al. (2017) described its potential including the achievement of interdisciplinary learning outcomes; inter-generational information flows; creating emotional connections with the marine environment; positive behavioural changes; science literacy and critical thinking; development through civic engagement; environmental stewardship; and ultimately environmental science agency—an ability to use science learning and participation as the grounds on which to act towards environmental sustainability. For this potential, youth-based citizen science represents an appealing tool in beach and dune management because it could address the disconnect between socio-cultural values ascribed to sandy shores and effective public participation in their conservation over large scales.

Projects focussing either partly or fully on youth-based citizen science for beach and dune management (Table 11.2) mainly revolve around protocol-based monitoring of beach litter, intertidal species and climate-change events;

beach profiling; and dune restoration. All these projects have also incorporated a strong educational component, both before and after participation. Overall, these initiatives had several positive effects, from increased knowledge and development of independent inquiry to the acquisition of skills and collection of valid data that could be used for management. From the project summaries (Table 11.2), eight key characteristics that effective youth-based beach and dune citizen science ought to possess are identified and described below.

1. Stakeholder cooperation: Citizen science ought to be underpinned by meaningful collaborations between scientists and formal education institutions to achieve the outcomes of conservation research and management as well as conservation learning and action.

2. Real science: Involvement in all aspects of science from data collection, entry, analysis and interpretation to the dissemination of scientific findings and the development of new hypotheses is understood to substantially enhance validation, pride and environmental science agency among youth.

3. Project ownership: Formal institutions are encouraged to have pupils commit to both long-term and short-term projects. The type of project may be selected depending on the scope of the problem to be investigated, local priorities and needs, the efforts required for data collection, and the time available for pupils to participate. Regardless of the nature of the citizen science project, however, it is critical to ensure an understanding and ownership of a particular protocol and action process.

4. Familiar places: Pre-existing relationships between youth and places they study in citizen science are important. These relationships tie well with the development of a sense of place, ownership and stewardship among youth.

5. Education and motivation: Preparatory educational and informative sessions are critical in stimulating interest and generating motivation to act to solve a particular problem. The message that all beings are tightly connected with different ecosystems, including coastal areas, is urgent and strategic in order to trigger various processes related to conservation and action.

6. Priority issues: The youth's attention should be drawn to different issues of priority and relevance for beach and dune management, which highlight ecosystems' complexity, connectivity and vulnerability.

7. Citizen science monitoring: The effects of participation could be monitored by integrating citizen science into school-based assessments, and by using pre- and post-participation research.

8. Versatility: Projects should possess characteristics, which enable the engagement of youth coming from different socio-cultural backgrounds, and include both coastal communities and inland communities. For instance, technological advances, create opportunities to garner youth access to beach and dune ecosystems remotely, using virtual reality or satellite imagery. Projects could also use technology to create connections between school pupils from completely different settings (e.g., mountain versus coast) through the exchange of photographic and video materials. Although technology can make remote communities feel included in beach-based citizen science, real-life experiences of beach and dune environments remain pivotal in creating profound links between youth and the coast and stimulating positive attitudes towards stewardship.

Using Education and Citizen Science to Connect People to Beach Values

There is a plethora of social and cultural values associated with sandy beach and dune ecosystems (Table 11.1). Given the myriad of threats to beaches, especially in the face of global change, the rich values humans assign to sandy shores provide an excellent opportunity to popularize these ecosystems and create a culture of beach and dune stewardship. Education and citizen science are key avenues through which existing values can translate into conservation actions. They offer opportunities for people to be in direct contact with sandy beach and dune ecosystems; gain knowledge of these ecosystems and the threats they face; be motivated to act in favour of beach and dune protection; and ultimately, be stewards and adopt roles of advocacy for coastal conservation. The available research on education and citizen science revolving either partly or fully around sandy beaches and dunes shows that there are measurable positive effects on the stakeholders involved, especially the youth (e.g,, Eastman et al. 2014, Hartley et al. 2015, Bryant 2018). Consequently school pupils, facilitated by the education system and by scientists, are a particularly important driver for change. Stimulating their curiosity while raising awareness, imparting scientific knowledge, and making them feel like they are part of the solution to global problems affecting the coasts can result in the achievement of several conservation goals set for sandy beaches and dunes. Over and above the benefits of engaging in citizen science for the participants, the resulting culture of sandy beach stewardship can foster that it could play a key role in supporting formal management and conservation actions taken by decision-makers and local or national authorities.

CONCLUDING REMARKS

The need to avoid, mitigate and restore the contemporary impacts to sandy shores calls for a change in the way society perceives, manages and conserves beaches. Furthermore, finding long-term, pragmatic and effective solutions is becoming increasingly urgent because climate change is exacerbating the consequences of the legacy of poor decision-making in coastal zones, particularly on inappropriately developed urban shores. In response, a two-part action plan where formal action by scientists and decision-makers was proposed to provide the high-level framework for sustainable development of the natural environment (part one) that more deliberately includes requirements of the unique biodiversity and critical ecological processes across the ecotone. This will require a shift in perception from the historical utilitarian model of beach use to a new ecocentric approach to beach and dune management. Understanding is the first step towards real appreciation and the key to resolving many of the beach management challenges. The supporting role of education and citizen science (part two) is thus expected to be instrumental in helping people transition to the new management and conservation paradigm. To this end, it is important to engage a broad range of stakeholder groups in understanding the ecological value of sandy beaches, thereby encouraging acceptance of their responsibility in sustainable use of beaches and dunes by playing an active role in stewardship. Importantly, the data generated by the citizen scientists can also be included in the formal management and conservation processes, and sites of ecological importance can be used as stages for key environmental education initiatives. Thus, the two parts of the action plan are highly complementary and explicitly not mutually exclusive.

Increasing our understanding of the dynamic relationships between the human and environmental dimensions of the sandy beach social-ecological system can contribute to increasing management effectiveness by guiding policy decisions for sustainable development, particularly around tourism and recreation, resource extraction and land-use planning. Consequently, it is encouraging to note that integrative studies are becoming more frequent and relevant in beach research, blurring the lines between socio-economic and natural sciences. There is also a key niche to fill in bridging the gap between science and society. Scientists have a role to play in that process, but it may also require specialist intermediaries to take the lead in actively bringing beach science information to lay audiences through environmental education and citizen science initiatives, particularly among school groups. These initiatives deliberately need to transfer knowledge on elements of sandy beach and dune ecology that can help to promote a culture of stewardship of this invaluable link between land and sea that, albeit for different reasons, one holds in such high regard. Ultimately, integration is the key to safeguard beaches and their unique biodiversity, and to maintain the many benefits they confer for generations to come through sustainable development. This integration is of coastal land and sea as the planning domain; of the social, economic and ecological dimensions of the sandy beach social-ecological system; of governance structures, decisions and actions; and of science and society.

Acknowledgements

We extend our gratitude to several colleagues for discussions and advice that contributed to this work. In particular, SL and CC wish to thank Martina Milanese and Thomas Schlacher.

REFERENCES

Able, K.W., M.J. Wuenschel, T.M. Grothues, J.M. Vasslides and P.M. Rowe. 2013. Do surf zones in New Jersey provide "nursery" habitat for southern fishes? Environ. Biol. Fish. 96(5): 661–675.

Ashbullby, K.J., S. Pahl, P. Webley and M.P. White. 2013. The beach as a setting for families' health promotion: a qualitative study with parents and children living in coastal regions in Southwest England. Health Place 23: 138–147.

Augé, A.A., B.L. Chilvers, R. Mathieu and A.B. Moore. 2012. On-land habitat preferences of female New Zealand sea lions at Sandy Bay, Auckland Islands. Mar. Mam. Sci. 28(3): 620–637.

Baker, G.J. and E. Readman. 2019. Closer to blue. pp. 377–400. *In*: G. Fauville, D.L. Payne, M.E. Marrero, A. Lantz-Andersson and F. Crouch [eds]. Exemplary Practices in Marine Science Education. Springer, Cham.

Ballard, H.L., C.G. Dixon and E.M. Harris. 2017. Youth-focused citizen science: examining the role of environmental science learning and agency for conservation. Biol. Conserv. 208: 65–75.

Barbier, E.B., S.D. Hacker, C. Kennedy, E.W. Koch, A.C. Stier and B.R. Silliman. 2011. The value of estuarine and coastal ecosystem services. Ecol. Monogr. 81(2): 169–193.

Barragán, J.M. and M. de Andrés. 2015. Analysis and trends of the world´s coastal cities and agglomerations. Ocean. Coast. Manage. 114: 11–20.

Bear, M. 2016. Perspectives in marine citizen science. J. Microbiol. Biol. Educ. 17(1): 56.

Beaumont, N.J., L. Jones, A. Garbutt, J.D. Hansom and M. Toberman. 2014. The value of carbon sequestration and storage in coastal habitats. Estuar. Coast. Shelf Sci. 137: 32–40.

Bell, S.L., C. Phoenix, R. Lovell and B.W. Wheeler. 2015. Seeking everyday wellbeing: the coast as a therapeutic landscape. Soc. Sci. Med. 142: 56–67.

Bender, M.A., T.R. Knutson, R.E. Tuleya, J.J. Sirutis, G.A. Vecchi, S.T. Garner, et al. 2010. Modelled impact of anthropogenic warming on the frequency of intense Atlantic hurricanes. Science 327(5964): 454–458.

Berry, A.J., S. Fahey and N. Meyers. 2014. Boulderdash and beachwalls—The erosion of sandy beach ecosystem resilience. Ocean. Coast. Manage. 96: 104–111.

Bonney, R., C.B. Cooper, J. Dickinson, S. Kelling, T. Phillips, K.V. Rosenberg, et al. 2009. Citizen science: a developing tool for expanding science knowledge and scientific literacy. BioScience 59(11): 977–984.

Botton, M.L. and R.E. Loveland. 2011. Temporal and spatial patterns of organic carbon are linked to egg deposition by beach spawning horseshoe crabs. Hydrobiologia 658: 77–85.

Botts, E.A., G. Pence, S. Holness, K. Sink, A. Skowno, A. Driver, et al. 2019. Practical actions for applied systematic conservation planning. Conserv. Biol. 33(6): 1235–1246.

Bravo, M., M. de los Ángeles Gallardo, G. Luna-Jorquera, P. Núñez, N. Vásquez and M. Thiel. 2009. Anthropogenic debris on beaches in the SE Pacific (Chile): results from a national survey supported by volunteers. Mar. Pollut. Bull. 58(11): 1718–1726.

Brockmann, H.J. 1990. Mating behavior of horseshoe crabs, *Limulus polyphemus*. Behaviour. 114(1–4): 206–220.

Brossard, D., B. Lewenstein and R. Bonney. 2005. Scientific knowledge and attitude change: the impact of a citizen science project. Int. J. Sci. Educ. 27(9): 1099–1121.

Bruun, P. 1962. Sea-level rise as a cause of shore erosion. J. Waterway. Div-ASCE 88: 117–130.

Bryant, M.F. 2018. Self-determined Exploration of the Outdoors and what Students can Teach Themselves: a Report on the Beach Investigators Summer Program. Master's thesis, Western Washington University, USA.

Burger, J., M. Gochfeld, L. Niles, N. Tsipoura, D. Mizrahi, A. Dey, et al. 2017. Stakeholder contributions to assessment, monitoring, and conservation of threatened species: black skimmer and red knot as case studies. Environ. Monit. Assess. 189(2): 60.

Cambers, G. and P. Diamond. 2010. Sandwatch: Adapting to Climate Change and Educating for Sustainable Development. UNESCO.

Campbell, E.E. and G.C. Bate. 1988. The estimation of annual primary production in a high energy surf-zone. Bot. Mar. 31: 337–343.

Campbell, E.E. and G.C. Bate. 1991. Ground water in the Alexandria dune field and its potential influence on the adjacent surf zone. Water SA 17: 155–160.

Castle, Z., S. Fletcher and E. McKinley. 2010. Coastal and marine education in schools: constraints and opportunities created by the curriculum, schools, and teachers in England. Ocean Year. 24(1): 425–444.

Cava, F., S. Schoedinger, C. Strang and P. Tuddenham. 2005. Science Content and Standards for Ocean Literacy: A Report on Ocean Literacy. COSEE, NOAA, COE and NMEA Report.

Cigliano, J.A., R. Meyer, H.L. Ballard, A. Freitag, T.B. Phillips and A. Wasser. 2015. Making marine and coastal citizen science matter. Ocean Coast. Manage. 115: 77–87.

Collins, D. and R. Kearns. 2010. 'Pulling up the tent pegs?' The significance and changing status of coastal campgrounds in New Zealand. Tourism Geogr. 12(1): 53–76.

Collins, D. and R. Kearns. 2013. Place attachment and community activism at the coast: the case of Ngunguru, Northland. New Zeal. Geogr. 69(1): 39–51.

Convention on Biological Diversity. 2011. Strategic Plan for Biodiversity 2011–2020 and the Aichi Targets: "Living in Harmony with Nature". Montreal, Quebec, Canada: UNEP.

Coupland, G.T., C.M. Duarte and D.I. Walker. 2007. High metabolic rates in beach cast communities. Ecosystems 10(8): 1341–1350.

Defeo, O., A. McLachlan, D.S. Schoeman, T. Schlacher, J. Dugan, A. Jones, et al. 2009. Threats to sandy beach ecosystems: a review. Estuar. Coast. Shelf Sci. 81: 1–12.

de Vos D., R. Nel, D. Schoeman, L.R. Harris and D. du Preez. 2019. Effect of introduced *Casuarina* trees on the vulnerability of sea turtle nesting beaches to erosion. Estuar. Coast. Shelf Sci. 223(31): 147–158.

Devictor, V., R.J. Whittaker and C. Beltrame. 2010. Beyond scarcity: citizen science programmes as useful tools for conservation biogeography. Divers. Distrib. 16(3): 354–362.

Dickinson, J.L., J. Shirk, D. Bonter, R. Bonney, R.L. Crain, J. Martin, et al. 2012. The current state of citizen science as a tool for ecological research and public engagement. Front. Ecol. Environ. 10(6): 291–297.

Drius, M., M.L. Carranza, A. Stanisci and L. Jones. 2016. The role of Italian coastal dunes as carbon sinks and diversity sources. A multi-service perspective. Appl. Geogr. 75: 27–136.

Domínguez-Tejo, E., G. Metternicht, E.L. Johnston and L. Hedge. 2018. Exploring the social dimension of sandy beaches through predictive modelling. J. Environ. Manage. 214: 379–407.

Dugan, J.E., D.M. Hubbard, M.D. McCrary and M.O. Pierson. 2003. The response of macrofauna communities and shorebirds to macrophyte wrack subsidies on exposed sandy beaches of southern California. Estuar. Coast. Shelf Sci. 58S: 25–40.

Dugan, J.E., D.M. Hubbard, I. Rodil, D.L. Revell and S. Schroeter. 2008. Ecological effects of coastal armoring on sandy beaches. Mar. Ecol. 29: 160–170.

Dugan, J.E., O. Defeo, E. Jaramillo, A.P. Jones, M. Lastra, R. Nel, et al. 2010. Give beach ecosystems their day in the sun. Science 329: 1146.

Dugan J.E., D.M. Hubbard, H.M. Page and J.P. Schimel. 2011. Marine macrophyte wrack inputs and dissolved nutrients in beach sands. Estuar. Coast. 34: 839–50.

Earp, H.S. and A. Liconti. 2020. Science for the future: the use of citizen science in marine research and conservation. pp. 1–19. *In*: S. Jungblut, V. Liebich and M. Bode-Dalby [eds]. YOUMARES 9-The Oceans: Our Research, Our Future. Springer, Cham.

Eastman, L., V. Hidalgo-Ruz, V. Macaya-Caquilpán, P. Nuñez and M. Thiel. 2014. The potential for young citizen scientist projects: a case study of Chilean schoolchildren collecting data on marine litter. J.I.C.Z.M. / R.G.C.I. 14(4): 569–579.

Erlandson, J.M. and S.M. Fitzpatrick. 2006. Oceans, islands, and coasts: current perspectives on the role of the sea in human prehistory. J. Island Coast. Archaeol. 1(1): 5–32.

Everard, M., L. Jones and B. Watts. 2010. Have we neglected the societal importance of sand dunes? An ecosystem services perspective. Aquat. Conserv. 20(4): 476–487.

Fanini, L., M. El Gtari, A. Ghlala and F. Scapini. 2007. From researchers to primary school: dissemination of scientific research results on the beach. An experience of environmental education at Nefza, Tunisia. Oceanologia 49(1).

Fanini, L., W. Plaiti and N. Papageorgiou. 2019. Environmental education: constraints and potential as seen by sandy beach researchers. Estuar. Coast. Shelf S. 218: 173–178.

Fauville, G., C. Strang, M.A. Cannady and Y.F. Chen. 2019. Development of the International Ocean Literacy Survey: measuring knowledge across the world. Environ. Educ. Res. 25(2): 238–263.

Feagin R.A., D.J. Sherman and W.E. Grant. 2005. Coastal erosion, global sea-level rise, and the loss of sand dune plant habitats. Front. Ecol. Environ. 3(7): 359–364.

Ferreira, M.A., L. Soares and F. Andrade. 2012. Educating citizens about their coastal environments: beach profiling in the Coastwatch project. J. Coast. Conserv. 16(4): 567–574.

Ford, C. 2009. ‘What Power What Grandeur What Sublimity!’: romanticism and the appeal of Sydney beaches in the nineteenth century. pp. 20–34. *In*: R. Hosking, S. Hosking, R. Pannell and N. Bierbaum [eds]. Something Rich and Strange: Sea Changes, Beaches and the Littoral in the Antipodes. Wakefield Press, Adelaide, South Australia.

Foster-Smith, J. and S.M. Evans. 2003. The value of marine ecological data collected by volunteers. Biol. Conserv. 113(2): 199–213.

Freiwald, J., R. Meyer, J.E. Caselle, C.A. Blanchette, K. Hovel, D. Neilson, et al. 2018. Citizen science monitoring of marine protected areas: case studies and recommendations for integration into monitoring programs. Mar. Ecol. 39: e12470.

Guest, H., H.K. Lotze and D. Wallace. 2015. Youth and the sea: ocean literacy in Nova Scotia, Canada. Mar. Policy 58: 98–107.

Harris, L., R. Nel and D. Schoeman. 2011. Mapping beach morphodynamics remotely: a novel application tested on South African sandy shores. Estuar. Coast. Shelf Sci. 92: 78–89.

Harris, L.R. 2012. An ecosystem-based spatial conservation plan for the South African sandy beaches. Ph.D. thesis. Nelson Mandela Metropolitan University, South Africa.

Harris, L., E.E. Campbell, R. Nel and D. Schoeman. 2014a. Rich diversity, strong endemism, but poor protection: addressing the neglect of sandy beach ecosystems in coastal conservation planning. Divers, Distrib. 20(10): 1120–1135.

Harris, L.R., M.E. Watts, R. Nel, D.S. Schoeman and H.P. Possingham. 2014b. Using multivariate statistics to explore trade-offs among spatial planning scenarios. J. Appl. Ecol. 51 (6): 1504–1514.

Harris, L., R. Nel, S. Holness, K. Sink and D. Schoeman. 2014c. Setting conservation targets for sandy beach ecosystems. Estuar. Coast. Shelf Sci. 150: 45–57.

Harris, L.R., R. Nel, S. Holness and D.S. Schoeman. 2015. Quantifying cumulative threats to sandy beach ecosystems: a tool to guide ecosystem-based management beyond coastal reserves. Ocean Coast. Manage. 110: 12–24.

Harris, L.R., C.J. Poole, M. van der Bank, K.J. Sink, D. De Vos, J.L. Raw, et al. 2019a. Benefits of coastal biodiversity. pp. 27–49. *In*: L.R. Harris, K.J. Sink, A.L. Skowno and L. Van Niekerk [eds]. South African National Biodiversity Assessment 2018: Technical Report. Volume 5: Coast. http://hdl.handle.net/20.500.12143/6374: South African National Biodiversity Institute. Pretoria.

Harris, L.R., M. Bessinger, A. Dayaram, S. Holness, S. Kirkman, T.-C. Livingstone, et al. 2019b. Advancing land-sea integration for ecologically meaningful coastal conservation and management. Biol. Conserv. 237: 81–89.

Hart, R.A. 2013. Children's Participation: The Theory and Practice of Involving Young Citizens in Community Development and Environmental Care. United Nations Children's Fund, Earthscan, London, New York.

Hartley, B.L., R.C. Thompson and S. Pahl. 2015. Marine litter education boosts children's understanding and self-reported actions. Mar. Pollut. Bull. 90(1–2): 209–217.

Hidalgo-Ruz, V. and M. Thiel. 2013. Distribution and abundance of small plastic debris on beaches in the SE Pacific (Chile): a study supported by a citizen science project. Mar. Environ. Res. 87: 12–18.

Hind-Ozan, E.J., G.T. Pecl and C.A. Ward-Paige. 2017. Communication and trust-building with the broader public through coastal and marine citizen science. pp. 261–278. *In*: J.A. Cigliano and H.L. Ballard [eds]. Citizen Science for Coastal and Marine Conservation. Routledge, London.

Honorato-Zimmer, D., K. Kruse, K. Knickmeier, A. Weinmann, I.A. Hinojosa and M. Thiel. 2019. Inter-hemispherical shoreline surveys of anthropogenic marine debris–a binational citizen science project with schoolchildren. Mar. Pollut. Bull. 138: 464–473.

Houston, J.R. 2013. The economic value of beaches—a 2013 update. Shore & Beach. 81(1): 3–10.

Hubbard, D.M., J.E. Dugan, N.K. Schooler and S.M. Viola. 2014. Local extirpations and regional declines of endemic upper beach invertebrates in southern California. Estuar. Coast. Shelf Sci. 150: 67–75.

Hyder, K., B. Townhill, L.G. Anderson, J. Delany and J.K. Pinnegar. 2015. Can citizen science contribute to the evidence-base that underpins marine policy? Mar. Policy 59: 112–120.

IPCC. 2014. Climate Change 2014: Synthesis Report. Contribution of Working Groups I, II and III to the Fifth Assessment Report of the Intergovernmental Panel on Climate Change, edited by Core Writing.

James, R.J. 2000. From beaches to beach environments: linking the ecology, human-use and management of beaches in Australia. Ocean. Coast. Manage. 43: 495–514.

Jaramillo E., J.E. Dugan, D.M. Hubbard, D. Melnick, M. Manzano, C. Duarte, et al. 2012. Ecological Implications of Extreme Events: Footprints of the 2010 Earthquake along the Chilean Coast. PLoS One. 7(5): e35348.

Kearns, R. and D. Collins. 2012. Feeling for the coast: the place of emotion in resistance to residential development. Soc. Cult. Geogr. 13(8): 937–955.

Koop, K. and C. Griffiths. 1982. The relative significance of bacteria, meio- and macrofauna on an exposed sandy beach. Mar Biol. 66: 295–300.

Koop, K., R.C. Newell and M.I. Lucas. 1982. Microbial regeneration of nutrients from the decomposition of macrophyte debris on the shore. Mar. Ecol. Prog. Ser. 9(1): 91–96.

La Mar, B.P. 2006. Marine-Based Ecological Education: Marine Discovery Centres, Millennium Kids, Environmental Citizenship, and a Vision for an Eco-Camp. Ph.D. thesis, Murdoch University, Australia.

La Cock, G.D. and J.R. Burkinshaw. 1996. Management implications of development resulting in disruption of a headland bypass dunefield and its associated river, Cape St Francis, South Africa. Landscape Urban Plan. 34(3–4): 373–381.

Lambert, J. 2006. High school marine science and scientific literacy: the promise of an integrated science course. Int. J. Sci. Educ. 28(6): 633–654.

Lasiak, T.A. 1983. The impact of surf zone fish communities on faunal assemblages associated with sandy beaches. In: A. McLachlan and T. Erasmus [eds.]. Sandy Beaches as Ecosystems. Developments in Hydrobiology. The Hague: Dr W Junk.

Lastra, M., H.M. Page, J.E. Dugan, D.M. Hubbard and I.F. Rodil. 2008. Processing of allochthonous macrophyte subsidies by sandy beach consumers: estimates of feeding rates and impacts on food resources. Mar. Biol. 154: 163–174.

Lastra, M., J. López and I.F. Rodil. 2018. Warming intensify CO_2 flux and nutrient release from algal wrack subsidies on sandy beaches. Glob. Change Biol. 24: 3766–3779.

Lawson, E.T. 2016. Re-thinking relationships between environmental attitudes and values for effective coastal natural resource management in Ghana. Local Environ. 21(7): 898–917.

Le Gouvello, D.Z.M., R. Nel, L.R. Harris and K. Bezuidenhout. 2017a. The response of sandy beach meiofauna to nutrients from sea turtle eggs. J. Exp. Mar. Biol. Ecol. 487: 94–105.

Le Gouvello, D.Z.M., R. Nel, L.R. Harris, K. Bezuidenhout and S. Woodborne. 2017b. Identifying potential pathways for turtle-derived nutrients cycling through beach ecosystems. Mar. Ecol. Prog. Ser. 583: 49–62.

Linke, S., M. Watts, R. Stewart and H.P. Possingham. 2011. Using multivariate analysis to deliver conservation planning products that align with practitioner needs. Ecography 34(2): 203–207.

Liu, J., N. Liu, Y. Zhang, Z. Qu and J. Yu. 2019. Evaluation of the non-use value of beach tourism resources: a case study of Qingdao coastal scenic area, China. Ocean Coast. Manage. 168: 63–71.

Lucrezi, S. and M.F. Van der Walt. 2016. Beachgoers' perceptions of sandy beach conditions: demographic and attitudinal influences, and the implications for beach ecosystem management. J. Coast. Conserv. 20(1): 81–96.

Lucrezi, S., M. Saayman and P. Van der Merwee. 2016. An assessment tool for sandy beaches: a case study for integrating beach description, human dimension, and economic factors to identify priority management issues. Ocean. Coast. Manage. 121: 1–22.

Lucrezi, S., M.H. Esfehani, E. Ferretti and C. Cerrano. 2019. The effects of stakeholder education and capacity building in marine protected areas: a case study from southern Mozambique. Mar. Policy 108: 103645.

Maguire, G.S., K.K. Miller, M.A. Weston and K. Young. 2011. Being beside the seaside: beach use and preferences among coastal residents of south-eastern Australia. Ocean Coast. Manage. 54(10): 781–788.

Maguire, G.S., J. Rimmer and M.A. Weston. 2013. Stakeholder perceptions of threatened species and their management on urban beaches. Animals 3(4): 1002–1020.

Margules, C.R. and R.L. Pressey. 2000. Systematic conservation planning. Nature 405(6783): 243–253.

Martin, V.Y. 2017. Citizen science as a means for increasing public engagement in science: presumption or possibility?. Sci. Commun. 39(2): 142–168.

McLachlan, A. 1982. A model for the estimation of water filtration and nutrient regeneration by exposed sandy beaches. Mar. Environ. Res. 6(1): 37–47.

McLachlan, A. and T. Erasmus. 1983. Sandy beaches as ecosystems. p. 757. *In*: A. McLachlan and T. Erasmus [eds]. Sandy Beaches as Ecosystems. Developments in Hydrobiology. The Hague: Dr W Junk.

McLachlan, A. and M. Burns. 1992. Headland bypass dunes on the South African coast: 100 years of mismanagement. pp. 71–79. *In*: R.W.G. Carter, T.G.F. Curtis and M.J. Sheehy-Skeffington [eds]. Coastal Dunes. A.A. Balkema, The Netherlands.

McLachlan, A. and O. Defeo. 2018. The Ecology of Sandy Shores, 3rd Ed. Academic Press, Burlington, MA, USA.

Nel, R., E.E. Campbell, L. Harris, L. Hauser, D.S. Schoeman, A. McLachlan, et al. 2014. The status of sandy beach science: past trends, progress, and possible futures. Estuar. Coast. Shelf Sci. 150: 1–10.

Nordstrom, K.F. 2000. Beaches and Dunes on Developed Coasts. Cambridge University Press, UK.

Peterson, C.H. and M. Bishop. 2005. Assessing the environmental impacts of beach nourishment. Bioscience. 55: 887–896.

Piwowarczyk, J. 2010. Consilience on the beach: inquiry-based educational programs on sandy beaches. pp. 214–216. *In*: M. Fichaut and V. Tosello [eds]. International Marine Data and Information Systems Conference, IMDIS 2010. Muséum National d'Histoire Naturelle, Paris, France.

Riedinger, K. and A. Taylor. 2019. Leveraging parent chaperones to support youths' learning during an out-of-school field trip to a marine science field station. pp. 59–80. *In*: G. Fauville, D.L. Payne, M.E. Marrero, A. Lantz-Andersson and F. Crouch [eds]. Exemplary Practices in Marine Science Education. Springer, Cham.

Rodella, I., F. Madau, M. Mazzanti, C. Corbau, D. Carboni, K. Utizi, et al. 2019. Willingness to pay for management and preservation of natural, semi-urban and urban beaches in Italy. Ocean Coast. Manage. 172: 93–104.

Rodil, I.F., E. Jaramillo, D.M. Hubbard, J.E. Dugan, D. Melnick and C. Velasquez. 2015. Responses of dune plant communities to continental uplift from a major earthquake: sudden releases from coastal squeeze. PLoS One. 10(5): e0124334.

Rodil, I.F., M. Lastra, L. López, A.P. Mucha, J.P. Fernandes, S.V. Fernandes, et al. 2019. Sandy beaches as biogeochemical hotspots: the metabolic role of macroalgal wrack on low productive shores. Ecosystems. 22: 49–63.

Rodríguez-Revelo, N., I. Espejel, C.A. García, L. Ojeda-Revah and M.A.S. Vázquez. 2018. Environmental services of beaches and coastal sand dunes as a tool for their conservation. pp. 75–100. *In*: C.M. Botero, O. Cervantes and C.W. Finkl [eds]. Beach Management Tools-Concepts, Methodologies and Case Studies. Springer, Cham.

Rolfe, J. and D. Gregg. 2012. Valuing beach recreation across a regional area: the Great Barrier Reef in Australia. Ocean Coast. Manage. 69: 282–290.

SANBI. 2013. Life: the State of South Africa's Biodiversity 2012. Pretoria: South African National Biodiversity Institute.

SANBI and UNEP-WCMC. 2016. Mapping biodiversity priorities: a practical, science-based approach to national biodiversity assessment and prioritisation to inform strategy and action planning. Cambridge, UK: UNEP-WCMC.

Scapini, F. and L. Fanini. 2009. The role of scientists in providing formal and informal information for the definition of guidelines, regulations or management plans for sandy beaches. pp. 87–94. *In*: A. Bayed [ed.]. Sandy Beaches and Coastal Zone Management —Proceedings of the Fifth International Symposium on Sandy Beaches, 19th–23rd October 2009, Rabat, Morocco.

Schlacher, T.A., D.S. Schoeman, J. Dugan, L. Lastra, A. Jones, F. Scapini, et al. 2008. Sandy beach ecosystems: key features, management challenges, climate change impacts, and sampling issues. Mar. Ecol. 29: 70–90.

Schlacher, T.A., S. Strydom and R.M. Connolly. 2013. Multiple scavengers respond rapidly to pulsed carrion resources at the land-ocean interface. Acta Oecol. 48: 7–12.

Schlacher, T.A., D.S. Schoeman, A.R. Jones, J.E. Dugan, D.M. Hubbard, O. Defeo, et al. 2014a. Metrics to assess ecological condition, change, and impacts in sandy beach ecosystems. J. Environ. Manage. 114: 322–335.

Schlacher, T.A., A.R. Jones, J.E. Dugan, M.A. Weston, L. Harris, D.S. Schoeman, et al. 2014b. Open-coast sandy beaches and coastal dunes. pp. 37–92. *In*: B. Maslo and J.L. Lockwood [eds]. Conservation biology 19: Coastal conservation. Cambridge University Press, Cambridge.

Schlacher, T.A., S. Lucrezi, R.M. Connolly, C.H. Peterson, B.L. Gilby, B. Maslo, et al. 2016. Human threats to sandy beaches: a meta-analysis of ghost crabs illustrates global anthropogenic impacts. Estuar. Coast. Shelf Sci. 169: 56–73.

Shirk, J.L., H.L. Ballard, C.C. Wilderman, T. Phillips, A. Wiggins, R. Jordan, et al. 2012. Public participation in scientific research: a framework for deliberate design. Ecol. Soc. 17(2).

Silvertown, J. 2009. A new dawn for citizen science. Trends Ecol. Evol. 24(9): 467–471.

Small, C. and R.J. Nicholls. 2003. A global analysis of human settlement in coastal zones. J. Coastal Res. 19(3): 584–599.

Steven, R., C. Morrison and J.G. Castley. 2014. Bird watching and avitourism: a global review of research into its participant markets, distribution and impacts, highlighting future research priorities to inform sustainable avitourism management. J. Sustain. Tour. 23(8–9): 1257–1276.

Storrier, K.L. and D.J. McGlashan. 2006. Development and management of a coastal litter campaign: the voluntary coastal partnership approach. Mar. Policy 30(2): 189–196.

Stronge, W.B. 2005. Economic Value of Beaches. pp. 401–462. *In*: M.L. Schwartz [ed.]. Encyclopedia of Coastal Science. Encyclopedia of Earth Science Series. Springer, Dordrecht.

Taylor, B. 2007. Surfing into spirituality and a new, aquatic nature religion. J. Am. Acad. Relig. 75(4): 923–951.

Thiel, M., M.A. Penna-Díaz, G. Luna-Jorquera, S. Salas, J. Sellanes and W. Stotz. 2014. Citizen scientists and marine research: volunteer participants, their contributions, and projection for the future. Oceanogr. Mar. Biol. Annu. Rev. 52: 257–314.

Thiel, M., S. Hong, J.R. Jambeck, M. Gatta-Rosemary, D. Honorato-Zimmer, T. Kiessling, et al. 2017. Marine litter–bringing together citizen scientists from around the world. pp. 104–131. *In*: J.A. Cigliano and H.L. Ballard [eds]. Citizen Science for Coastal and Marine Conservation. Routledge, London.

Thompson, R. 2007. Cultural models and shoreline social conflict. Coast. Manage. 35(2–3): 211–237.

Tinley, K.L. 1985. Coastal Dunes of South Africa. A report for the Committee for Nature Conservation Research, National Programme for Ecosystem Research. In South African National Scientific Programmes Report No 109. Pretoria: CSIR.

Tunstall, S.M. and E.C. Penning-Rowsell. 1998. The English beach: experiences and values. Geogr. J. 164(3): 319–332.

UN General Assembly. 2019. A/RES/73/284: United Nations Decade on Ecosystem Restoration (2021–2030). Resolution adopted by the General Assembly on 1 March 2019. Available at: https://undocs.org/A/RES/73/284.

United Nations. 2015. Transforming our world: The 2030 Agenda for Sustainable Development. A/RES/70/1. United Nations, New York.

van Buskirk, J. and L.B. Crowder. 1994. Life-History Variation in Marine Turtles. Copeia 1994(1): 66–81.

Vann-Sander, S., J. Clifton and E. Harvey. 2016. Can citizen science work? Perceptions of the role and utility of citizen science in a marine policy and management context. Mar. Policy 72: 82–93.

Vivian, E.V.C. and T.A. Schlacher. 2015. Intrinsic and utilitarian valuing on K'gari-Fraser Island: a philosophical exploration of the modern disjunction between ecological and cultural valuing. Australas. J. Env. Man. 22(2): 149–162.

Voyer, M., N. Gollan, K. Barclay and W. Gladstone. 2015. 'It's part of me'; understanding the values, images and principles of coastal users and their influence on the social acceptability of MPAs. Mar. Policy 52: 93–102.

Wakita, K., Z. Shen, T. Oishi, N. Yagi, H. Kurokura and K. Furuya. 2014. Human utility of marine ecosystem services and behavioural intentions for marine conservation in Japan. Mar. Policy 46: 53–60.

Ward, E.J., K.N. Marshall, T. Ross, A. Sedgley, T. Hass, S.F. Pearson, et al. 2015. Using citizen-science data to identify local hotspots of seabird occurrence. PeerJ 3: e704.

Weiss, E. and B. Chi. 2019. ¡Youth & The Ocean! (¡YO!): partnering high school and graduate students for youth-driven research experiences. pp. 27–58. *In*: G. Fauville, D.L. Payne, M.E. Marrero, A. Lantz-Andersson and F. Crouch [eds]. Exemplary Practices in Marine Science Education. Springer, Cham.

West-Newman, C.L. 2008. Beach crisis: law and love of place. Space Cult. 11(2): 160–175.

Wolch, J. and J. Zhang. 2004. Beach recreation, cultural diversity and attitudes towards nature. J. Leis. Res. 36(3): 414–443.

Index

About the Editors

Dr. Sílvia C. Gonçalves is a Portuguese researcher, with a PhD in Biology (specialty of Ecology) by the University of Coimbra, Portugal. She is presently Coordinating Professor at the School of Tourism and Maritime Technology (ESTM), of the Polytechnic of Leiria (Portugal), lecturing mainly in the scientific areas of ecology, biological diversity and marine environment. She is also a Senior Researcher at MARE - Marine and Environmental Sciences Centre and focusses her research on marine ecology, in the following domains: (1) bio-ecology of crustacean populations on sandy beaches and their potential as ecological indicators of human impacts; (2) structure and function of macrobenthic communities; (3) monitoring and evaluation of environmental quality in coastal ecosystems, especially on sandy beaches and coastal lagoons. Her scientific research activities also include environmental biotechnology (e.g., bioremediation of trace metals by halophyte plants) and aquaculture (e.g., maintenance and reproduction of sea urchins and fish). Her scientific production includes book chapters in international books, research papers in international scientific periodicals with referee, publications in international congresses with referee, oral and poster presentations on international scientific meetings and the authorship of one documentary of scientific outreach about marine life in coastal areas. She is also referee of several international scientific periodicals from leading publishers (e.g., Elsevier, Springer Verlag, Taylor & Francis) and Associate Editor of the journal Frontiers in Marine Science.

Dr. Susana M. F. Ferreira is a Portuguese researcher, with a PhD in Biology (specialization in Ecology) from the University of Coimbra (Portugal), since 2005. She is currently an Adjunct Professor at the Polytechnic of Leiria (Portugal), where she lectures biology, zoology, histology, embryology, population dynamics, and is a Coordination Commission member of the Higher Professional Technical Course in Laboratory Analysis. She is also Senior Researcher at MARE - Marine and Environmental Sciences Centre (PLeiria), Collaborator at CFE - Centre for Functional Ecology (University of Coimbra), and member of the ORBEA - Body Responsible for the Animal Welfare of MARE-PLeiria. Her scientific activity includes areas of ecology, biotechnology and aquaculture. The domain of her specialization relates with research on biological and ecological processes in aquatic ecosystems: environmental impact studies, with emphasis on eutrophication processes; population dynamics, biology and secondary production of benthic organisms; community structure and functioning of coastal systems; influence of trematode parasites on estuarine crustaceans and molluscs. Her fascination for the sea and marine life encouraged her to pursue several nautical and underwater activities (local skipper, VHF-A operator, PADI advanced open water and enriched air diver, plus digital underwater photographer). She is currently developing research on trace metals and phytoremediation in coastal habitats; stock assessment of aquatic organisms and their use in aquaculture and biotechnology (namely, sea anemones, polychaetes, sea urchins and fish); animal welfare; and ocean literacy. Her scientific production counts on an international book, chapters in international books, articles in international journals with scientific referee, abstracts in international congress proceedings with scientific referee, oral and panel communications, an invention of a feed for rearing omnivorous fish patented in Europe, plus the direction and authorship of a documentary on marine life.